CLIMATE CHANGE AND THE HEALTH OF NATIONS

CLIMATE CHANGE AND THE HEALTH OF NATIONS

FAMINES, FEVERS, AND THE FATE OF POPULATIONS

ANTHONY J. MCMICHAEL
WITH
ALISTAIR WOODWARD
AND CAMERON MUIR

OXFORD
UNIVERSITY PRESS

OXFORD
UNIVERSITY PRESS

Oxford University Press is a department of the University of Oxford. It furthers
the University's objective of excellence in research, scholarship, and education
by publishing worldwide. Oxford is a registered trade mark of Oxford University
Press in the UK and certain other countries.

Published in the United States of America by Oxford University Press
198 Madison Avenue, New York, NY 10016, United States of America.

© Oxford University Press 2017

Library of Congress Cataloging-in-Publication Data
Names: McMichael, A. J. (Anthony J.), author. | Woodward, Alistair, author. |
Muir, Cameron.
Title: Climate change and the health of nations : famines, fevers, and the fate of
populations / Anthony J. McMichael with Alistair Woodward and Cameron Muir.
Description: New York, NY, United States of America : Oxford University
Press, [2017] | Includes bibliographical references.
Identifiers: LCCN 2016034533 | ISBN 9780190262952
Subjects: LCSH: Human beings—Effect of climate on. |
Climatic changes—Health aspects.
Classification: LCC GF71 .M46 2017 | DDC 304.2/5—dc23
LC record available at https://lccn.loc.gov/2016034533

3 5 7 9 8 6 4
Printed by Sheridan Books, Inc., United States of America

To the grandchildren of our world

CONTENTS

⬥

LIST OF ILLUSTRATIONS

PREFACE

———⟫◆⟪———

This book is about how climate has played upon the health and fates of populations throughout the 200,000-year odyssey of the human species. You may ask: "Why write a book about the past when today's big challenge is to avert future climate change?" My answer is that knowledge about past human experiences of natural climate shifts alerts us to the risks that humankind faces. The landscapes of past millennia are littered with famines and fevers, often provoked by natural changes in climate systems. The future prospect may be darker if, as now seems likely, unusually large and rapid climate change, accompanied by extreme weather variability, lies ahead.

"Histories are written to help explain how we got where we are,"[1] and I argue that history can also help explain where we may be heading. Earth's biography offers us a natural historical analog of future climate change—albeit one that cannot clearly predict the consequences of the more extreme forms of climate change that may occur later this century.

But historical insights may help jolt us out of policy-paralyzing complacency and deferral of responsibility—the "moral economy of inaction."[2]

Naive assumptions that incremental technical adjustments and adaptations will allow us to pursue business as usual reflect a deep ignorance about the complex nature of the real world. Yet I believe that there is still cause for optimism. Global information flows are heightening awareness of the causes, processes, and consequences of climate change. Young researchers, willing to look beyond the conventional bounds of their disciplines, are now engaging with this topic, responding to challenges, and knowing that their work is contributing to the global common good at this epochal time. As people's knowledge about, and experiences of, climate change accumulate, as climate science strengthens, and with a reality check garnered from history's archives, we humans may yet take extraordinary international action. Many national governments now see that their future economic security is threatened by climate change—and that responding to it provides social and economic opportunities.

Despite the economic stranglehold exerted by the fossil fuel industry, for example, renewable energy systems are now evolving rapidly.

As an environmental epidemiologist[3] and medical graduate, my interest in this topic began when writing an Australian national newspaper column, Spaceship Earth, in the early 1970s—a time of rising concerns about "limits to growth" and the environmental hazards associated with industrial expansion.[4] There was much concern about direct chemical toxicity and the spread of synthetic pesticides through nature's food chains, but concern also was growing that human intrusion on the natural environment was endangering living organisms and their ecosystems.

The risks posed by climate change to human health and physical survival are now better known. Since the early 1990s, I have contributed to the assessment of health risk by the UN Intergovernmental Panel on Climate Change (IPCC), and have done likewise in Asian and Pacific countries for the World Health Organization. My knowledge of this field has benefited greatly from working with colleagues who are experts on diverse aspects of climate change, the physical and social sciences, and population health.

My hope for this book is that better insights into the influence of climate change on human populations and their societies will help motivate more effective changes in how we live and how we care for the planet. Exposure to unusual climatic fluctuations and shifts over the past two million years accelerated the evolution of a powerful, flexible, and creative human brain. Having long used that brain to satisfy the basic urges to own, acquire, control, and consume, we must now apply our collective brainpower to halting the unusual change in climatic behavior that our creative and acquisitive actions have caused.

Tony McMichael, Australian National University,
Canberra, July 2014

My husband, Tony McMichael, worked on this book in between other commitments from around 2011 to September 2014. The book stems from Tony's keen interest in the long history of health and disease and upon his pioneering work, stretching over more than 20 years, on the impact of climate change on human populations. He wanted this book to contribute to a better understanding of the intertwined relationship between humans and their environment, and to how human populations are impacted by, and impact upon, the life support systems of our planet. A full manuscript was submitted to Oxford University Press in early 2014 and was subsequently accepted subject to attention to reviewer comments and a reduction in length. Tony was looking forward happily to completing the book, but he died unexpectedly on September 26, 2014.

The book has been revised and edited, therefore, by Alistair Woodward and Cameron Muir. Alistair, a longstanding colleague of Tony's and his successor as a coordinating lead author for the Intergovernmental Panel on Climate Change, applied his population health perspective. Cameron brought his knowledge of environmental history and overall editing expertise.

My family, as well as Tony's many friends and colleagues, miss him sorely. We are happy that this, his last book, has been published and joins his many other publications in expanding our understanding of the intersections between environments and human well-being and the grave risks that we humans face in a warming world.

Judith Healy, Australian National University,
Canberra, January 1, 2017

ACKNOWLEDGMENTS

—————⋗◆⋖—————

Tony McMichael enjoyed discussing aspects of this wide-ranging book with many people during its gestation, and some people read and commented upon early drafts of the whole book or particular chapters. Tony did not keep an up-to-date list, however, of the many people whom he consulted, since he knew who they were, what suggestions they had made, and the lines of inquiry along which they directed him. We are embarrassed, therefore, that the following acknowledgments are very partial and hope that the many people not mentioned will understand.

Tony appreciated the support given him by the Australian National University and by his colleagues at the National Centre for Epidemiology and Population Health, whose director he was from 2001 to 2007, then as an NHMRC Australia Fellow and latterly as emeritus professor. The Rockefeller Foundation Bellagio Center offered a stimulating environment for researching and presenting his ideas during a residency in 2012. Tony also appreciated the steadfast assistance of Andrew Schuller in shepherding the book through to a publisher—as did we during the later stages of the book. We are also grateful to Laura Ford for checking and completing the references.

Many other people assisted in various ways, and as acknowledged, the following is a very incomplete (in family name alphabetic order) roll call: Phillip Baker, John Brooke, Colin Butler, Andrew Glikson,

xx   ACKNOWLEDGMENTS

Billy Griffiths, Andy Haines, Adrian Hayes, Clive Hilliker, Allen Isaacmann, Allan Kearns, Tord Kjellstrom, Sari Kovats, Philip McMichael, John McNeill, Jonathan Patz, Eric Richards, Kirk Smith, Robin Weiss, and Steve Zavestoski.

Judith Healy, Australian National University,
Canberra, January 1, 2017

I

Introduction

TRENDS IN GLOBAL GREENHOUSE emissions during the first two decades of this twenty-first century are leading us to a much hotter world by 2100, perhaps 3°C–4°C above the late-twentieth-century average temperature[1,2] and hotter than at any time in the last 20–30 million years. Further, the *rate* of heating would be about 30 times faster than when Earth emerged from the most recent ice age, between 17,000 and 12,000 years ago. At that speed, environmental changes may outstrip the capacity of many species to evolve and adapt.

Having once relied on fires in caves, humans in the late eighteenth and nineteenth centuries increasingly began to burn fossil fuels to release vastly more energy—and, inadvertently, vastly more carbon dioxide. About 600 billion metric tons of that invisible, stable, and odorless gas have been emitted since 1750, about two-thirds of which will persist in the atmosphere for centuries. The resulting 40 percent increase in atmospheric carbon dioxide concentration is the main cause of human-driven climate change. We have wrapped another heat-retaining blanket around the planet, causing warming of Earth's surface at a rate that far outpaces nature's rhythms.

Humans have lived in climatically congenial times for the past 11,000 years of the Holocene geological epoch compared with the rigors of the preceding ice age. Figure 1.1 shows the world's estimated average surface temperature over that era, and the right-hand side of the graph shows the likely global warming by 2100 averaged across many published modeled projections. The difference between the peak temperature of 7,000 years ago and the nadir of the Little Ice Age 350 years ago is 0.7°C. By early in this twenty-first century, the global average temperature had edged higher than for the past 11,000 years—by 0.6°C in six decades.[3,4] If the world's temperature were to rise by 3°C–4°C within just three generations, our descendants might struggle to remain healthy, raise families, and survive within stable societies. I am certainly not the first to say this . . .

A 4°C temperature increase probably means a global carrying capacity below 1 billion people. (Hans Joachim Schellnhuber, Director, Potsdam Institute for Climate Impact Research, at the Copenhagen Climate Change Conference in 2009)[5]

In such a 4°C world, the limits for human adaptation are likely to be exceeded in many parts of the world, while the limits for adaptation for natural systems would largely be exceeded throughout the world. (Rachel Warren in *Philosophical Transactions of the Royal Society* 2011)[6]

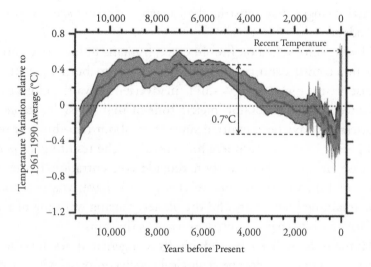

FIGURE 1.1 Reconstructed global temperature variations during the 11,000 years of the Holocene. Marcott, et al., "A Reconstruction of Regional and Global Temperature for the Past 11,300 Years."[7]
Permission received from AAAS.

Piecing together an untold story

No one has yet told the full story of the historical interplay between climate change, human health, disease, and survival. Scholars have written about parts of the story, in some parts of the world, but a comprehensive account was not possible before recent advances in understanding of the climate system and the high-tech and detailed reconstructions of past climates.[8,9,10] Also, more information is emerging about population health, from sources such as high-resolution identification of food traces in pottery fragments and genotyping of bacterial DNA in human skeletal samples.

Ecologists are using past situations of natural climate change to learn how species and ecosystems responded.[11] Modern technologies, particularly those of molecular biology, are enabling us to learn from past human experiences about aspects of our species' biological vulnerability, such as the consequence for the immune systems of gene exchanges via interbreeding between *Homo sapiens* and our Neanderthal and Denisovan cousins around 40,000 years ago.

The field of climate science has expanded enormously, and scholars have produced a large amount of literature on the science of past and present climate change and future climate projections. This includes the pivotal synthesizing work of the Intergovernmental Panel on Climate Change (IPCC), a large international scientific body established by the United Nations in 1988 in response to the anticipated damages from human-caused climate change.[12] The IPCC's flagship series of five Assessment Reports on the science and impacts of climate change, published during 1990–2014, are a key source of information.

Even so, much less has been written about the likely risks to *humans* and their communities—especially the risks to population health, survival, and social stability.[13,14] In this book, I aim to weave historical threads together to gain a better understanding of how climate change affects human populations. Unlike the other momentous human-induced global environmental changes that now press on the living world, such as stratospheric ozone depletion, there is a historical analog of today's human-driven climate change. The world's climate has always varied naturally, a fluctuating and sometimes turbulent platform for the evolution of life and for the emergence of human cultures and societies, and the health of their populations.

Some argue that it is the fact that the future may be like the past that makes science possible, while the fact that the future may *not* be like the past makes science necessary.[15] Although climate change this century will differ greatly in speed and extent from natural climate changes over past epochs, thus making new modes of science necessary, I believe insights from studying past climatic experiences will help us anticipate the scale and nature of hazards that may lie ahead.

Historian Geoffrey Parker concludes from his research on the global crisis of the seventeenth century that Europe's dire climate at that time strongly influenced the health and survival of communities and their social and political stability. He wrote:[16]

> "We have only two ways to anticipate the impact of a future catastrophic climate change ... Either we 'fast-forward' the tape of history and predict what might happen on the basis of current trends; or we 'rewind the tape' and learn from what happened during global catastrophes in the past. Although many experts (mainly climatologists, sociologists, and political scientists) have tried the former, *few have systematically attempted the latter* [my emphasis] ... [from which] we may learn some valuable lessons for dealing with the climate challenges that undoubtedly await us and our children."

The history of *natural* climate changes over the past six or seven millennia should shed light on how those changes influenced the health and survival of human societies. Interesting questions abound. Did climate change during the approaching depth of the glacial period around 23,000 years ago B.C.E. (Before Current Era) hasten the demise of our cousins, the Neanderthals? How did the prolonged drying of the Sahara 5,500 years ago, or of southern Mesopotamia 4,000 years ago, affect food yields and hence the health and the viability of those societies? How did climatic fluctuations influence the outbreaks of the first two bubonic plague pandemics in the sixth and fourteenth centuries C.E. (Current Era) respectively?[17] Did the five dynastic collapses in

China over the past millennium mostly occur after periods of unusually adverse climatic conditions?

Other exogenous environmental factors, such as regional differences in soil fertility, water supplies, numbers of plants and animal species amenable to domestication, and access to trade routes, influenced the long-term development, wealth, technical sophistication, and power of societies.[18] The same can be said of the natural shifts, cycles, and fluctuations in the climate. Examples include:

- The 500-year Roman Warm, particularly the temporary northward excursion of the Mediterranean climate zone in Europe that facilitated the Roman Empire's expansive phase by, for instance, boosting agricultural productivity.
- The wild rodent expansion in Central Asia around 1300 C.E. associated with climatic change, setting in motion the sequence of events leading to outbreaks of bubonic plague in China and subsequent epidemics in Europe in the 1340s.
- The influence early in seventeenth-century Europe of the unusually cold and wet weather on food crises, starvation, deaths, epidemics, social disorder, conflict, and warfare, most notably the chaotic Thirty Years' War.

But are the past and future comparable? Does today's global connectedness, for example, offset the risks that once applied to regionally isolated societies? Have rich urban-industrial populations buffered themselves against the main risks of climate change? The answers may surprise us unpleasantly. The OECD warns, for example, that modern hyper-integrated IT-dependent urban populations have become *increasingly* vulnerable to the potential havoc caused by major external environmental stressors.[19] The partial collapse of New York's infrastructure caused by Superstorm Sandy in late 2012 underscores that point.

Avoiding the shoals of "climatic determinism"

Changes in climate rarely act alone. Typically, climate change acts as a *contributor* or an *amplifier*. It may be the dominant causal factor in a surge of deaths during an extreme heat wave; it may be a contributory factor to a decades-long downturn in regional crop yields with downstream effects on nutritional health and mortality; it may be a predisposing factor in a community's decision to migrate.

Contending academic disciplines see the world differently; the debate over "environmental determinism" is a prime example. During the nineteenth and early twentieth centuries, there was strong support for the idea that "a society's physical environment can control its cultural development,"[20] especially among anthropologists and archaeologists. But many social scientists and historians were hostile to this view. In the mid-1950s, Swedish economic historian Gustaf Utterström broke ranks, arguing that the cold and adverse climate in the sixteenth and seventeenth centuries accounted for much of Scandinavia's demographic and economic setback at that time.[21] The French historian Fernand Braudel also acknowledged that changes in climate had played an important role, particularly as part of what he called the *longue durée* changes in population profiles, environment, and climate—that is, the great undercurrents that shaped the fates of civilizations, their cultures, ideologies, and power structures.[22]

Other scholars who argued that social historians should take climate into account include Emmanuel Le Roy Ladurie in his seminal book *Times of Feast, Times of Famine*, published in 1972.[23] Subsequently, several contemporary historians have argued that their discipline should be foregrounding climate as an influence on human affairs.[24,25] For example, John Brooke writes, "Over the very long term, the history of a volatile and changing earth has driven biological and human evolution: it has been a rough journey, and we are products of that journey."[26]

Still others have raised questions about the "environmentalist's paradox."[27] Why has human life expectancy continued to increase while our erosion of much of the planet's resource base and life support system has been increasing? Well, there is no real paradox; it is mostly a problem of inappropriate time horizons. The example of land use is helpful. During the past five millennia, human populations have cut down about half the world's temperate and tropical forests, transformed nearly half the non-ice non-desert landscape to croplands or pasture, and built almost one million dams, many of which impede natural river flow.[28] Some of these changes clearly have been associated with major benefits for public health, including increased food yields. Over time, many negative impacts of land use change have also occurred. For example, dams and irrigation projects have boosted the occurrence of vector-borne infectious diseases, including malaria and schistosomiasis in parts of Africa and South Asia.[29,30] But the full impacts of large-scale human colonization of the planet, such as loss of species and disruption of the global nitrogen cycle, have not yet become apparent.[31]

Pathways by which climatic changes affect human health

Climate change does not exist, or act, in isolation. Some health impacts of climate change occur directly, from extreme climate-related exposures (see Figure 1.2). For example, heat waves and

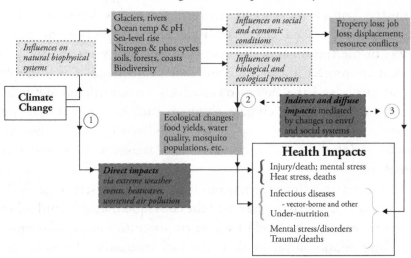

FIGURE 1.2 Schematic diagram of the main paths, direct and indirect, by which climatic changes affect human health. McMichael, "Globalization, Climate Change, and Human Health."

Source: Author.[32]

Permission received from Massachusetts Medical Society.

floods directly kill thousands of people every year. Most climate-related health risks, however, are mediated by less direct pathways and therefore are modulated by a host of environmental and social factors—and are likely to have much greater adverse consequences for health and survival than the direct-acting risks. Indirect health effects arise via impacts on food yields, water flows, patterns and ranges of infectious disease occurrence, stresses on housing and settlements, impoverishment of the vulnerable, and the movement, sometimes displacement, of groups and populations. In the Horn of Africa, for example, recent surface warming of the western Indian Ocean by up to 2°C, and the associated shifts in patterns of ocean water circulation and easterly moisture-laden winds, have contributed to the emergence of long-term regional droughts and lethal food shortages.[33,34,35] Climate changes may also undermine the integrity of natural and human-built protection against natural disasters (for instance, coral reefs, mangroves, and storm water systems in big cities).

It is important to identify factors that aggravate or diminish the impacts of climate change, since this knowledge may help us understand the magnitude and distribution of health risks in the future, and also points to practical steps that may be taken to protect those who are most vulnerable. For example, food yields and the downstream effects on nutritional status are sensitive not only to climate extremes such as drought and heat waves, but are affected also by local biodiversity, soil quality, the vigor and health of farmworkers, and stable, functioning food markets.[36]

Changes in climate, environment, and ecosystems generally occur on a large spatial scale, and the consequent health (and other) risks impinge at the level of whole communities or populations.

These relationships, in their internal interactive dynamics and their scale, are therefore essentially ecological, so we must put aside our usual preoccupation with how specific discrete environmental exposures and personal behaviors can affect an *individual's* health. Climate change and its accompanying environmental changes require an ecological reframing of our understanding of how *groups and populations* are affected by and respond to these large-scale external influences.

Climate and the evolution of Homo sapiens

The world's climate is naturally restless; sometimes it changes dramatically, sometimes subtly. Natural influences on the climate system span many tens of thousands of years to periods of months. Over long time spans, successful species and ecosystems have evolved in response to changes in prevailing climatic-environmental conditions; but great extinctions and replacements have occurred sporadically because of climatic convulsions. Human societies depend on many climate-sensitive natural systems for their biological products (such as food, fiber, timber, and medicines) and for their environmental and stabilizing functions. Our well-being is closely linked, both directly and indirectly, to climatic conditions.

For 3.8 billion years, changes in the world's climate have been a driving force in the evolution of life on Earth. Around 5–10 million years ago, tectonic movements and global cooling and drying transformed the eastern and southern African landscape, and as forests receded, our primate ancestors ventured down from the trees. Under pressure to find new sources of food and ways of acquiring it, our ancestors evolved into upright-walking animals able to gather and, later, hunt for food in the open savannah and woodland. As the *Homo* genus emerged around 2.5 million years ago, legs lengthened, jaws and teeth changed, thumb and finger became able to grasp, large bowels shrank, and brains grew larger and more complex.

As human anatomy continued to evolve, so too did cultural and tool-making habits. Our own species, anatomically modern *Homo sapiens*, emerged around 200,000 years ago. Fire and cooking had already been domesticated by earlier *Homo* species; stone blades and spear tips were being refined; cave painting, music making, and funeral rituals appeared within the past 40,000 years.

From around 17,000 years ago, as ice sheets receded and the world warmed by around 6°C over the next 6,000 years, life began to change for *Homo sapiens*, by now an accomplished toolmaker with a large brain and well-developed speech.

Taking a longer view, the most enduring influence of the climate on our species has been on the biological evolution of human body form (size, shape, skin pigmentation, and so on) and metabolism. These Darwinian adaptations typically evolved via natural selection over many thousands of years. For example, skin color lightened, the better to absorb the solar energy needed to synthesize Vitamin D as early farming communities from the eastern Mediterranean region gradually spread north, over about 5,000 years, into Europe's cooler climes.

Climatic changes also influenced the evolution of many organisms that infect humans. The specialized human body louse, for example, now the carrier of deadly typhus infection, evolved from the lineage of ape-infesting lice as sustained global cooling set in around 70,000 years ago, early in the most recent ice age.[37] Humans lacked thick ape-like body hair, but as temperatures fell our ancestors began wearing thicker fur garments, which, from a louse's viewpoint, made for an acceptable body-hair substitute. Via natural selection, a variant strain of lice evolved that thrived in this newly available niche, and so arose a successful new species that remains an unwelcome companion of *Homo sapiens*.

The past 11,000 years of the Holocene are merely the latest phase in the total climate experience of humans and predecessor hominid species. Climate change is therefore ingrained in our genes, bones, and many cultural practices. Relatively stable climatic conditions have prevailed (so far) during the Holocene, with century-to-century average global temperatures varying by no more than plus or minus 1°C. The stability in the climate conferred many advantages on human societies and their food production systems, and settled agrarian living slowly emerged around much of the inhabited world. Then came villages, towns, wind and water power, cities, trade, larger-scale warfare, empires, sea power, accelerating population growth, and industrialization.

During the Holocene, the influence of climate on human biology and health has occurred on relatively short timescales. Later chapters will describe the effects of year-to-year and decade-to-decade variations in weather and climate on nutrition, epidemics, physical trauma, community destruction and displacement, impoverishment, and warfare.

Past climates: Trends, cycles, convulsions

The extent, speed, and form of human-induced change in the world's climate do not replicate any past documented phase of natural climate change. Yet the past offers indicative estimates of the risks to health and survival, particularly the risks borne by low-income populations living in rural conditions that differ little in many ways from those of past agrarian-based societies.

Figure 1.3 shows the average temperatures for the northern hemisphere (where more complete climatic information is available). Brief temperature dips of 2°C–3°C were mostly due to atmospheric shrouding associated with major volcanic eruptions that typically lasted 5–10 years. Occasional extreme El Niño events have caused severe droughts and hotter (1°C–2°C) conditions, particularly in Australia, China, and South Asia, lasting for a decade or more.

Note that the longer-term natural variations, such as the Holocene Climatic Optimum (when El Niño conditions dominated much of the world's climate), the briefer Roman Warm, and the Little Ice Age, are much smaller than the human-driven warming of 4°C (the dashed line) that may occur by 2100.[38]

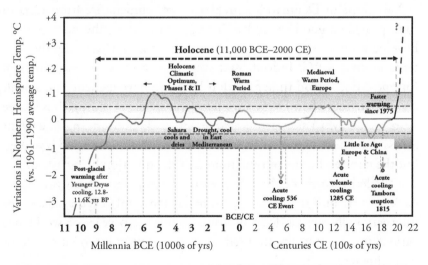

FIGURE 1.3 Variations in average Northern Hemisphere temperature during the past 12,000 years. Time on left (B.C.E.) is compressed tenfold relative to the 2000 years C.E. McMichael, "Insights from Past Millennia into Climatic Impacts on Human Health and Survival."[39]

Measuring past climates

The global "climate" is a mosaic of regional conditions. The many natural changes in climate caused by geological and cosmological forces, on many timescales, have always occurred unevenly around the world.[40] Over the past 4,000 years, the estimated temperature trends in China and Europe differed, particularly early on when China was apparently about 1°C warmer than Europe. From around 850 to 1100 C.E., the northern hemisphere was generally warmer than the southern hemisphere. In contrast, in South America and Australasia a warm period occurred from around 1160 to 1370 C.E.—while colder regional climates were emerging in the Arctic, Europe, and Asia between 1200 and 1500.

To understand what the past experiences of climate change might mean for future human populations, we should ask, first, what are the most appropriate measures of past excursions in temperature, rainfall, and other aspects of climate, and second, what units of time are most relevant?

Changes in *average* northern hemisphere temperatures have occurred over hundreds of thousands of years. But this broad-brush picture masks both regional and shorter-term temperature variations.

Variations in Denmark's annual average temperature from 1870 to 1990 provide a good example. During 1933 to 1941, the first six years hovered around 8.3°C, whereas the years 1939–1941 were tightly clustered around 6.3°C. Indeed, 1941 was the century's coldest winter in Europe—not a good time for Adolf Hitler to have launched an assault on Russia. Yet, when Denmark's temperature chart during those same years is redrawn using a rolling average, a slight warming is evident across that entire nine-year segment of time.[41] Neither version is wrong, but the choice of units of time must suit the question being asked.

Detailed records of directly measured temperatures do not reach back very far into the past; thermometers have only been available for internationally standardized use over the past 150 years or so. Hence various proxy measures are used to piece together information about past climates.

These include written records in literate societies and estimates of temperature and rainfall from tree-ring widths, cave speleothems (e.g., stalactites), corals, lake and seabed sediments, and ice cores. Ice-core measurements of oxygen isotope ratios can reach back hundreds of thousands and, at lower resolution, millions of years. Microscopic analyses of chemical impurities and growth bands in tiny planktonic shells can provide estimates extending back a remarkable 100 million years.[42] At a finer scale, modern proxy measurement techniques have made it possible to reconstruct month-by-month rainfall in the Mesoamerican Mayan civilization reaching back 1,500 years. Paleoclimatology is now able to fill in vast gaps in our knowledge, although not for every part of the world.[43] The two regions with reasonably continuous written records of temperature extending back beyond the past two centuries are Europe and China.

Information about annual and seasonal weather patterns was not kept systematically in most of Europe until fourteenth-century parish-based records emerged.[44] In China, systematic observational records of climate and weather began about the same time.

Human ingenuity has added colorful depth to this scouring of the past for climatic information. Scholars have obtained valuable information from nineteenth-century gentleman-naturalists' diaries, parish records, monastic documents, and medieval Arabic writings.

The Anthropocene: Oversized footprints

Late in the eighteenth century, industry was about to take a mighty step forward. James Watt's refinements of steam engine technology dramatically increased engine efficiency, power, and potential applications. The subsequent harnessing of heat from coal combustion to steam power opened up new vistas of industrial production and transport. The rich dividends of empire would further supplement this material gain. The combined impact of a large and growing population, human technological ingenuity, an energy bonanza, and intensified exploitation and disruption of the natural world was signaling a shift from the Holocene, and by some accounts the emergence over the last 200 years of the *Anthropocene*, the Age of Humans.[45,46] And with it came the risks that arise from disruption of the workings of the biosphere.

This new geological epoch, extraordinarily, is dominated by just one species.[47] Human numbers have multiplied, forests have been cleared, farmland irrigated, coal burned, and carbon dioxide released in, literally, industrial quantities. Humans and their livestock now account for almost 98 percent of the total vertebrate biomass on Earth.[48] This shift has

ancient roots, argues paleogeologist Andrew Glikson; human domination was kindled by the mastery of fire from around 1.5 million years ago,
culminating now in "a global oxidation event on a geological scale, a rise
in entropy [disorder] in nature and the sixth mass extinction of species."[49]
Humans first exploited fossil fuel for cave-warming and cooking, then for
hunting megafauna, clearing forests, raiding and warfare, early smelting,
and eventually coal-powered industry and oil-powered transport.

Darwinian evolution selects for randomly variant offspring best
able to survive and reproduce in the prevailing, though ever-change-
able, environment. But the mastery of fire and flint and the associated
development of a large brain conferred extra human power over the
environment and other species. The logic of Darwinian evolution was
thus fractured by an animal species that could substantially reshape
and disrupt its environment, both deliberately and inadvertently.

Human footprints, population growth, and deficit budgets

In his presciently titled book *Man as a Geological Agent*, written almost
a century ago, Robert Sherlock notes "the possibility of a considerable increase in the amount of carbon dioxide in the atmosphere as
the result of the burning of fuel, and the probable effect on climate of
such an increase, if it occurs."[50] The effect, he suggests, "is likely to be
in some degree inimical to the higher animals, and therefore to favor
lower forms of life in the 'struggle for existence'; and also to raise the
average temperature of the earth." Humankind is now using up Earth's
capacity to supply, regenerate, restore, and absorb our effluent much
faster than the planet can keep up with. To an increasing extent we are
living off nature's capital rather than doing as all other species must
do—live off nature's dividends.[51] The evidence of damage and loss is
now extensive: depleted fisheries, rapidly vanishing species, half-empty
aquifers, degraded soils, vast losses of topsoil from land to sea as river
silt, and shrinking forests. For example, land clearance and its consequences are complex. A long-standing feature of human impact over
the last several thousand years, it may well have caused marginal increases in greenhouse warming well before the industrial revolution.[52]

The topic of population growth is complex and politically sensitive.
Growth rates, and the environmental impact of different populations,
occurs unevenly between regions of the world. Public discussion of

climatic and environmental concerns typically skirt nervously around the "population" variable in the equation. Who has the right, and who has the will, to propose a coordinated international strategy to limit the number of humans pressing on the natural world and on its climate? The Royal Society in 2012 stated that "rapid and widespread changes in the world's human population, coupled with unprecedented levels of consumption, present profound challenges to human health and well-being, and the natural environment."[53] Here, unusually, was a statement about human effects on the natural environment that explicitly invoked the problem of population growth and the consequences for population health.

During the twentieth century, the world population quadrupled to 6 billion, a rapid surge without precedent. The level of per-person economic activity, globally averaged, grew at a similar rate last century, driven primarily by the burgeoning production and consumption demands of the world's wealthy countries. This Great Acceleration[54] is now turbocharged by the development of large middle-income countries such as China, India, Brazil, and Mexico.

Ten thousand years ago, as early farming emerged, the estimated total population was around five million. Numbers grew to a total of nearly 200 million at the B.C.E./C.E. junction about 2,000 years ago. This number then doubled to almost 500 million during the next 1,500 years and has escalated 15-fold in the subsequent 500 years to 7.5 billion by 2014. By 2100, according to the UN's medium-scenario projection, our great(-great)-grandchildren will have 11.2 billion contemporaries.[55]

The oversized *global environmental footprint* of humankind is thus growing even larger.[56,57] Total human demands first exceeded Earth's "biocapacity" around 1980. Today that demand requires the equivalent biocapacity of 1.5 Planet Earths to feed, provide materials, regenerate, self-replenish, and absorb wastes.

Most of the recent growth in per-person energy use comes from the huge energy subsidy extracted from fossil fuels. While early farmers each used around 20 megajoules of energy (physical labor) daily, the average North American now operates daily on at least 1,000 megajoules, and the current global average is around one quarter of that.[58,59] Today, with 1,400 times as many people on Earth as in the early

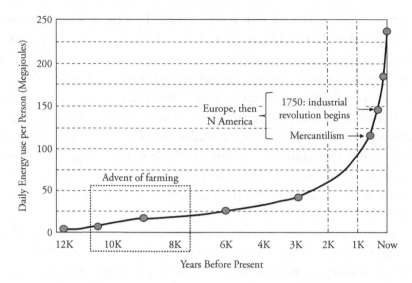

FIGURE 1.4 Changes in globally averaged daily energy use per person during the Holocene. Per-person energy use has increased around 16-fold from the early farming period through to the present. Based on data from several sources. Smil, *Energy in World History*; Ponting, *A New Green History of the World*; Boyden, *Biology of Civilisation*.[60,61,62]

farming days, total global energy use has increased almost 20,000-fold (Figure 1.4).

Earth's operating system

Systems science refers to a mode of scientific thought and inquiry that tackles the nature and workings of complex systems as a whole, seeking to understand how the fully assembled complex works.[63] The fact that most of the natural world is made up of geophysical and ecological systems has important consequences for scientific research methods and for gaining insights from modeling a system's dynamic workings. Real-world processes are much more complex and changeable than has been assumed within the imposing edifice of reductionist science. System behavior is less amenable to exact description and measurement, and behavior under future unfamiliar conditions cannot be confidently estimated. Yet crucial policy decisions such as climate change abatement must accommodate this complexity and uncertainty.

The recent emergence of Earth System Science draws heavily on systems theory, developed from studying the self-stabilizing behavior of biological organisms.[64] Ecologists soon embraced the idea,[65] and systems theory now is widely applied to studying and managing many real-world processes. The relationships between climate, agriculture, and population health offers a simple example. Climatic conditions affect harvest yields, and harvest yields affect food prices and human nutrition. A decline in harvests due to climatic stress leads to the use of more nitrogenous fertilizer, which results in increased soil emission of nitrous oxide, a major greenhouse gas. Warming increases a little and yields decline a bit further. Additional forested land may then be cleared to increase crop acreage, but that releases a pulse of carbon dioxide into the atmosphere from burned or decomposing trees. Meanwhile, an undernourished rural farming population works less efficiently, and yields fall yet further—and health and livelihoods are jeopardized.

Life-supports: The ecological framework

The systems that support and protect life include the climate system, the ultraviolet-filtering ozone layer in the stratosphere, the circulation and productivity of oceans, the growth of plants, the vitality of forests, the hydrological cycle and its multi-function wetlands, and the regeneration of soils.

Human population health has always depended on the continued functioning and productivity of these systems—systems and relationships that are best understood within what Stephen Boyden calls a *biohistorical* context.[66]

Humankind has done well for itself over the past 150 years, although with very unevenly shared benefits. For example, life expectancy in much of the world (particularly the wealthier parts) has doubled, but in some regions (such as sub-Saharan Africa) improvements have been modest and subject to the historic threats of war, famine, and epidemic infection. The gain in human population health globally has been made at the expense of the environment, and we now face an unexpected and daunting bill.[67] The techno-optimists may protest that we are smart enough to capture vast new amounts of solar energy. We could then use this energy to desalinate seawater, to power huge new factories that synthesize artificial protein, grow genetically modified plants, and supply all our transport and building needs. Might not a techno-garden future be more secure and microbe-free than nature? Well ... perhaps. But, feasibility aside, time alone is not on the side of developing that type of high-tech Brave New World. As global environmental conditions change rapidly, scientists and policymakers must find solutions in this century if Earth's life-supporting capacity is to be sustained.

It is the complexity of those human-impact paths of climate change, the scale of the impacts and their protracted likely future, that make two things clear. First, this is a *mega*-problem of disordered human ecology, of a disordered relationship with the natural world upon which our health, cultures, and societies depend. Second, its extraordinary dimensions and hydra-headed threats to human populations necessitate an extraordinary response by communities and their governments. We now *also* face an unfamiliar and qualitatively different type of hazard arising from changes to the biosphere's complex life-supporting systems. In 2009, a pioneering study tackled the nature of these threats and framed its conclusions in terms of crossing nine critical "planetary boundaries."[68] Three changes were deemed to have already transgressed the probable safe level for humankind's future: loss of biodiversity, disruption of the global nitrogen cycle, and global climate change. (The other systemic changes are depletion of stratospheric ozone, aerosol accumulation in the lower atmosphere, interference with the phosphorus cycle [part of a boundary with the nitrogen cycle], acidification of oceans, landuse change and degradation of arable land, over-consumption of freshwater supplies, and spread of chemical pollutants.) To ensure operating space on the planet for future generations, the global population must reduce its excessive pressures on the global environment. Yet sufficient resource and energy "space" must be available to enable low-income countries to achieve satisfactory material and social development.

This crossing of safe planetary boundaries comes at an awkward time in relation to the now-urgent programs of achieving development and reducing population health problems in the world's low-income countries.

In summary, climate change is part of the wider-ranging *syndrome* of human-induced changes to the Earth system that are the dominant environmental signature of the Anthropocene. Although climate change sometimes acts on its own, more often its impact is influenced or supplemented by concurrent stresses from other systemic environmental changes. For example, the effect of water shortage on crop growth varies with temperature and with soil nitrogen levels. Loss of biodiversity, nutrient overload of the nitrogen cycle, and ocean acidification are examples of large-scale disturbances that act in concert with climate change to undermine the ecological foundations of human health and well-being.

The Goldilocks Zone and the Faustian bargain

Humans, like all life-forms, thrive within a particular climatic range, a *Goldilocks Zone* in which climatic conditions are "just right." This is the range of climatic and environmental conditions within which they have evolved biologically and, in the case of humans, culturally also. Outside that range, biological impairment, dysfunction, and disorder occur. Crops fail if temperatures are too hot and dry, or too cold and wet. Mosquito survival and malaria transmission show similar responses. This "just-right" graph for the insect and microbial vectors that carry diseases that afflict humans is typically U-shaped, as shown in the two accompanying graphs (Figure 1.5) from the Pacific Islands.

If climatic conditions move well outside the range to which human bodies, social institutions, and species and systems within the natural biophysical environment are attuned, stresses will accrue. Indeed if conditions change substantially, the biological health, productivity, and even viability of individual species and of ecosystems and ecological processes can change quite rapidly.

Humans, buffered by cultural and behavioral options, can often tolerate largish changes in climatic conditions. But many of the species and ecosystems we depend on are more vulnerable to climatic shifts. This just-right phenomenon is analogous to the astronomers' *Goldilocks Zone*, the narrow "not too hot and not too cold, but just

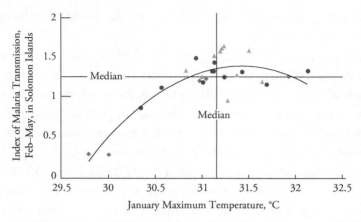

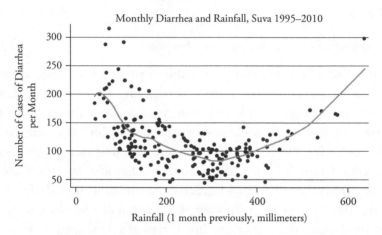

Note: Triangle indicates El Niño years, diamond is La Niña and dots are non-ENSO years.

FIGURE 1.5 Relationships between temperature and malaria cases in the Solomon Islands, and rainfall and diarrhea cases in Suva, Fiji. Abawi et al., "Relationship between El Nino-Southern Oscillation and the Incidence of Malaria in the Solomon Islands"; McIver et al., "Climate Change and Health In Fiji: Environmental Epidemiology of Infectious Diseases and Potential for Climate- Based Early Warning Systems."[69]
Permission received from Professor Yaha Abawi and from Dr. Lachlan McIver.

right" zone in which planets can support life via the existence of surface water.[70] The climatic comfort zone that sustains each particular society's food yields, water supplies, stability of infectious disease occurrence, and other basic needs is limited to a surprisingly small range of temperature and annual rainfall. Either cooling or warming of more than 1°–2°C relative to the long-term regional average and either excessive or deficient rainfall can disrupt those fine-tuned natural systems and threaten the mainspring supports for human health and survival.

In Goethe's masterpiece, *Faust*, the story of a man who offers his soul to the devil in exchange for a lifetime of favors that will confer knowledge, wealth, and power, Mephistopheles (the Devil) declares triumphantly:[71]

> *Now is he mine, without a saving clause,*
> *For fate has put a spirit in his breast*
> *That drives him madly on without a pause,*
> *And whose precipitate and rash behest*
> *O'erleaps the joys of Earth and natural laws.*

Those last two lines sound uncomfortably familiar. Viewed within this frame, human-induced climate change is a late-stage symptom of the playing out of that ancient bargain; the industrialized world has indeed *overleaped* the limits of the natural world, and is moving away from the Goldilocks Zone.

Living in a warmer and climatically less erratic Holocene world, the early agrarians' ways of achieving food sufficiency and more secure and comfortable lives gradually, though unforeseeably, led to accelerating increases in population growth, acquisition of property, accrual of wealth, co-option of energy sources, and the power of technologies.

Our species has become a geological force in its own right, and the Faustian bargain appears to be turning sour as overheated conditions loom menacingly. It is time we took up the option of renegotiating the deal—this time on terms acceptable to Nature.

2

A Restless Climate

INSTRUCTIONS ACCOMPANYING NEW DOMESTIC devices are tedious. So, uninstructed, we plunge into assembling the device . . . and the likely result is all too familiar. By analogy, a basic understanding of the climate system and the forces influencing it will shed more light on later chapters. The climate system has many interacting parts, encompassing the linkages between atmosphere, oceans, land, and ice surfaces. The atmosphere and oceans are the prime global distributors of that part of incoming solar energy that Earth retains in the form of heat, much of which is then re-expressed as water vapor, wind, and ocean currents.

As part of the Earth system, the world's climate is always changing. The internal dynamics of the climate system are complex and region-ally distinctive, and include, on shorter timescales, chaotic behavior. Meanwhile, larger forces are at work. As continents coalesce and then drift apart; as massive mountain ranges get pressed skyward; as Earth's elliptical orbit around the sun alternates between greater and lesser ro-tundity; as both the tilt and wobble of the planet's axis vary; as solar sunspots come and go; as fluctuations occur in the great ocean-based regional climatic cycles (the Pacific's El Niño Southern Oscillation, the Indian Ocean Dipole Oscillation, the North Atlantic Inter-Decadal Oscillation, and others); and as volcanic eruptions enshroud the lower atmosphere—so the world's climate varies on time-scales ranging from tens of millions of years to just several years.

The main engine of the climate system, the atmosphere, is made up of many local circulation subsystems, often interacting with (or "coupled" with) the oceans. At a regional scale, different combinations of local circulation systems account for changes in climate, such as the decline in rainfall in Mesopotamia during the third millennium B.C.E., or the southward encroachment of the Arctic polar vortex that imposed the Big Freeze on much of the northeastern USA in the 2013–2014 winter. Although the complex details of climate dynamics are not needed for the purpose of this book, familiarity with major subsystems—such as the El Niño Southern Oscillation, the North Atlantic Oscillation, the seasonal monsoons, and the prevailing winds in different latitude zones—sharpens our understanding of what our forebears were dealing with, what we will have to deal with, and why the human impacts do not occur evenly around the world. Familiarity with the long-term planetary orbital and rotational influences on climate, and shorter-term influences of solar activity and volcanic activity, will help as well.

What distinguishes "climate" from "weather"? A region's climate is the long-term average of its short-term weather patterns. It is said that climate is what you expect and weather is what you get: we may expect a hot summer, and get a day with afternoon thunderstorms. Weather tends to involve atmospheric changes over days or months and includes rainfall, temperatures, wind speed, and air pressure. Climatologists often define a region's climate as the average of these weather patterns over a 30-year period. Changes in climate can be studied between generations, or over many thousands of years.

In the beginning . . .

Earth's climate has been on a roller-coaster ride since the planet first coalesced from cosmic debris 4.6 billion years ago. Mighty geological rifts and ruptures along with the more orderly cosmological cycles have ensured a restless global climate. From violent beginnings with an atmosphere dominated by volcanic eruptions, storms, and lightning, the global temperature has swung hugely and erratically; alternating periods of great warmth and icy glaciation have come and gone.

For the first two billion years, the atmosphere contained very little oxygen. Rudimentary self-replicating proto-life emerged around 3.8 billion years ago as novel molecules formed from simple unoxidized carbon compounds in the air and oceans. Among the random assemblage of molecules were, by chance, simple nucleic acids, the building blocks of genes. They may then have linked with amino acid molecules sprayed around by comet impacts,[1] creating a molecular milieu conducive to template-based self-replication.[2] Charles Darwin's "primordial filaments" were setting something momentous in motion.

As one thing led evolutionarily to another, primitive single-celled algae-like organisms eventually emerged as *anaerobes* sustained by energy acquired from reactive chemical sources such as iron and sulfur compounds. Unceasing selection pressure in favor of an ever more energy-efficient metabolism led to cellular organisms able to use solar energy for rudimentary photosynthesis. This was a signal event: these photosynthesizing organisms began emitting oxygen as a "waste" gas into the atmosphere. The *Great Oxygenation Event* began 2–3 billion years ago.[3] Anaerobes, faced with an oxygen toxicity crisis, receded, and the oxygen-powered aerobes began taking over. The substantial amount of methane in the atmosphere was oxidized to water and carbon dioxide, and without the powerful global heating effect of unchallenged methane, the Earth cooled dramatically.

The time-scales of geological change and biological evolution are vast. A billion years ago, the only aerobic organisms were single-celled algae and primitive bacteria. But multicellular venturers were on the horizon, and by 700 million years ago, as rising oxygen levels in watery environments enabled the evolution of oxygen-capturing muscle-powering proteins, multicellular creatures with moving parts appeared on life's stage.[4] Then, 540 million years ago, came the spectacular Cambrian Explosion, spawning a profusion of (to us) bizarre organisms.

By 400 million years ago, enough atmospheric oxygen (O_2) had been converted into ozone (O_3) to provide a stratospheric shield that minimized biologically damaging solar ultraviolet rays. It was now safe to come out of the water onto land. First came the plants, followed by fish-like vertebrate animals that ate the plants. Land surfaces were becoming carbon-rich, and 100 million years later much of that carbon was entombed in sedimentary layers, destined to become great stores of fossil solar energy: "bottled sunshine." Around 250 million years ago, small reptilian vertebrates began to thrive in warming conditions, and the dinosaur dynasty was soon founded. The rest of the timeline is fairly familiar to us.

Cosmological and geological influences on climate

Around 100 million years ago, the world of dinosaurs was 8°C–10°C hotter than now—not a place that primates like us could inhabit. The subsequent 65 million years of cooling constitute merely a tiny extension of the climate's long and sinuous historical journey.

Natural changes in Earth's climate have occurred on widely differing time spans ranging, at the high end, from hundreds of millions of years due to the momentous consequences of tectonic upheavals when continents collide to tens of thousands of years in response to varying combinations of cycles in Earth's orbital and axial geometry in relation to the sun. Climatic changes on progressively shorter time scales include those influenced by changes in the oceans' slow-moving conveyor-belt distribution of heated deep-ocean water from the equatorial Pacific regions[5], variations in solar flare activity (a relatively weak influence),[6] multiyear influences of regional circulatory oscillations such as the El Niño Southern Oscillation (ENSO), and most acutely, volcanic activity.

Some geologists and paleoclimatologists argue that the recent rise in world temperature is merely business as usual for a climatically ever-restless planet; it is not the work of humans.

But this "either-or" argument is spurious. The relevant question is: what *additional* changes to the climate are humans now superimposing on the ever-present natural background changes?

The workings of the climate system

The climate system, via atmosphere and oceans, redistributes solar energy: the sun provides and the climate distributes. The pioneering English physicist John Tyndall wrote in 1872: "The Earth and its atmosphere constitute a vast distilling apparatus in which the equatorial ocean plays the part of the boiler and the chill regions of the poles the part of the condenser."[7]

As the sun warms our planet, some of the initially absorbed solar energy is reradiated outward from Earth's surface as infrared radiation. As it passes back through the atmosphere, it must run the gauntlet of the lower atmosphere's naturally occurring "greenhouse" gases, predominantly carbon dioxide, which capture some of that energy—which then manifests as heat at Earth's surface. Of the total retained heat, 90 percent is absorbed by the oceans, around 7 percent by land and ice sheets, and just 3 percent by the lower atmosphere.

The equatorial regions are more exposed to incoming solar radiation than are the polar regions. The resulting temperature gradient between equator and poles drives the redistribution of heat-energy by the atmosphere's circulation system and the ocean's deep currents. This circulation system has many interacting, or "coupled," parts. In addition to the many shorter-term regional circulatory oscillations, larger cycles operate in the background with durations that range from several centuries to several millennia, such as the 200-year DeVries/Suess cycle and the 2,300-year Hallstatt cycle. Deeper in the background and longer in time span are the three influential Milankovitch planetary axial and orbital cycles (Figure 2.1). Since *all* these cycles are active concurrently, the prevailing regional climatic conditions at any one time depend on the particular transient configuration of the cycles.[8]

The Milankovitch cycles were identified and described geometrically and mathematically by Serbian astronomer Milutin Milanković

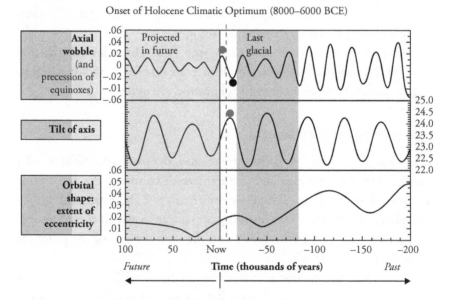

FIGURE 2.1 The Milankovitch cycles in Earth's orbital and axial geometry. Adapted from Wikimedia Commons graphs, with annotations and dashed lines by author.[9]

almost a century ago (the spelling of the cycles has been Anglicized). The three cycles involve:

- The shape of Earth's annual orbit around the sun, which varies from more to less elliptical on an approximately 100,000-year cycle. Currently, the orbit is at its most circular, which minimizes differences between the two hemispheres in season-specific temperature.
- Earth's axis of rotation relative to its orbital plane, which varies from more to less oblique on an approximately 41,000-year cycle.
- The 23,000-year cycle as Earth wobbles around its current axial tilt.

Many combinations of these three cycles are possible, as with the many combinations of the hour, minute, and second hands of a clock. The ever-changing combinations of Milankovitch cycles influence Earth's climate on a scale of thousands to hundreds of thousands of years. For example, reflecting the wobble/precession cycle, currently the northern *winter* occurs when Earth, on its annual circuit, is at its closest to the sun. Yet 13,000 years ago, the northern *summer* occurred when Earth was closest to the sun; summers were then warmer and winters colder.

The dark dot in Figure 2.1 shows the wobble index bottoming around 13,000 years ago. Such changes, Milanković argued, influenced the great fluctuations of ice-sheets and glaciations. Different combinations create different climatic conditions: e.g., the onset of the northern hemisphere's Holocene Climatic Optimum 8,000 years ago occurred when axial tilt was maximal (24°), amplified by a high positive "wobble" (lighter-shaded dots)—and peak summer sun exposure occurred at high northern latitude.

Climate system: Circulation patterns, cycles, oscillations, and winds

As Earth's surface spins towards the east, its passengers "travel" with it, leaving the sun behind to set in the west. People at the wide-girthed equator travel much faster than those in the polar regions. Sitting in Canberra, Australia, at 34°S, while writing this book, I am traveling, though imperceptibly, at around 1,200 kilometers per hour.

The atmosphere, while also spinning, does not quite keep pace with Earth's spinning surface. As it lags behind, the surface travelers typically experience the air as "blowing" westward. At least that is the case in the

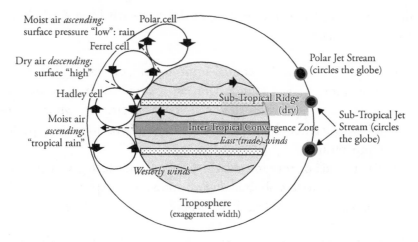

FIGURE 2.2 Schematic representation of the Hadley (three-cell) Circulation, plus main winds, and high- and low-pressure regions. These are the result of the combination of Earth's rotation and the equator-to-pole distribution of heat and moisture. The jet streams are fast-flowing winds high in Earth's atmosphere that influence the movement of cold and warm air masses that can then affect weather patterns. *Source:* Author.

faster-traveling equatorial region, where the stronger westward-heading winds, the *easterlies*, are the well-known *trade winds* (Figure 2.2).

But all is not as it might seem; at higher latitudes, the surface travelers encounter *westerlies*. That seems paradoxical: how can the lower atmosphere at 40°–50° latitude, north and south, travel east *faster* than the eastward-rotating land surface beneath? The explanation lies in the combined effect of the atmospheric redistribution of heat from equator to poles and the fact that Earth-bound people at 40°–50° latitude are traveling east more slowly than those at the larger-girth equator. So the heat-transporting air that is heading toward the poles arrives at around 40° latitude with its original equatorial faster-moving eastward momentum, and this is now *faster* than the eastward motion of the land surface at 40° latitude. So mid-latitude surface travelers experience a westerly wind blowing toward the east.

The architecture of that crucial latitudinal redistribution of heat from equator to poles is fascinating; it is called the Hadley Circulation.[10] The redistribution occurs via three successive cycling atmospheric cells (Figure 2.2): the Hadley, Ferrel, and Polar cells. The action of the cells resembles three cogwheels that mesh and turn. This Hadley circulation explains much that is relevant in later chapters, including the Inter-Tropical Convergence Zone (ITCZ) and the Sub-Tropical Ridges.

The air streams in the two adjoining (northern and southern hemispheres) Hadley cells each take up warm moisture from the tropical ocean surface as they traverse the bottom, surface-level, of their cyclical path, each heading towards the equator. They then converge at the ITCZ, where their moisture-laden air rises, cools, and releases copious rain. That now drier and higher air then travels away from the equator before descending at 30° latitude. Meanwhile, the Ferrel cell has been cycling in mirror-image fashion to the Hadley cell. The poleward stream of surface air sheds its ocean-derived moisture as it rises at the 60° juncture with the Polar cell, then the now drier and higher Ferrel air returns to 30° latitude and descends alongside the Hadley cell's dry air column. This convergent descent at 30° creates a meteorological "high-pressure zone" as the two columns of dry air press downward, hence the band of drier and desert landscape that circles the world at around 30° latitude in each hemisphere. Those bands are the Sub-Tropical Ridges; they each resemble a "girdle round the earth," Puck's fleet-footed flight path in Shakespeare's *Midsummer Night's Dream*.

Oscillating and seasonal systems

Spanning each of the world's main oceans is at least one major gradient of atmospheric pressure. Over periods of a half to several decades, each gradient "oscillates" as the ocean-atmosphere interaction varies. The main oscillations are the Pacific Decadal Oscillation and its widely influential component the El Niño Southern Oscillation (ENSO), the North Atlantic Oscillation (NAO) and Atlantic Multi-Decadal Oscillation, and the Indian Ocean Dipole (Figure 2.3).

The Indian Ocean Dipole (IOD) exerts considerable influence on the frequency of droughts and floods in countries on the rim of this ocean. Warming conditions in the western Indian Ocean early this century affected rainfall in Somalia and hence the severe drought and food crisis. Australia—for long a recognized front-line victim of the extreme ENSO phases—is also hostage to the phases of the IOD.

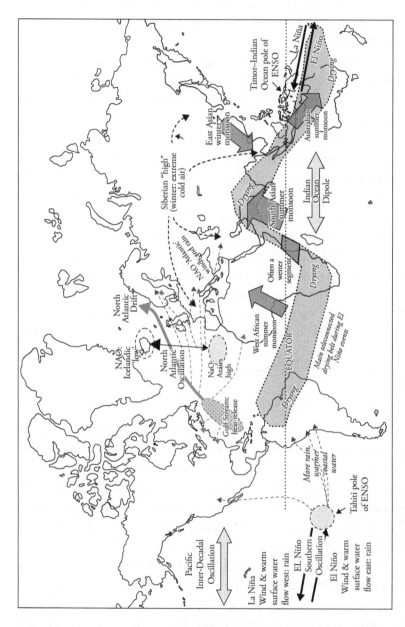

FIGURE 2.3 Map of world climate oscillations

Source: Author.

Indeed, under projected climate change conditions, that influence is likely to triple the frequency of droughts, bushfires, and floods by later this century.[11]

In 1890, Peruvian geographer Federico Pezet provided the first formal description of El Niño events, though these had been recognized (and named) many years previously by Peruvian fishermen and sailors.[12] The parent system, the El Niño Southern Oscillation, was named by British meteorologist Gilbert Walker in the early 1900s. Working in India with reams of archived meteorological records, Walker sought to impose order on the seemingly chaotic year-to-year vagaries of the South Asian monsoon. What became apparent was a larger climatic fluctuation that originated in the Pacific Ocean, the El Niño Southern Oscillation (ENSO), with its two extremes—the El Niño and La Niña phases.

ENSO and the North Atlantic Oscillation each affect climates far afield. These two great ocean-atmosphere oscillations, discussed in more detail below, have influenced many historical fluctuations in climatic conditions, with consequent impacts on human affairs and well-being.

Much is also going on within the heat-circulating engine of the oceans. The ocean-based North Atlantic Thermohaline Circulation is best known because of the bounty of warmth and moisture it confers on Europe via the Gulf Stream. In happy consequence, the west of Europe, the British Isles, and southwest Scandinavia enjoy conditions about 5°C warmer than Newfoundland, which lies on the same latitude on the other side of the Atlantic. Originating from the planet's giant solar panel, the Pacific Ocean, slow-moving deepwater currents redistribute heat around the world's oceans. One such current branches around the tip of South Africa into the Atlantic Ocean and surfaces in the Gulf of Mexico. That current of warm water then heads northeast toward the Arctic Ocean, releasing heat and moisture on the way—the celebrated Gulf Stream. The surface water thus becomes saltier, and further north, under the influence of the less dense meltwater from the Greenland ice-sheet, it sinks and begins its return journey south; hence the name "thermo-haline (heat-salt) circulation."

Another important climatic influence comes from the southern and northern "annular modes," the westerly wind belts that encircle the planet at latitudes of around 45°–60°. Early sailors crisscrossing the Atlantic knew to travel west with the low-latitude easterly trade winds and to return eastward with the mid-latitude westerlies. In the southern hemisphere, these westerly winds, sweeping unimpeded around the Southern Ocean, have long been called the Roaring Forties. In the imagination of nineteenth-century Europeans, they were the most famous of the southern winds;[13] French historian Jean-René Vanney described these great southern winds in aural terms, writing that in the forties latitude they roared, in the fifties they screamed, and in the sixties they whistled.[14] The changeable latitude of these belts influences the strength and track of cold fronts and storm systems, and hence rainfall. The Southern Annular Mode gets a clear run around the Southern Ocean, and its undulating shifts in latitude affect the low-pressure storm tracks that bring rain to southern Australia, New Zealand, and South America.

As the world warms, these annular airstreams are gradually displaced toward the poles. This is an increasing worry in southeast and southwest Australia, where germination of the wheat crop depends on the "normal" winter rains. In recent decades, there has been a 10–20 percent reduction in southern winter on-land rainfall. As this great southern girdle of wind and weather moves poleward, tightening its grip around Antarctica, the mid-continental subtropical dry zone is stretched further south, curtailing late autumn rainfall in southern Australia, southeastern Africa, and southern Chile.[15] Although rainfall is declining in some areas, note that the world's total annual rainfall rises as warming increases the evaporation of surface water—although some then remains in the warmer atmosphere as water vapor. The hydrological cycle will thus intensify with global warming, humidity will rise, and extreme rainfall events will increase.[16]

Meanwhile, always a Cinderella to temperature and rainfall in public discussion of climate, the world's wind systems are a central component of climate change. Changes in wind strength and direction reflect the natural cyclical changes in geographic temperature gradients as the angle of incoming solar radiation varies between less and more oblique in summer and winter, respectively. The monsoon winds in particular, and the moisture they bring, are critical to the annual rhythms of human life, agriculture, and nutrition in many countries.

El Niño and the North Atlantic Oscillation: Brief biographies

The Pacific-based El Niño Southern Oscillation is the great global engine of short-term fluctuations in regional climates. It affects most of the Pacific Rim countries and, more impressively, much of the low- to mid-latitude world.[17] The oscillating system is driven by a seesawing of the atmospheric pressure gradient across the Pacific Ocean between, approximately, Tahiti and Timor-Indonesia, typically on a five- to seven-year cycle.

During an El Niño phase, an atmospheric "low" off the coast of Peru allows warm surface water to heap up on that eastern side of the Pacific, producing moister air that dumps rain on land. Meanwhile, the higher pressure and cooler surface water in the western Pacific causes temporary lowering of the sea level, diminished rainfall, and cooler regional temperatures. During the La Niña phase, the process is essentially reversed. The temperature gradient influences wind strength and the flow of surface ocean waters in the equatorial Pacific—from west to east during the El Niño phase, and from east to west during the La Niña phase.

In early 1992, a cargo of many thousands of rubber duckies fell off a cargo ship from Hong Kong at around 45°N in the mid-Pacific Ocean.[18,19] Initially carried east by the sub-Arctic current and the westerly winds for a year or two, an estimated three-quarters of the duckies then veered south toward equatorial waters, before changing course again and sailing back westward. This reversal was probably partly due to the westward-flowing warm tropical surface current of the 1995–1997 La Niña. Oceanographers, interested to learn from this unplanned experimental study of ocean currents, offered rewards to beachcombers who handed in a washed-up rubber duckie. Such non-experimental studies typically must deal with a bit of background "noise" and, in accord with human nature, a number of claimants emerged with "beached" rubber duckies which, on inspection, looked to have been buffeted by no more than a few ripples in the domestic bathtub.

The near-global geographic reach of ENSO events reflects the huge uptake of solar heat by the expansive Pacific Ocean. The cycle regularly influences patterns of rainfall, floods, droughts, warming, and cooling in over 50 countries, in turn affecting food yields, water supplies, physical safety, and outbreaks of some mosquito-borne infections.

During the La Niña phase, warmer coastal waters, evaporation, cloud formation, and, hence, heavy rainfall impinge on eastern Australia and the western Pacific Islands. Meanwhile, on the Peruvian coast the outgoing westward flow of sea-surface waters draws nutrient-rich cold water to the surface from the deeper north-running Humboldt Current, nourishing Peru's prolific anchovy fisheries. A commercial fishing bonanza ensues. When the cycle reverses during an El Niño phase, the warm surface waters flow east, accompanied by moistened winds. Torrential rain then falls on the western slopes of the Andes, boosting food yields and local water security. A less welcome result, though, is that the piling up of warmer coastal surface waters causes the Peruvian fisheries to falter and, with occasional very strong El Niños, to fail.

Further afield, and extending far to the west, El Niño events lead to hotter, drier weather and drought conditions that often extend across a vast subtropical swathe of the globe, from eastern Australia through Southeast Asia to India, southern Africa, and on to the northeast of South America. But all is not drying and warming during an El Niño event; in equatorial eastern Africa and the south of North America, the rainfall increases and winters are cooler. During strong El Niño events, the eastern Mediterranean region and southern Europe are also affected, becoming cooler. In Egypt, the ancient records of the annual Nile floods show that during the past two millennia the flood levels have often mirrored the comings and goings of El Niño conditions, which reduce rainfall in the Horn of Africa and in highland Ethiopia and thus diminish the flow of the Blue Nile, which has its source in those highlands.

In contrast to ENSO, the atmospheric pressure gradient that defines the North Atlantic Oscillation lies on a north-south, meridional, axis. Its northern pole hovers over the Icelandic region at about 60°N and its southern pole lies over the Azores, west of Portugal, at 30°N. Its more usual *positive* state entails a strong pressure gradient between an Azores high and an Icelandic low. Unlike ENSO, the NAO neither oscillates with any regularity nor truly reverses; instead, a substantial weakening of this gradient constitutes the *negative* phase.

A *positive* NAO generates stronger westerly winds and cool summers and milder wet winters in Western and Central Europe, the eastern Mediterranean, and western Russia. This was important in the early rise of agrarian societies in Mesopotamia and Egypt; in prolonged positive mode, the NAO accounted substantially for the protracted warming of the Holocene Climatic Optimum. Along coastal southern Europe, however, rainfall declines during a positive North Atlantic Oscillation, while winters become milder in much of North America. In *negative* mode, the NAO causes west winds to weaken and Central and Western Europe to experience colder winters. Furthermore, the winds now track further south, and bring increased rainfall in southern Europe and North Africa.

Occasionally, when the temperature gradient between equator and polar region weakens, the increased Icelandic atmospheric pressure intensifies and the meandering subpolar jet stream slows. This can shut off the usual flow of warmer Gulf Stream air from the high-latitude North Atlantic into Europe and Scandinavia, allowing colder arctic air to be drawn south into Central and Western Europe. Lethally cold winters can result—as happened in early 2012, when several hundred deaths from extreme cold occurred among homeless people in Central Europe.

The influence of the North Atlantic Oscillation on the range and strength of the west winds is particularly important in the Middle East's winter months, when conditions are usually cool and wet, whereas summers are hot and dry. The viability of pastures and farming in this part of the world, where water sources are scarce, depends on the NAO-generated winter rainfall in Turkey, since this rainfall sustains various south-flowing rivers, particularly the great Tigris-Euphrates river system, and the adjoining agricultural lands. The NAO played a key role in extended periods of global cooling, drying, and climatic instability that occurred in the seventh, fourth, and second millennia B.C.E., and during the Little Ice Age in the past millennium.[20]

Regional monsoon winds

Daily land and sea breezes offer a simple analogy to seasonal monsoon winds. During the day, the land warms faster than the sea, while at night it cools more quickly; water has a higher thermal mass than do rock and soil. Hence, afternoon sea breezes occur as air flows from sea to land to fill the "space" vacated by the warmed air that rises from the day's well-warmed land surface. Conversely, overnight the land cools rapidly while warmed air rises above the still-warmed sea surface, and land breezes therefore flow from land to sea.

A summer monsoon is a sea breeze writ large. The static land surface becomes steadily warmer during summer months, while ocean currents limit the ocean's warming. Hence the air flow is from sea to land. Monsoons are often thought of as extreme rain events when the skies turn to sheets of water. In fact, they are major moisture-bearing *wind* events, responding to these seasonal temperature gradients between land and sea. The two largest monsoon systems, each a crucial source of seasonal rainfall for populous pastoral-agrarian regions, are the South Asian and West African summer monsoons. There are many other localized monsoons in the tropical and subtropical regions, occurring at differing seasonal junctures.

The history of regional sea trade with equatorial Singapore nicely illustrates this seasonal reversal of winds. Each year in the early nineteenth century, a flotilla of brightly painted large junks would arrive from the ports of southern China during December and January, bringing goods for Chinese New Year. During those months, the regional winds were blowing southward from the cold land mass of southern China toward the relatively warmer South China Sea and Bay of Bengal—the equivalent of a land breeze. The wind-dependent junks then waited for a month or two for the winds to reverse and carry them back home.

The South Asian summer monsoon originates from the Arabian Sea and Indian Ocean and blows to the northeast, carrying moist air over Iran, Pakistan, and much of India and Bangladesh. This happens during the summer months of July–September and contributes three-quarters of the total annual rainfall in much of the region; it is crucial for agriculture, livelihoods, and lives in Bangladesh, India, Nepal, and Pakistan. With warming, this monsoon system is likely to weaken and arrive a little later, and its rainfall will become more erratic.[21,22]

The profile of the West African summer monsoon resembles that of the South Asian monsoon. During the period from June to September, it blows from the southwest, originating in the Gulf of Guinea. It carries moisture inland around 10°–20°N, including much of the rain-dependent Sahel. Africa experiences marked summer-versus-winter contrast in wind and rain as the Inter-Tropical Convergence Zone oscillates north-south over the course of each year.

Prolonged shifts in that monsoon system were critical in determining the outcome of territorial rivalries in West Africa six to seven centuries ago, when the region had long been dominated by the powerful Kingdom of Mali. During the period 800–1700 C.E., the West African Sahel region (the "Mali region") experienced several different climate periods.[23] Although we cannot be certain, it seems that the first few centuries from 800 C.E. were relatively wet, followed by a progressively drier period during 1100–1450 and then reversion to a wetter period for several centuries.[24] Three considerations are important for understanding the relationship between climate and human populations in West Africa. First, rainfall, not temperature, is the critical determinant of food yields. Second, the main staples, millet and sorghum, are hardy crops well suited to dry conditions, though slow and low in yield. Third, the southern border of the savannah is the northern border of the tsetse fly habitat, the blood-feeding vector of the infectious trypanosome that causes sleeping sickness. If horses, cattle, and the humans that depended on them transgressed this southern climatic barrier, the risk of death from sleeping sickness was high. When that climatic barrier drifted 100–200 kilometers further south under weakened monsoon conditions, woodland and forest populations benefited by a boost in horsepower and hence military mobility and might. A further boost accrued from similar southward contraction of the malaria transmission zone.

During those nine centuries, West Africa's Sahelian empires of Ghana, Mali, and Songhai came and went, with considerable overlap of territory and some overlap in time as they thrived or declined. After the Ghana Empire (which bore little geographical resemblance to modern-day Ghana) collapsed in 1230 C.E., its successor, the Mali Empire, expanded and flourished until the mid-fourteenth century and then weakened until its demise in the late sixteenth century.

In the early fourteenth century, the Mali rulers built a grand palace in the strategically located city of Timbuktu, just north of the River Niger and on the cross-Sahara trade-route network. Meanwhile, the Sahara Desert was extending southward. With the desert came the

Arabic Tuareg camel pastoralists from the north, demanding national autonomy. In 1483, Timbuktu fell into Tuareg hands, consolidating the rising Islamic Songhai Empire based at nearby Gao, a little further east on the Niger. Benefiting from enhanced north-south trading routes during the wetter fifteenth and sixteenth centuries, the Songhai Empire endured until 1591.

Great floods occurred in southern Mali during that third, wet, period from late in the fifteenth century, including some that inundated Timbuktu. These floods reflected the failure of the summer monsoon rainfall to reach above latitude 12°N, instead dumping the rain at around 11°N in the headwaters of the River Niger in Guinea near the western coast. While episodes of severe flooding caused deaths and displacement in the southern Mali region, drought and famines increased in the north and continued throughout the seventeenth century, causing many northern Sahelian people to migrate south into moister climes.[25] Recurrent drought periods of up to two decades have recurred often in the Sahel over the past 500 years, but earlier ones did not cause as much drying and food shortage as have the droughts in recent decades.[26] Indeed, changes in Sahelian climate conditions since 1970 have caused the most dramatic long-term decline in rainfall recorded anywhere in the world over the past two centuries.

Cyclones (hurricanes), storms, and droughts

Tropical cyclones are a familiar phenomenon of the increasingly energetic climate system. Absorbing energy from warm tropical waters by the uptake and then condensation of evaporated surface water, a cyclone can generate gale-force winds that often extend hundreds of kilometers from the center ("eye") of the cyclone. A small increase in sea-surface temperature can greatly influence the intensity of the cyclone's wind speed and rainfall. For example, Cyclone (or "typhoon") Haiyan, which devastated the Philippines in late 2013, drew its energy from equatorial waters that were just 1–2 degrees warmer than usual.

Severe storms over land, including tornadoes and hailstorms, will be more likely in a warmer world with its heightened energy levels in the lower atmosphere. For example, in April 1999 in Australia, a costly storm occurred in Sydney's suburbs. It produced extraordinary hailstones of up to nine centimeters in diameter (tennis-ball size) and within one hour caused insurance losses of around $1.7 billion.[27]

And, finally, droughts: a topic that recurs throughout this book. Droughts are well understood in lay terms: no rain, drier conditions, and a threat to food harvests and river flows. They come and go, whereas aridity is a long-term state. Droughts come in three forms that are not mutually exclusive:

- *Meteorological* droughts, with below-normal rainfall and (often) warmer temperatures. Defined by threshold criteria, they do not necessarily affect food yield.
- *Agricultural* droughts, due to reduced rainfall (total or seasonal) and increased evaporation; these reduce crop and livestock yields.
- *Hydrological* droughts, when river flows or water storage falls below critical long-term levels. These droughts often evolve slowly, but may then resolve rapidly when "the dry" breaks.

Our interest, historically, will focus on agricultural droughts and their impacts on food yields and prices, and hence human health and social and political stability. Under global warming, climate models forecast more frequent extremely hot years, more variable rainfall, and shifts in some rainfall systems.[28] This may well increase the frequency and intensity of regional droughts and the area of land affected, although drought modeling remains an uncertain area.[29,30] The more permanent desertification of areas can happen surprisingly quickly. While the drying of the Sahara from around 6,000 years ago occurred in unhurried fashion over several centuries, regional changes in climate in the Dead Sea region in the early millennia of the Holocene led to desertification within decades.[31]

Emerging evidence suggests that longer-term drying is under way in southern Spain, not far north of the Sahara. Unwelcome surprises may be afoot in several populated and food-producing regions in the temperate and subtropical zones. The mega-droughts that occurred in the American southwest during the twelfth to fourteenth centuries adversely affected vegetation, water flows, and the pueblo-dwelling Anasazi peoples and their neighbors. That area is now a very fast-growing region of the United States, with a burgeoning demand for water—and if a mega-drought recurred in this region it would have disastrous environmental, economic, and social consequences.

The greenhouse effect

Incoming solar energy powers the Earth System, including winds, ocean currents, evaporation and rainfall, and plant photosynthesis, which produces organic material. Much of that organically captured energy was entombed in Earth's crust around 250 million years ago, when trillions of tonnes of dead vegetation, from forest giants to algae, were covered over by sediment to form what we refer to as (carbon-rich) fossil fuels—coal, oil, and gas.

Much of the incoming solar energy initially physically absorbed by Earth's surface as heat is subsequently reradiated out toward space as long-wave infrared radiation. But not all of it manages to escape Earth's clutches. Some is intercepted by *greenhouse gas* molecules in the lower atmosphere (troposphere) and then dispersed as heat, warming the lower atmosphere and Earth's surface.[32] The naturally occurring greenhouse gases include carbon dioxide (CO_2), water vapor, smaller amounts of nitrous oxide, and several potent short-lived "heat-and-run" compounds, particularly methane and ozone. Carbon dioxide is the most important greenhouse gas, despite accounting for less than one-thousandth of the total volume of air. Before the industrial revolution, its concentration was 270 parts per million (ppm); now it is over 400 ppm and rising—and is higher than at any previous time in the past 15 million years (about when our great ape lineage emerged).

Human activity has increased the atmospheric concentration of all these naturally occurring gases and has added several more, including industrial halocarbon gases and fine particulate black carbon (soot). The potency and longevity of these atmospheric additives vary greatly. The inert gas carbon dioxide is by far the longest-lasting and strongest determinant of long-term temperature change at Earth's surface, and is therefore a useful single-factor index of the lower atmosphere's global warming potential.

We Earthlings have been lucky.[33] The troposphere, courtesy of the heat-trapping by carbon dioxide and water vapor, acts as a blanket over the planet's surface. That natural greenhouse warming in today's world raises the surface temperature by 32°C. If Earth had no atmosphere, the average surface temperature would be *minus* 18°C and the planet would be a giant iceball. In contrast, Venus has a dense atmosphere of carbon dioxide and a surface temperature of 440°C. Earth's natural greenhouse effect, yielding a temperature compatible with the existence of surface water, makes the difference between life and no life—at least for life forms that we can realistically imagine. Now, though, human-kind's escalating emission of CO_2, compounded by the gradual decline in the "sink" capacity of the ocean to continue absorbing yet more of this excess CO_2, is the main cause of human-induced climate change.

The tight "Inconvenient Truth" relationship between time-trends in CO_2 levels and in temperature over the past million years has caused widespread dispute: which causes which? In fact, it is not a simple one-way matter of either causing or being caused by temperature change. The relationship is actually two-way: time trends in CO_2 and temperature mutually influence one another, and their relative dominance varies systematically over the different stages of each glacial-interglacial cycle.[34] Every 100,000 years (approximately), the Milankovitch cycles have initiated glaciation (a new ice age) by triggering the formation of ice-sheets, which reflect away more of the incoming solar energy. As the cooling oceans absorb more CO_2, the greenhouse effect is diminished, reinforcing the cooling. Meanwhile, as the configuration of Milankovitch cycles changes and the ice-sheets recede, the cooling processes weaken and the glacial episode ends.[35] Oceans then warm and release CO_2 back to the atmosphere (likewise, a neglected glass of cold beer goes flat as it warms), and this accelerates the postglacial warming. And so a new equilibrium state, a warmer interglacial, is temporarily established while the Milankovitch cycles regroup.

So here we are today, somewhere in the middle of an interglacial, but now adding a major new component—human-generated warming—to the mix. By burning carbon-rich fossil material to release its buried treasure, ancient stored solar energy, we have increased the amount of oxidized carbon (CO_2) circulating in the biosphere. This has been further supplemented by land-clearing and its release of terrestrial carbon from vegetation and soil. Since 1800 C.E., fossil fuel combustion and land-clearing have released over 500 billion tonnes of carbon, which, if expressed as a too-simple average over two centuries, equals 2–3 billion tonnes per year. In fact, for the past two decades our total annual carbon emissions have been around 8–9 billion tonnes, and rising. That figure is 200 times greater than the amount naturally released by land and seabed volcanic activity. That surely qualifies us for founder-membership of the Anthropocene.

Climate change science: What do we know?

To be uncertain is disagreeable; but to be certain is absurd
VOLTAIRE

For our ancestors, the sky was the realm of the gods, inaccessible to mere humans. Two hundred years ago, scientists would have scorned the idea that, by the late twentieth century, humans would have begun to change Earth's atmosphere and climate. Yet after 8,000 generations of *Homo sapiens*, today's generation is witnessing the now definite onset of human-induced changes in the composition, dynamics, and workings of the lower atmosphere.

Pioneers, climate science, and the Intergovernmental Panel on Climate Change

The first glimmerings of scientific insight into human influence on the climate date back to the 1820s, when Jean Baptiste Fourier, versatile French mathematician and physicist, first estimated the surface-warming effect of Earth's atmosphere.[36] In 1859, British physicist John Tyndall discovered experimentally that both carbon dioxide and water vapor absorb infrared radiation.[37] This radiation energy, when absorbed by atmospheric carbon dioxide and water vapor, warms the lower atmosphere and keeps Earth's surface warmer than it would otherwise be. Tyndall concluded that without this natural heat-trapping effect of the atmosphere, "the warmth of our fields and gardens would pour itself unrequited into space, and the sun would rise upon an island held fast in the iron grip of frost."[38]

The year 1859 was a bumper one for British science, being also the year in which Charles Darwin's *The Origin of Species* was published. But the connection between those two names runs much deeper than the calendar. John Tyndall was a leading light in the X Club, a mysteriously named group of scientists and thinkers who met and exchanged ideas regularly. They sought to promote scientific rationality over institutionalized religious doctrine, and they shared a close alignment with Darwin's ideas.[39] Indeed, Darwin's two strongest public advocates, Joseph Hooker (botanist) and Thomas Huxley (biologist and polymath), were founding members of this small but politically influential group.

Ironically, 1859 was also a bumper year for the future of cheap and densely stored energy. The first successful commercial oil well was drilled in Pennsylvania, in the eastern United States, and within four decades, rudimentary cars would be on the roads. Indeed, in 1896, near the Crystal Palace in London, Bridget Driscoll was the first pedestrian to be struck down and killed, by a car "speeding" at around seven kilometers/hour. The coroner said that he hoped such a dreadful thing would never happen again.

Late in the nineteenth century, Swedish scientist Svante Arrhenius estimated the likely global surface temperature rise for a doubling in atmospheric carbon dioxide concentration. His eventual best estimate was around 5°C, about double the figure of 2°C–3°C estimated by modern climate science.[40] English scientist-engineer Guy Callendar and several others were impressed by the greenhouse idea, and their research during the 1940s promoted new interest and gradual acceptance.[41,42]

In the 1960s, Charles Keeling identified, from continuous sampling of mountaintop air at his research site in Mauna Loa in Hawaii, a clear uptrend over time in atmospheric CO_2 concentration. Human activity was the suspected cause. These Mauna Loa data have iconic status in climate change science, with added luster conferred by former United States vice president Al Gore, who recounts that as a student in the 1970s, it was seeing Keeling's graph that galvanized his concern. In 2014, the Mauna Loa graph of annual CO_2 concentration reached 400 ppm, more than 40 percent higher than the preindustrial concentration of 275 ppm.

As a global warming trend became apparent during the latter 1970s, discussion about the world's climate trend focused increasingly on the rise in greenhouse gas emissions.[43] The UN Climate Conference, convened in 1985, concluded that further emissions of greenhouse gases would most probably cause significant future global warming. And so the radical step was taken to establish, within the UN framework, the Intergovernmental Panel on Climate Change (IPCC). This would be a science-based body of formal review and assessment of published research, drawing on international expertise from domains such as insurance, land management, and energy systems, in order to advise national governments on the science of climate change, its risks and impacts—while not recommending actual policies. This comprehensive and on-going assessment process, based on a series of major reports, began in 1988. A new model of formal international assessment of climate change science had been launched; so, too, had a high-flying target for climate change doubters and deniers.

Taking the world's temperature

Meaningful analyses of temperature trends must, say scientists, span several decades to minimize background "noise" from short-term, mostly natural, fluctuations. Earth's year-to-year temperature is influenced by the El Niño cycle, volcanic eruptions, and solar activity. Spurious conclusions about Earth's temperature trends can be drawn by selectively picking two single-year points, as shown by the two circles in Figure 2.4.

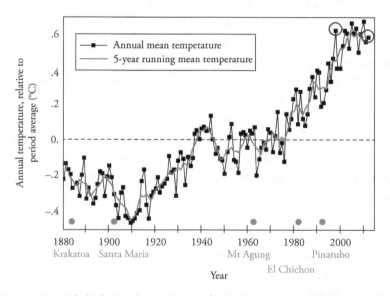

FIGURE 2.4 Global land-ocean surface temperature trend during 1880–2012, shown as deviations from the 1951–1980 average global temperature. Major volcanoes are shown as dots. The two circled years at top right illustrate how an inappropriate singling out of two years can be used to demonstrate either warming or cooling—in this case, the latter. Adapted by author from United States NASA, "Global Land-Ocean Temperature Index."[44]

Earth's average temperature has increased by around 0.7°C since the 1970s. As warming continues, that surface temperature will not increase in simple linear fashion—as is evident from the 130-year period 1880–2010 (Figure 2.4). There was, for example, an intervening period of slight cooling during 1945–1975 coinciding with reversals in the two main multidecadal oscillations in heat distribution (crudely characterized as release or storage) in the Pacific and North Atlantic Oceans.[45] The rapid postwar expansion of urban-industrial air pollution, obstructing incoming solar radiation, may also have contributed. The pause in surface warming since around 2000 (Figure 2.4)—a ready focus of controversy and confusion—is thought to have been mostly due to the influence of natural cooling phases in both the Pacific and Atlantic Decadal Oscillations, supplemented by stronger east-west winds, when more of the solar heat absorbed by Earth goes into the deeper ocean layers.[46,47,48]

The much more important question is how to explain the longer-term temperature rise since the mid-1970s. Analyses of the recorded information on time trends in global greenhouse gas concentration, along with information on volcanic eruptions, solar activity, and atmospheric aerosol-pollutant concentrations, show that the dominant influence is the rise in atmospheric greenhouse gases—for which humans are responsible. There are also corroborative "fingerprints" of greenhouse-induced warming, such as the greater warming at high latitudes versus low latitudes, over land versus sea, and in winter versus summer.

Impacts on the living world

Meanwhile, the biological world is also sending an overall signal. Literally hundreds of studies, worldwide, have noted recent climate-related changes in plants (flowering time, growth rates, changes in geographic range) and in animals (nesting, breeding, feeding, migrating).[49,50] Spring now arrives approximately two weeks earlier in temperate zones than it did several decades ago.[51] This has wide-ranging consequences for birds, bugs, bears, and buds—for example, the uncoupling of the timing of births and hatchings and the seasonal appearance of food sources. In bats, the earlier onset of spring results in an increased proportion of female offspring.[52] Many insect and spider species are changing their geographic range in order to keep within their climatic comfort zone. In response to warming in the United Kingdom over the past quarter century, spiders, butterflies, beetles, and grasshoppers have moved to higher altitudes at the average rate of eleven meters per decade, and to more northern latitudes at the rate of seventeen kilometers per decade.[53] Of course, no single observation of this kind proves that warming is causing these biotic changes, but the overall pattern from hundreds of observations is difficult to explain in any other way.

If average surface warming of 3°C–4°C occurs by the end of the century,[54,55] this will threaten the range and viability of many plant and animal species, especially as warming at high latitudes and high altitudes would be even greater. Most plant and many larger animal species could not keep up with that rate of warming, whether by migrating poleward or up mountain slopes. And where do you go after reaching the summit? In South Africa, there is great concern that the distinctive native fynbos vegetation in the southern region will disappear as it attempts to migrate further south—until confronted by the Southern Ocean. An estimated one-tenth of mammals in the Americas will be unable to outpace the rate of climate change caused by a mid to high scenario of future greenhouse gas emissions.[56] Many primate species, including tamarins, spider monkeys, marmosets, and howler monkeys, some of them already endangered, will be hard-pressed to reach a safe climatic haven.

One further important aspect of climate change, becoming evident around the world, is that warming influences the frequency, severity, and character of most types of extreme weather events.[57,58] The warming of oceans and atmosphere loads the dice in favor of warmer, wilder, and wetter extremes of weather. Recent unusually severe heat waves, now occurring every decade or so, will become commonplace by midcentury.[59] In Australia and Bangladesh, for example, the average annual number of very hot days has been increasing noticeably over the past four decades. In the western tropical Atlantic, the intensity, though not frequency, of cyclones (hurricanes) has increased a little.[60]

Projected warming this century

In the early 2000s, the uneasy consensus view between scientists and governments was that an average global temperature rise greater than 2°C above the 1990s level would take the planet into dangerous territory.[61] A decade later it has become increasingly probable, indeed near certain, that we will exceed that temperature as emissions have continued to rise.

Carbon dioxide is not a lone villain. Human-generated increases in other greenhouse pollutants (especially methane, nitrous oxide, and particulate black carbon[62]) have increased the total atmospheric *carbon dioxide–equivalent* (CO_2-e) concentration to around 440 ppm. These climate-active pollutants have added an extra one-third to the warming potential of carbon dioxide on their own.[63] The amount of that total mix that is already loaded into the atmosphere may suffice to push the global average temperature up by 2°C—perhaps more if there is no counteractive influence. In fact, the cooling effect of human-generated aerosols (predominantly sulfate particulates and fine dust) has been attenuating the total greenhouse warming effect by about one-third.[64] Early this century, the surge in sulfur emissions from China's and India's coal-fired power stations contributed hugely to this greenhouse masking effect by other air pollutants;[65] many other developing countries are also contributing.

Ironically, since industrial-urban air pollution causes almost one million deaths worldwide each year, there is pressure on governments to clean up their local air—thus reducing both the directly caused extra deaths *and* the greenhouse-masking effect of the air pollution.

As an approximate guide, an atmospheric carbon dioxide–equivalent level of 500 ppm during this century would carry a one-in-two risk of a global 3°C–4°C rise, while 600 ppm would imply a 4°C–5°C rise, and 700 ppm would cause even greater warming. Currently we appear to be heading for around 500–600 ppm by this century's end[66]—and a global temperature that has not existed for at least the last 20 million years. Beyond 2100, under so-called high-end emission scenarios (that essentially envisage a "business as usual" future), emissions and temperatures continue to rise. That would not just be dangerous territory; it would be alien territory, possibly, in large part, uninhabitable by humans.

Those estimates of future warming, in response to different plausible scenarios of emissions, come from carefully constructed and tested climate models. Estimating likely future temperatures many decades into the future cannot, of course, be done by gazing into crystal balls.

Instead, scientists must use models that draw on actual prior observed relationships and on understanding of how the climate system responds to particular physical and chemical changes in atmospheric composition—and which they then test against the measured past reality. For example, using the 1992 Mount Pinatubo volcanic eruption to test their climate model, NASA scientists found that it accurately predicted the observed global cooling caused by the brief shrouding by volcanic aerosols. Like all models, climate models simplify reality and make some assumptions. Indeed, our own mental models do likewise every day: using prior experience and observations ("data"), our brain quickly estimates the risk of dashing across the road before an oncoming bus—but we cannot easily take account of oil slicks on the road or whether the bus driver is likely to accelerate. As scientific knowledge accrues and as models evolve, so their predictive capacity improves.[67]

Feedback loops

Various feedback loops will occur in the future, especially if or when climatic conditions move beyond the bounds of recorded experience. There are both amplifying (or reinforcing) feedbacks and dampening (or restabilizing) feedbacks, and there are fast and slow feedbacks. On balance, feedbacks are likely to accelerate rather than slow the warming process.

First, then, the *amplifying* feedbacks. One of the best known is the "fast" amplifying feedback when surface warming causes Arctic sea-ice to melt. This ice loss diminishes Earth's albedo (reflectance), allowing more of the sun's incoming energy to be absorbed by the now darker Arctic Ocean surface—and as the ocean water warms further, the melting accelerates. On a longer timescale, as the oceans absorb more carbon dioxide from the air, their "sink" capacity gradually becomes saturated, CO_2 absorption slows, and greenhouse warming therefore increases. This is a "slow" amplifying feedback.

A more threatening feedback may result from ongoing warming causing the release of the vast stores of methane currently embedded in frozen tundra, permafrost, and polar seabed sediment. The amount of potentially mobile carbon locked up in these frozen northern landscapes is estimated to be at least twice the amount of carbon already in the world's atmosphere. Once triggered, the amplified greenhouse-driven warming would probably continue for centuries. Meanwhile, at lower latitudes, warming will increase the biological respiration of organic carbon compounds (from leaf litter and other sources) in the world's soils, and thus accelerate the release of soil carbon into the atmosphere. Already the annual amount of carbon released, actually recycled, by soils is around eight times greater than the amount of newly liberated carbon from fossil fuel combustion.[68]

On the other side of the feedback ledger are those that *dampen*, or restabilize. An example comes, again, from the world's great CO_2 sink. The oceans absorb around one-quarter of each additional unit of CO_2 released into the atmosphere. On land, when a region loses vegetation cover from clearing or drying, the now exposed bare land surface has an increased reflectance which reduces the region's surface warming. Many other, sometimes novel, feedbacks will also occur.

A final point: regional climates often behave in seemingly paradoxical ways. Explanations lie with the composite nature of the climate system and with shifts in the range and strength of various local circulation subsystems. Consider the following two examples.

By early this century, the West Antarctic, centered below South America and extending about halfway around the polar circumference, had *warmed* by 4°–5°C, whereas parts of East Antarctica had *cooled* by 2°–4°C.[69] Furthermore, while coastal sea-ice sheets were receding in the west they were increasing in the east. The likely explanations include geographic shifts in Antarctic snowfall associated with polar warming, higher-energy "Roaring Forties" westerlies around the Southern Ocean (increasing the upwelling of warmer deep waters), and an accelerated flow of fresh chilled meltwater from beneath the huge East Antarctic ice-sheet and into the sea. Global warming does not occur uniformly. Noticeable warming first started around the regions circling the Arctic and subtropical regions in both hemispheres—the largest accumulated warming to date is actually at the northern mid-latitudes. In some areas of the world, cooling has occurred. For example, from about 1910 to 1980, while the rest of the world was warming, some areas south of the equator, near the Andes, cooled until the mid-1990s.[70]

In early 2014, much of northeast America was subjected to massive snowfalls and potentially lethal sub-freezing temperatures of minus 20°–30°C, prolonged for several dire weeks. This was colder weather than most people in the region had ever experienced. How is this possible in a supposedly warming world? The relatively fast warming of the Arctic and loss of sea ice in recent decades, and the consequent weakening of the Arctic jet stream's function as a "meteorological fence," enables unusually large masses of freezing polar vortex air to breach those normal boundaries, escape southward, and cause protracted extremes of cold and snow[71,72]—a process that seems to have beset the north of Europe and America increasingly often as this century progresses.

Concluding remarks

The restlessness of the climate system reflects its role in redistributing the solar energy retained at Earth's surface and redistributing moisture. The climate, as a complex system, cannot be studied in piecemeal fashion. Nor can simple linear forecasts be made.

The concentration of carbon dioxide, the dominant greenhouse gas, is already in the range likely to cause a global 2°C rise within a half century. Meanwhile, annual greenhouse emissions continue to rise, foreshadowing a probable 3°–4°C rise by 2100. And in the political background, the economic and social costs of deferring effective action rise inexorably.

The next chapter explores ways in which climatic conditions affect human health and survival.

3

Climatic Choreography of Health and Disease

EVER SINCE HUMANS FIRST looked to the skies for relief, the changing mood of the climate has been assumed to be beyond human control, other than through supplication, ceremony, and ritual sacrifice. In secular modern times, there has been little interest in studying climatic influences on patterns of disease and survival. After all, we can curb cigarette smoking, but we cannot change the climate. Or so we thought. Now, though, there is new interest in understanding how human-driven climate change affects human health, in the present and into the future.

The risks to human health extend far beyond the well-known dangers from heat extremes, fires, floods, and mosquito proliferation, as signaled in Chapter 1. Changes in regional climates influence crop yields and livestock productivity and hence the occurrence of hunger, undernutrition, and stunted child development; they affect the ranges, seasonality, and rates of many infectious diseases. Heightened extremes of weather precipitate cholera outbreaks in impoverished crowded communities and, given the often destructive impacts of many events, can result in post-traumatic stress, long-term depression, and survivor guilt.[1,2]

The list goes on. There are both physical and mental health consequences of rural droughts and climate-exacerbated population displacement, migration, and resource conflicts. In the Canadian and Alaskan Arctic region, where 2°C warming has already occurred since 1950, the loss of coastal sea ice and permafrost is disrupting traditional Inuit hunting routines.[3] Without access to prey species such as seals and caribou, physical activity levels have decreased and the population's reliance on imported energy-dense processed foods has increased. Rising levels of obesity, cardiovascular disease, and type II diabetes have been the result.[4,5]

As climate change tightens its grip in coming decades, an increasing portion of all adverse health impacts is likely to result from indirect effects such as reduced food yields, depleted freshwater supplies, and loss of the physical protection provided by reefs, mangroves, and forests. In the past, moderately warmer periods in particular regions, spanning several centuries, often enhanced crop yields and population growth. But the world that lies ahead is on track to be much hotter by later this century than any period in recent times, far above the moderate warmth that may be associated with balmier living.

Climate change may initially confer some localized health gains, but the overall and longer-term outcome for population health everywhere is projected to be decidedly negative.[6] In at least the earlier stages of climate change, adverse health impacts are, and will be, greatest in poor and middle-income countries because of their relatively higher existing rates of disease (such as diarrheal disease and undernutrition), weaker health systems, and lower resilience.

A broad spectrum

Among the many paths by which a change in climate can influence health, those that are direct are clear enough: deaths from heat waves, heat stress and organ damage in very hot workplace settings, and exacerbation of urban air pollution by rising temperatures. Less well understood but more serious are the threats to population health and survival caused by disruptions of the biosphere's life-supporting system, especially as they affect food yields, water sufficiency, patterns of infectious diseases, and the stability of the physical environment. Furthermore, the economic and social consequences of environmental disruption and degradation often lead to job loss, impoverishment, migration, and perhaps violent conflict,[7] all of which are causes of injury, disease, undernutrition, misery, depression, and premature death.

Climate change is unlikely, in general, to spring surprises by causing *new* health disorders or diseases. However, novel infectious diseases could appear if cross-species microbial traffic increases in response to changes in the density and migration of animals, birds, insect vectors, and humans. Bat populations, for example, often change their roosting colony sites because of changes in climatic and feeding conditions. Bats harbor a myriad of coevolved viruses, some of which have found that our species is a congenial host for their replication, leading to diseases such as that caused by the Nipah virus.[8,9] The first documented outbreak of this infection occurred in Malaysia in 1998 during a severe El Niño event when the forests had been damaged by extreme fires, heat, and smoke that reduced wild-fruit yields.[10] Forest-feeding bats sought alternative sustenance from fruit orchards adjoining commercial pig farms, where they duly infected the omnivorous pigs via half-eaten fruit. The pigs then infected their human handlers, resulting in over 100 deaths.

Climate change often acts as a *risk multiplier*, amplifying existing health problems such as diarrheal disease, workplace heat-related disorders, and child stunting. The effects stretch beyond specific organ diseases; they include nutritional deficiencies, disorders of the immune system, physical trauma, and post-disaster mental trauma.

Heat extremes and health

Humans survive in very different climates, primarily via cultural adaptations: physical, social, and behavioral. Given time, biological evolution also plays a significant part. Hence the Inuit, Bedouin Arabs, and Ethiopians have adapted anatomically and physiologically, as well as culturally, to living in very different climate regimes. But there are limits. Just as car engines may stall in the depths of winter and overheat on very hot days, so there are limits to human physiological capacity to cope with extremes of cold and heat.

The graph of daily death rate against daily temperature in most temperate and prosperous countries is approximately U-shaped (Figure 3.1). There is a central comfort zone, with rising death rates on either side as temperatures get colder or hotter. Brisbane and Melbourne, two major Australian cities, each show such a relationship. However, because Brisbane residents live in a warmer northern climate their comfort zone is 2°C higher than that in cooler southerly Melbourne.[11] There are many variations on that relationship because of the local influences of topography, housing quality, age structure, and adequacy of emergency services.[12]

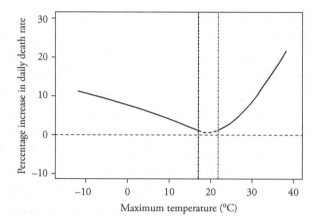

FIGURE 3.1 Graph of daily death rate in relation to maximum daily temperature, New York City. The comfort zone lies between 17°C and 22°C. The steeper rise in death rate with increasing heat versus increasing cold typifies the temperate-zone relationship. Li, et al., "Projections of Seasonal Patterns in Temperature-related Deaths for Manhattan, New York." [13]
Permission received from Nature Publishing.

The risk of death typically rises more steeply on the hot side as physiological coping capacities fail. In situations of extreme heat, maintaining the body's normal core temperature of 37°C relies principally on the evaporative cooling effect of sweat—and this will be less effective in a warmer and hence more humid world.[14] (Dogs, lacking hairless skin, must use their moist flapping tongue as an evaporative cooler.) The cause of the increase in death rate on the other side of the graph, as the daily temperature gets colder, appears more complex. It may be substantially due to nontemperature factors, including winter outbreaks of influenza, dietary changes affecting heart function, and health-impairing reductions in sunlight exposure and Vitamin D levels.[15,16,17] As temperatures rise during this century, deaths and other health adversity from increased heat exposure are projected to progressively overshadow any health gains from reduced cold exposure.[18]

As warming progresses, very hot episodes will occur more often—a trend that has emerged more clearly during this second decade. The frequency of extreme heat waves in temperate countries is likely to increase five- to ten-fold by 2050, and unless urban environments are radically renovated and climate-proofed, the consequent health risks will be heightened by the aging of populations and their clustering in unshaded suburbs reliant on air-conditioning. The conglomeration of concrete, asphalt, and masonry turns many urban environments into heat traps, or "urban heat islands," amplifying heat-wave temperatures by several degrees Celsius relative to outlying suburbs and adjoining countryside. Further, the huge thermal mass of densely settled inner urban environments retains the extra heat overnight; no relief for would-be sleepers.

The resulting health risks are greatest in the elderly, those with advanced underlying chronic diseases, and those most directly exposed—the homeless, people living in low-grade housing, and those working in excessively hot workplaces. Most heat-related deaths are due to heart attacks, strokes, and respiratory failure. As the living and working environments get hotter, underlying conditions such as high blood pressure, coronary artery disease, and chronic lung disease translate more readily into hospitalization and death. Marathon running provides an analog between 1981 and 2012, 11 runners died in the annual London Marathon from cardiovascular stress due to dehydration and serious core-body overheating. Due to high temperatures and humidity, only 70 percent of the elite competitors in the 2016 U.S. Olympic Team Trials Marathon in Los Angeles finished; it was the warmest day in Trials history.

In August 2003, Europe was gripped by an extreme ten-day heat wave that caused over 50,000 extra deaths[19]—particularly in France, where around one-third of the grain harvest was lost. In Paris, the death toll during this episode was almost 10 times higher than had occurred during a more typical summer heat wave a month earlier.[20] The August heat wave was hotter and longer, and was accompanied by increased exposures to airborne ozone, a known health hazard that forms faster at higher temperatures. Three years later, in California, an extreme two-week heat wave caused over 600 extra deaths and several thousand emergency calls and hospitalizations, and cost the state $5.4 billion.[21] Vast costs may await the next generation.

While heat waves get the headlines, the uptrend in average temperature has impacts, too. For example, the increase in average temperature in Stockholm between 1980 and 2009 caused an estimated several hundred more premature deaths than if warming had not occurred.[22] Among four million chronically ill people living in 135 United States cities, followed over time, death rates were consistently higher in years with above-average swings in summer temperature, particularly affecting those with diabetes, chronic lung disease, and heart disorders, as well as those more vulnerable because of poverty and living in urban regions lacking green space.[23]

Most public discussion about heat extremes refers to risks faced by the general community. Yet even greater extremes of heat exposure do and will occur in many occupational settings, posing special risks to health, behavior, and work capacity. Outdoor workers in construction, roadwork, or agriculture, and indoor workers in the hot and unventilated factory environments that are still common in tropical countries, are especially vulnerable.[24] Poorly paid cane-cutters in the hot and humid conditions of Nicaragua can lose up to 9 to 10 liters of water a day.[25] The health consequences include severe dehydration with damage to kidney function and eventual renal failure,[26] increases in physical accidents, and rash behavior. Furthermore, as heat stress rises, economic productivity falls.[27] In Southeast Asia in 2050, for example, more than half the afternoon work hours may be lost because of the need for rest breaks.[28] Indeed, projections indicate that much of the full-day workplace productivity around the world would be halved during the hotter months in a future world 4°C–6°C warmer than now.[29]

Humans do not suffer alone; plants and animals are also vulnerable to heat extremes. In such periods, birds may lose up to 5 percent of their body mass per hour and become dehydrated. In Western Australia in January 2009, a heat wave with air temperatures above 45°C killed thousands of birds, mostly zebra finches and budgerigars.[30] In past decades, heat waves, often combined with droughts, have caused mass deaths in Australian koalas, impaired reproduction in frogs, and brought cyanobacterial blooms in lakes and the spread of invasive species.[31,32,33,34]

Climate amplification of air pollution and aeroallergens

Pollution of the ground-level atmosphere has long been the trademark signature of industrialization and the proliferation of automobiles. Outdoor air pollution now causes over two million deaths each year, especially in the dense and highly polluted cities of East and South Asia.[35] The Chinese government was forced to close down local industry and inner-city traffic during the Beijing Olympic Games in 2008 in the hope of restoring breathable air for athletes and a blue sky for tourists.

Climatic conditions influence the formation and concentration of several outdoor air pollutants, especially ozone, which damages the respiratory and cardiovascular systems.[36,37] With wind assistance, the dispersal of newly formed ozone also damages crops in surrounding farmland. Nature, when overstimulated by temperature, humidity, and atmospheric carbon dioxide concentrations, increases the production and release of the pollens and spores that exacerbate respiratory allergies.[38,39] This presumably accounts for extensions of the *Betula* pollen season in Finland,[40] the intensified ragweed (*Ambrosia artemisifolia*) pollen season in the northern United States,[41] and the allergenic grass pollen season in Spain.[42]

The effects of climate change on rainfall patterns and wind speeds can amplify dust storms and wildfires and their vast palls of drifting dust and smoke. In Sydney, Australia, exposure to several dozen bushfire smoke episodes over a recent decade caused a 5–10 percent increase in nonaccidental deaths, especially heart and blood-vessel disease deaths. During the occasional blanketing of Sydney by severe dust storms, hospitalizations and deaths increase by around one-sixth.[43]

Weather extremes—is climate change the culprit?

Weather extremes and disasters are becoming more frequent in many regions and in some cases more damaging, as basic climate science has predicted.[44,45] Further, there has been no obvious trend in *geological* extreme events (earthquakes, tsunamis, and volcanic eruptions) for the past half century, whereas a strong uptrend has occurred in *meteorological-climatic* events: cyclones, floods, droughts, storms, and wildfires.[46] Climate scientists have become persuaded that this upturn is at least partly due to the background warming—that is, the greater energy content of the climate system.[47]

Can we estimate how much of the adverse health impact of such events is attributable to background climate change? Yes, we can, but only in terms of probabilities, although critics may not be satisfied by general statements such as "it is very likely that the recent increase in major flood events is due to climate change." For many people, the difficulty lies with the concept of the *probability* that more extreme events are occurring because of underlying human-induced warming. However, this is what is happening: natural extremes of weather will continue to occur, but an energy-laden climate system adds extra punch to many of those events. It is not an *either-or* issue. From now on, it is probable that the frequency or intensity of any category of weather extreme has been influenced by the warmer climate. Hence, for the next decade or two the human consequences of amplified extreme weather events are likely to be the most recognizable face of climate change's impact on deaths; injuries; diarrheal, skin, and respiratory infections; food shortages; and post-traumatic stress disorders.[48]

Climate, food yields, and water supplies

Food shortages, undernutrition, and starvation have been the great killers throughout agrarian history—literally the grim reaper.[49] Many of those crises and famines were triggered or amplified by natural climatic changes. Yields of food and flows of freshwater are highly dependent on temperature, rainfall, and other climatic conditions, and irrigation systems and water-flow management are crucial to food security in many parts of the world. Much of sub-Saharan Africa, where most agriculture still depends on rainfall, is already prone to sporadic food shortages and hence to illnesses, deaths, and civil conflict. That region's general vulnerability to future climate change is therefore very high.

But the causal chain gets more complex. Since food yields are affected by many other factors, including soil degradation, groundwater shortage, pest infestations, and the diversion of harvests into biofuel production, estimating what proportion of an observed change in food yield is due to regional climate change can only be approximate, even though the *fact* of that change occurring is of crucial importance to human health, social stability, and geopolitical security. As world food prices escalated during 2008, the World Food Program drew attention to the likely contribution of recent adverse climatic influences on food yields.[50] Subsequently, there has been a growing recognition that climate change and extreme weather events are affecting food yields in some regions[51,52,53] and that this threatens nutritional sufficiency, child development, general health and vitality, and, in the extreme, survival.[54] Yet most public discussion has been slow to connect the dots between climate change, food yields, and human health. Many economists, politicians, and food corporations still view agricultural production primarily as an economic entity rather than as something that humans must find or produce to ensure their health and survival. "Food" has become increasingly marketized.

The general connection between climate-related changes in regional or local food availability and human nutritional needs and health outcomes is understood generically, but quantifying it—especially in relation to future circumstances—is a challenge. The work of one research group, at the London School of Hygiene and Tropical Medicine, illustrates how we might grasp this nettle.[55] Using a multistep linked model, they estimate that for a medium greenhouse warming scenario there would be an additional 95,000 child deaths from undernutrition in 2030, compared to a world in which there was no climate change.[56] All such modeling must take account of the known or likely nonlinear relationships that characterize how plants and livestock respond to climate, many relationships being of "Goldilocks" form. Assumptions of straight-line relationships are often inappropriate. For example, when rainfall is either too low or too high, the concentration of trace elements and other minerals in soil tends to be low and plant growth is impaired. Dry soils cannot dissolve minerals from rocky formations, and very wet soils leach out much of the mineral content.[57] With these cautions noted, studies such as the assessment by the London School nevertheless provide policy-useful estimates of where trends and risks are heading—a cut above reliance on soothsayers or techno-triumphalist reassurances.

An impaired food supply follows upon reduced water sufficiency and quality. Freshwater is essential to health and survival, for many reasons beyond its fundamental influence on food yields. Water is essential for safe drinking, basic personal hygiene, sanitation, and wastewater management. Water in excess, as with flooding, will often contaminate local drinking water and vegetable gardens and exacerbate outbreaks of cholera, dysentery, and other diarrheal diseases, especially in poorer communities. Diarrheal diseases are ready targets of climate change, since currently a little over half of all diarrheal deaths are due to lack of adequate drinking water, sanitation, and hygiene.[58]

On a larger scale, cross-border tensions are likely if river flows recede as regional rainfall or mountain snowpack declines and if upstream water diversion reduces downstream supply. These adversarial situations can easily translate into hunger, conflict, displacement, and their many health consequences. In South Asia and Southeast Asia, many of the great rivers that flow off the northern slopes of the Himalayas are destined to have lower flows as mountain glaciers melt.[59] As national water shortages escalate, these rivers are liable to be dammed as they flow off the Tibetan plateau, disrupting downstream irrigation, river ecosystems, and fish yields. The long-standing seasonal rhythm in the Mekong River flow is crucial to the two-way inflow and outflow hydraulics and ecology of the remarkable Tonle Sap lake system in Cambodia,[60] the main source of the nation's freshwater fish catch.

Finally, the influence of climate on cholera has a further, intriguing dimension. Outbreaks in coastal Bangladeshi communities often occur during episodes of warming of coastal water.[61] The warmer water and its nutrient eutrophication by nitrogen- and phosphorus-enriched agricultural and wastewater runoff stimulate algal blooms. Algae serve as hosts for quiescent cholera bacteria in the natural watery ecosystem, and during algal blooms those bacteria are ingested by algae-feeding fish. As the cholera bacteria pass up the food chain, the larger infected fish then become hazardous food for coastal communities.

Infectious diseases

From personal experience, most of us know that bacterial food-poisoning is more likely to strike in hot summer months. Indeed, as for food-borne bacteria, so for the microbial world in general; proliferation typically increases in a warmer world. Some infections spread from person to person; others use the intermediary services of a "vector" organism such as mosquito, tick, or flea; and zoonotic diseases (*zoo* meaning "animal") spread from animal to human, either directly or via a vector, and usually without subsequent person-to-person transmission. In all such cases, temperature, rainfall, and humidity can affect the growth and proliferation of the infecting microbe, the pathogen itself, and where applicable the biology, longevity, and biting behavior of the vector organism. Cold-blooded insect vectors are very sensitive to temperature, and as they become dehydrated in dry conditions they must feed more often.

Among the vector-borne infectious diseases (VBDs) related to climate, malaria has iconic, if still debated, status. More on this topic later in the chapter, but in brief, when the temperature is warm enough for the malarial plasmodium to develop and multiply inside the mosquito gut, a further 1°C rise can halve the time required for maturation and hence readiness for onward transmission via the mosquito's bite. In principle, even light warming can speed up the cycle of malaria transmission and make a local outbreak of disease more likely.[62] Warming may cause a large increase in mosquito numbers and hence in the rate of transmission of other VBDs, including dengue (or "breakbone") fever, which is distinguished by high temperatures, rashes, and painful head and body aches. Dengue now occurs widely in tropical and subtropical regions. It is estimated that about 390 million people are infected with dengue virus each year, and roughly a quarter of these cases are symptomatic.[63] There has been a 30-fold increase in dengue since the 1960s in response to the combined influences of population growth, urbanization, increased international trade and travel, and climate change.[64,65]

In the case of zoonotic infections, abundance of the pathogen's natural host species is often influenced by climatic change. Variations in rainfall, temperature, and vegetation affect the density of reservoir-hosts, such as deer and forest mice for tick-borne Lyme disease in Europe and northeast North America, kangaroos and wallabies for Ross River Virus in Australia, and burrow-dwelling wild rodents for flea-transmitted bubonic plague in Asia and some other parts of the world. Without those natural hosts, the prevalence of infection in the wild declines, as does transmission to humans.

Many of the infections that spread between individuals without vector assistance are also sensitive to climatic conditions. Influenza viruses survive better in cooler air, and most flu epidemics occur in winter months. Rates of gastroenteritis (diarrheal disease) typically increase during warmer periods when *Salmonella* and other bacteria in food multiply more rapidly, as evident from recent multinational time-trends in salmonella gastroenteritis.[66,67] Diarrheal diseases, the curse of many travelers, look set to become more notorious as the world warms (and as antibiotic defenses become less effective). For example, in the northern summer of 2004 an outbreak of gastroenteritis from seafood contaminated with *Vibrio parahaemolyticus* bacteria occurred on a cruise ship off the Alaskan coast. Over the preceding decade, the summer-season coastal water had gradually become warmer, and in 2004 it reached a sustained temperature that enabled the bacteria to multiply in the oyster beds throughout the summer months.[68] Warming also contributes to ciguatera poisoning from eating reef fish with unusually high loads of a toxin associated with higher sea-surface temperatures, especially evident in Pacific Island communities.[69,70]

At lower latitudes, far from the Arctic, the common housefly *Musca domestica*—the culprit responsible for much fecal contamination of food—flourishes in warmer conditions. Studies in Botswana, in southern Africa, forecast substantial increases in housefly numbers in response to projected climatic changes in a medium- to low-emissions scenario.[71] Diarrheal disease would consequently rise. But not all gastrointestinal pathogens proliferate in warmer conditions. Infections with norovirus and rotavirus (an important childhood diarrheal infection) peak in winter and in spring, respectively. If that is a pure temperature effect, then those diseases may recede in a warmer world.

We come now to a key question: Has climate change yet affected infectious disease patterns? While the evidence for some diseases is equivocal, the geographic ranges of others have certainly changed alongside regional warming. Examples include malaria extending to higher altitudes in the eastern African highlands, discussed below, and changing patterns of tick-borne encephalitis (TBE) and Lyme disease.[72] In Sweden during the 1980s and 1990s, the ticks that transmit TBE extended progressively northward as winter temperatures rose,[73] although the sequence of warmth and spread is somewhat unclear.[74,75] In the Baltic Sea, where seawater has warmed since the early 1980s, the concentration of *Vibrio* bacteria has increased, as has the frequency of *Vibrio* infection outbreaks in coastal communities around the Baltic coast.[76]

Humans of course are not the only species at climatically altered risk of infectious disease. Blue-tongue virus, a deadly infection of livestock, and its midge vector have been extending northward in Europe in association with warmer temperatures.[77] More generally, both plant and livestock species will become more vulnerable to infection as climate change extends the range and increases the proliferation of many pathogens.

As understanding about these relationships grows, better biologically based models of disease transmission can be constructed to forecast likely changes in ranges and rates. In both Canada and Europe, for example, tick-borne Lyme disease is likely to spread further north.

Similar projections for central-eastern China indicate that warming during the coming half century will extend the transmission of the ancient disease schistosomiasis (called the "God of Plague" by Chairman Mao). Water snails, the natural intermediate host for schistosomiasis, cannot survive north of the midwinter freezing zone. The local reservoir-host animal for schistosomiasis (*Schistosoma japonica*) is the water buffalo, which excretes the early-stage pathogen in urine, from where it infects a water snail, develops to its next free-swimming adult life-stage, and subsequently reaches humans via skin contact. This ancient disease, which causes damage to internal organs, swollen bellies, wasting, and often death, remains a serious disease in rural China, although receding as tractors replace water buffalo.

Several aspects of this overview are important to bear in mind.

First, climatic conditions can influence infectious disease occurrence via both biological and social pathways. Both cooling and warming are important: cooling can slow pathogen proliferation but may increase opportunities for transmission, such as more time indoors for both humans and livestock, while warming can increase proliferation of bacteria, mosquitoes, and host species and influence social changes in contact patterns (e.g., swimming pool use) and behaviors such as outdoors eating.

Second, temperature is important, but so are rainfall, humidity, and winds. Colonial records from British India show that smallpox outbreaks occurred most often in hot and dry states.[78] In sub-Saharan Africa, the interyear pattern of summer rainfall affects rates of mosquito infection with West Nile virus and hence outbreaks in humans,[79] while in Venezuela malaria outbreaks tend to follow heavy rainfall events conducive to mosquito breeding and survival.[80] The role of wind in Kawasaki Disease in the north Pacific is particularly striking.

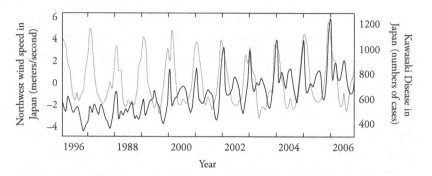

FIGURE 3.2 Westerly wind strength and rates of Kawasaki Disease in Japan, 1996–2006, by month. Based on Rodo et al. 2011.[81]

This mysterious infectious disease impairs the blood supply to heart muscle, mostly in children. Curiously, it occurs 20 times more frequently in Japan than in other countries, due possibly to the strong westerlies coming from southeast Russia and Mongolia.[82,83] These winds carry tiny dust particles to which infectious bacteria or viruses can attach for long-distance flights.[84] Overlaying the graph of month-by-month westerly wind strength on the monthly incidence of Kawasaki Disease cases in Japan shows an obvious correlation (Figure 3.2). This fits with observations that when another strong wind funnels northward from the southwest Pacific and, before veering east towards Canada, intersects with the incoming dust-bearing westerly wind, simultaneous outbreaks of Kawasaki Disease occur in Japan, Hawaii, and western Canada. One wind-stream seemingly infects the other.

As the climate continues to change, so too will wind patterns and the spread of microbes that are "blowing in the wind" or are carried aboard migrating birds.

Sea-level rise

During the postglacial thaw between 18,000 and 11,000 years ago, the seas rose by around 125 meters, the height of a 40-story building. The rate of rise early in this century has been around three centimeters per decade, having approximately doubled in the past hundred years because of the expansion of warming ocean water and, increasingly, the melting of land-based ice-sheets. Each now contributes about half the rise.

Because the redistribution of heat throughout the ocean's layers occurs slowly, even if we curb global warming in this century the seas will continue to rise for at least a thousand years.[85] The distant past provides a clue: around five million years ago, during the Pliocene when the world's prevailing temperature was about 2°C higher than today, the seas were around 10 meters higher than now.[86] Within a shorter time-frame, a possible warming of 4°C by 2100 could cause a sea-level rise of more than five meters over the next several centuries.[87]

Rising seas threaten food production, freshwater supplies, and physical safety in low-lying small island states and other coastal communities.[88] Rising seas combined with stronger winds will expose many coastal communities to severe storm surges and floods. India and neighboring Bangladesh, with densely populated, low-lying 7,000-km coastlines, are vulnerable. Many coastal rice fields are already threatened by inundation and salinization; a one-meter sea-level rise would flood almost half of Bangladesh's rice-growing area, jeopardizing food sufficiency and health. In June 2013, the low-lying Marshall Islands in the Pacific were hit by a disastrous king tide that displaced over 1,000 people, destroyed homes and crops, and left many people without livelihood. The Foreign Minister pleaded that his island state "will literally be wiped off the map sometime before the end of the century, given the appalling lack of effort by big emitters to reduce their greenhouse gas emissions. . . . It would mean that the Marshall Islands, its people and its culture would be lost forever."[89] Before the Paris climate conference in 2015, leaders of Pacific Island countries, whose islands face submersion or devastation, pleaded with other governments to take action.

Climate refugees: Displacement, migration, and health

Over millions of years, species have migrated in response to changes in climate, sometimes seasonally, sometimes on a longer-term basis. The annual long-distance migratory habits of many bird species and whales are well known. Over longer time-spans, the alpine tree line ascends or descends as the prevailing temperature rises or falls. Our preagrarian human forebears were no different; they followed the food sources and the freshwater supplies.

Human migration is as ancient as the comings and goings of ice-bound frontiers, droughts, floods and famines.

Homo erectus began migrating out of Africa around a million years ago, perhaps in response to the long-term global cooling and the onset of fluctuating millennium-long glacial periods with consequent reductions in food sources in cooler and more arid eastern Africa. Closer to modern times, around 5,000 years ago, eastern Saharan nomadic-pastoralist tribes migrated into the fertile Nile Valley as the Sahara began drying.

Climate change will contribute to displacing many millions of people over coming decades,[90] and this will affect the health of many of them via mechanisms such as nutritional deficits, infectious disease risks, and physical and mental trauma.[91,92] Most migrations will occur in the world's poorer regions, often confined *within* countries; but many will choose, or be forced, to cross national boundaries. One-fifth of the world's population lives in coastal areas at risk from sea-level rise and natural weather disasters; indeed, island populations in Tuvalu, Kiribati, the Maldives, and some parts of the Caribbean already face the prospect of wholesale displacement. According to the Asian Development Bank in 2012, increasingly frequent extreme weather events in Asia and the Pacific region since early this century have greatly expanded the numbers displaced by storms, floods, heat and cold extremes, drought, and sea-level rise.[93] In regions such as the Horn of Africa, thousands of displaced people have become long-stay refugees, living in precarious underfed and unhygienic conditions, their children deprived of education and prone to mental health disorders and despair.[94]

Displacement may entail forced movement of populations, planned resettlement, or intentional migration. Each poses its own profile of health risks.[95] Large-scale forced displacement increases the exposures to infectious disease outbreaks and food shortages. Planned resettlement typically entails adverse social outcomes, including loss of land, livelihood, and home, and leads to food insecurity and poor mental health. Displaced impoverished communities that inhabit improvised settlements in and around cities are often located where there are high risks of climate change impacts such as floods and landslides.

Mental health and psychological disorders: An unseen burden

Mental health, despite the large burden of illness and disability it represents in the general population, remains the Cinderella of public health research and prevention in most countries. The disturbed workings of an apparently normal brain are much less easy to measure than, say, blood pressure, and the diagnostic categories are more elusive. Post-traumatic stress disorders will rise in the aftermath of more frequent extreme floods, fires, cyclones, and other disasters, as will depression following the loss of family and friends, property, and community. Experience in Australia indicates that after a devastating weather event, up to one-fifth of people suffer the debilitating effects of extreme stress, emotional injury, and despair.[96]

Children, too, are vulnerable to anxieties and stress disorders. As public awareness of climate change and its consequences grows, many children will face special stresses from long-term insecurities and anxiety about climate change.[97] What will the future be like? Chronic anxiety in childhood causes baseline neuro-hormonal changes with increased blood levels of cortisol (the "stress hormone"). These hormonal and metabolic changes affect long-term disease processes, including blood pressure, heart function, and energy storage in response to insulin and hence the occurrence of type II diabetes. As ever, the influences of climate change on human health are much more complex than the relation between warmer conditions and physical changes such as melting sea-ice.

Climate change is, ultimately, no respecter of wealth, status, or national borders; its health impacts will affect populations everywhere, albeit with different intensity and consequence. Bangladesh is vulnerable on many fronts, including widespread poverty and food insecurity, a large coastal population, high rates of tropical infections, and threats upstream to river water flows from receding Himalayan glaciers. Richer countries also harbor within-population differences in vulnerability. For example, in the United States the health and mortality impacts of the 1995 heat wave in Chicago and the 2005 Hurricane Katrina in New Orleans differed markedly between ethnic and socioeconomic groups; the low-income and residentially disadvantaged were at greatest risk.

As climate change escalates, differences in vulnerability between rich and poor may diminish as protection thresholds for wealthier populations are breached. Poverty levels become less significant in the face of increasingly intense tropical cyclones and other weather disasters.[98]

The indiscriminate devastation of high-intensity Superstorm Sandy in and around New York in December 2012 underscored the point that climate change can be the Great Leveler.

Malaria and climate change

Malaria is an ancient, even venerable, foe. References survive in early records from Egypt, Mesopotamia, China, and India, and the disease was well recognized by the Greeks and the Romans. The disease owes its name to the swampy "bad air"—*mal' aria* in Italian—around the marshes, which was associated with fever. The main symptoms are fever, chills, shivering, sweating, and vomiting. The young, newly injected plasmodia transmitted by mosquitos multiply in the infected person's liver before dispersing and infecting red blood cells, their source of the micronutrient iron. If untreated, the infection can be life-threatening, disrupting blood supply to vital organs, including the brain. Malaria pathogen proteins have been detected in Egyptian mummies from 5,000 years ago.[99] The *Edwin Smith Surgical Papyrus*, an ancient medical text from that period, describes an annual pestilence, referred to as "the pest of the year," that struck this Nile Valley civilization.[100] This was presumably not an accolade but a description of an unwelcome annual occurrence—most probably, say experts, the malaria outbreaks during the annual flooding of the Nile. Molecular analyses have identified remnants of the malarial plasmodium in the opulently entombed mummy of King Tutankhamen. The boy-pharaoh died in 1324 B.C.E. at age 19, stricken with multiple lower-limb and spinal deformities attributed to his strongly inbred genetic inheritance; his royal-divine parents were apparently brother and sister. With a weakened immune system, his death may have been due to severe cerebral malaria, the lethal calling-card of *falciparum* malaria.[101]

During the second warm phase of the Holocene Climatic Optimum around 5,300–3,500 years ago, with more consistent and far-reaching moist westerly winds, malaria could readily have spread around the eastern Mediterranean. In Mesopotamia, ancient cuneiform clay-tablet texts from that time record the characteristic 48-hour fevers of malaria. Carvings of the Goddess of Epidemics on Mesopotamian stone statues point to the ravages of a new pestilence.

Malaria today is primarily a scourge of health and a brake on economic development in hotter and humid regions, particularly in sub-Saharan Africa, South America, South Asia, and Southeast Asia. Each year, 200–300 million cases occur, causing around 700,000 deaths, including about one-fifth of all child deaths in sub-Saharan Africa.[102] Recently, there has been both good and bad news. Successful interventions at community and household levels with insecticide spraying, mosquito-repellent bed-nets, and local surface water management have reduced malaria incidence. But meanwhile, pursuing its own survival interests, the plasmodium has evolved resistance to the major drugs used to protect and treat people. And so the evolutionary arms race goes on, extending a process that stretches back many millions of years (Box 3.1).

BOX 3.1 Malaria: A long evolutionary history

Charles Darwin's opponents might have asked of malaria: how could such a complex transmission pathway have evolved without benefit of Intelligent Design? Well, time is evolution's assistant, and countless dead ends in the evolution of plasmodium biology and transmission have long disappeared. The plasmodium's evolutionary origins extend back several hundred million years to a simple parasitic organism that survived without the mediation of a vector organism. Subsequent evolution, occurring as environments changed, has step by step yielded the current plasmodium-mosquito-human configuration.

Most long-standing infectious diseases in human populations came either from our ape/hominin lineage or from subsequent cross-species contacts with domesticated or cohabiting wild animals (rodents, birds, bats). Plasmodial genes indicate that human-adapted forms of malaria emerged at least 5,000–6,000 years ago. Indeed, *vivax* malaria may have entered the early *Homo* lineage from primate sources over a million years ago. The origin of the deadlier *falciparum* malaria is less certain, although its lethality suggests that it entered the human population more recently, probably in Central-Western Africa when forest-clearing agrarians came in contact with malaria-infected mosquitoes.[103]

Today's two main species of human malaria have different climate and biology profiles. The more lethal *Plasmodium falciparum* requires higher temperatures to develop and multiply within the mosquito, and is mostly confined to the warmer tropical regions. The less virulent *Plasmodium vivax* has evolved to survive in cooler climates where, during winter, there are no hovering mosquitoes available to transmit it to another human host. It quietly overwinters in the human liver, awaiting biochemical news of the external onset of warmer weather, whereupon it relaunches into the bloodstream. Viewed on the world map, malaria looks like a tropical disease. Yet in the nineteenth century, cold-adapted *vivax* malaria extended north in Europe, even reaching the Arctic Circle in Sweden and northwest Russia in summertime.

Malaria has been a persistent focus of malodorous debate about infectious disease and climate change. Is social and economic development the main determinant of malaria's range? Yet even if this is so—and there is supportive evidence for the proposition[104]—a climatic influence is certainly not precluded; indeed, that influence could become more important in a much warmer world. Mosquitoes are essential for the disease's transmission, and, within their "just-right" limits, they thrive best in warmer and more humid conditions, and the immature plasmodium develops increasingly faster in the mosquito at higher temperatures (Figure 3.3).

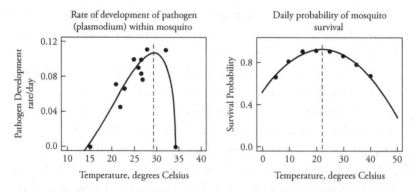

FIGURE 3.3 Bimodal responses to temperature change: The "Goldilocks Zones" for two critical biological components of malaria transmission in relation to temperature. Mordecai et al., "Optimal Temperature for Malaria Transmission is Dramatically Lower than Previously Predicted."[105]
Permission received from John Wiley & Sons via RightsLink.

First, though, something of malaria's recent history in relation to climatic changes in Europe.

Malaria: Recent history

The "summer fevers" that afflicted much of Europe during the past 2,000 years or more were mostly caused by *vivax* malaria. While the vivax plasmodium could survive the winters in Europe, the deadlier *falciparum* malaria only occasionally encroached on southern Europe during warmer periods. As populations and agriculture expanded in Europe with warmer temperatures in the eleventh and twelfth centuries, the malarial mosquito's range expanded, particularly northward. The disease became a Europe-wide scourge. Along the coasts of Holland and southeastern England, marshy areas were well known for their summer fevers, as were the lower reaches of France's Loire Valley.[106] Further south, until the mid-twentieth century, malaria has been a tyrannical oppressor for 2,000 years in the central Italian plains, debilitating many thousands of farmworkers.

During the Little Ice Age (ca. 1300–1850 C.E.), malaria receded in the north but still held much of its ground in Europe. In London in the 1660s, the third most common cause of death listed in John Graunt's Bills of Mortality was "agues and fevers." A large component of that disease category would have been malaria. The word "ague" was from an abbreviation of the Latin *febris acuta*, or acute fever. Agues and fevers were often referred to in literature, including by William Shakespeare, who lived during the coldest period of the Little Ice Age in Europe. It was widely believed that the sun spread this particular ague by drawing up telluric vapors from the marshes. Hotspur, in *Henry IV Part 1* (4.1), complains that:

> *Worse than the sun in March,*
> *This praise doth nourish agues.*

In Shakespeare's final play, *The Tempest* (2.2), published in 1611, Prospero's fractious slave Caliban curses his master:

> *All the infections that the sun sucks up*
> *From bogs, fens, flats, on Prosper fall . . .*

In chilly seventeenth-century London, soon after the nadir of the Little Ice Age, both Oliver Cromwell and, soon after, King Charles II contracted malaria. Cromwell, an ardent Protestant, refused treatment with the "Popish" quinine powder, newly discovered by Spanish Jesuit missionaries in Peru who observed its use as a traditional anti-fever medicine.

Cromwell duly died in 1658. But in 1679, Charles II, a practicing Catholic, accepted this quinine-based treatment, administered by the physician royal, Sir Robert Talbor. The king survived, and Sir Robert subsequently did very nicely from successful ministrations of this Peruvian wonder-drug to European nobility—he also cured the son of France's King Louis XIV—and from marketing his book *The English Remedy: Talbor's Wonderful Secret for Curing of Agues and Feavers* (1672).[107]

During the especially cold late sixteenth and seventeenth centuries of the Little Ice Age, when temperatures in parts of Europe were often, if briefly, 3°C lower than the northern hemisphere's average Holocene temperature,[108] endemic malaria receded in northern Europe.[109] The fact that *vivax* malaria maintained active annual transmission in much of Europe during this cooler time underscores the role of other nonclimatic factors. For example, farming families often shared crowded indoors accommodation with their sheep and cattle. So here, under the one roof, were warmth and abundant sustenance for mosquitoes, which, while preferring livestock blood, could top up on the human variety and thus maintain the community's infection rate.

During the nineteenth century, modifications to land use and to agricultural and animal husbandry practices accelerated the decline of malaria in Europe, bearing out the folk wisdom that "malaria flees before the plough."[110] The physical separation of the living quarters of families from livestock barns greatly reduced the risk, since mosquitoes preferentially followed the cows and sheep into their new quarters. Fortuitously, the blood of those animals is too salty for plasmodium survival, so, while the animals gave freely of their blood, they could not be infected. Hence, with separation of humans and animals, malaria dwindled further, and then further still following the draining of wetlands and marshes.

In the first half of the twentieth century, with new insight from Ronald Ross's identification in 1897 of mosquitoes as the malaria vector, and building on mosquito eradication experience with DDT, malaria was banished from many high-income countries. These included England and North America in the early 1950s and the Netherlands and Australia in the early 1960s.

Malaria's high international profile as a child-killer and economic burden in low-income countries makes it an obvious focus of climate change–related concern and research. While climatic conditions set the geographic and seasonal boundaries on where endemic (ever-present) malaria can occur, other factors determine where, within those boundaries, malaria actually occurs. A local climatic fluctuation might enable the disease to break its bounds and cause a brief epidemic in a neighboring, immunologically susceptible, population. But there are bigger questions, mentioned earlier. Might the range of malaria increase if the world gets warmer and, in many places, wetter? And at what point might some regions, such as parts of West Africa, become too hot or dry for mosquitoes?[111] And what critical thresholds would need to be passed for malaria to potentially re-establish itself in high-income countries?

Early studies in the 1990s in sub-Saharan Africa, South Asia, and South America confirmed that short-term changes in climate influenced the outbreak of malaria. In Rwanda, a period of local warming late last century was accompanied by an increase in malaria incidence,[112] and in parts of South Asia and Venezuela malaria epidemics occurred in association with El Niño events.[113,114] But what might the effects be of longer-term warming trends?

In the Kenyan, Ethiopian, and Colombian highlands, malaria has recently been edging higher up the slopes with regional warming trends.[115,116] In both Ethiopia and Colombia, year-to-year variation in the altitude ceiling of malaria has followed closely the variation in temperature. The large Kericho tea estate in the Western Kenyan highlands has been of particular interest in light of the laboratory-confirmed increase in malaria incidence that occurred between 1979 and 2009.[117,118] High-quality, long-run records like those from the tea estates are rare, and they show that the local temperature increased by a little over 0.2°C per decade. The key question is: were the two trends connected? Although the malaria rate in that region may have decreased,[119] reflecting better prevention and treatment in a known problem area, the more climate-relevant signal is that malaria cases have been reportedly occurring at higher altitude.

The extent to which climate change influences future patterns of malaria will depend on countries' levels of wealth, resources, and risk-management capacities.[120] Thousands of years ago, humans enabled malaria in Africa to extend its primate-host repertoire by felling forests and experimenting with agriculture. Are we now causing climatic and environmental changes that will facilitate malaria's persistence and possible spread? Most probably yes.

Interpreting climate and health relationships from the past

Changes in climate over the past 10,000 years have brought good times and bad times. Human life expectancies have fluctuated, fertility rates likewise. The historical record, however, is asymmetric; adversity receives more detailed treatment. A severe famine is very likely to have been recorded, while a period of congenial climate, good food yields, and no major epidemics is unlikely to have been explicitly recorded. A further obvious constraint is that the quality of information varies over the centuries. Written records extend back no more than about 5,000 years to the early Mesopotamian and Egyptian clay tablets; elsewhere, written records emerged much later (if at all).

The use of physical proxy measures has enabled rapid advances in reconstructing past climates, but proxy measures of past population health are in short supply. Most of the records and indicators relating to health and disease come from Europe, the Eastern Mediterranean, and China, and to a lesser extent from India and, much later, North America and Australia. Since most historical experience predates modern medical knowledge and its classification of specific diseases, much early evidence is coarse-grained and nonspecific, and mostly refers to undernutrition and starvation, infectious diseases ("pestilence" and "fevers"), and injury and death from extreme weather events and conflicts over natural resources. Those were for long the three dominant modes of serious illness and early death. Other information comes from archaeological and fossil evidence: skeletal remains may show the signs of undernutrition, stunting, and high mortality among children and young adults.

Records of major disease epidemics from Egypt and Eurasia extend back three to four millennia, but without reference to climatic conditions. More reliable are records of the great plagues of Athens (5th century B.C.E.) and Rome (the Antonine and Cyprian plagues in the second and third centuries C.E., respectively), and the Plague of Justinian affecting the Eastern Roman Empire (sixth century C.E.). Information about epidemics gradually sharpens after the medieval period in Europe and in China, becoming much more detailed in recent centuries.

The recent advances in molecular biological assays and the identification of infectious agents by DNA analysis of historical material offer means of resolving disputed diagnostic possibilities.[121]

Famines and fevers: Relations between nutrition and infection

Food shortages and fevers are not unrelated outcomes. Each can predispose to the other. Over the centuries, many episodes of serious undernutrition and starvation have been followed by outbreaks of infectious diseases. Lice-borne typhus caused most of the deaths during the Irish potato famine of the mid-nineteenth century. The great droughts of India in the late nineteenth century caused mass starvation and (curiously) a surge of malaria. The climatic extremes of the early fourteenth century in Europe and the resulting Great Famine presumably increased the susceptibility of the population to severe infections and death from the Black Death three decades later. An undernourished and weakened immune system often heightens susceptibility to infection.[122] Today, chronic childhood diarrheal disease in poor communities is a widespread cause of nutritional deficiencies and accounts for much of the stunted growth in young children.[123]

Sometimes, during social and economic disruption, epidemic disease causes insecurity. Food shortages resulted from the wholesale rural depopulation that followed each of the first two pandemics of bubonic plague: the Plague of Justinian, beginning in 542 C.E., and the Black Death.

Although social resources may suffice to prevent an epidemic when starvation-related deaths rise,[124] in general, populations that harbor endemic infections are more vulnerable to major epidemics following a famine. Pre-existing measles, for example, often breaks out as a killer disease following famines in much of sub-Saharan Africa today.[125] Nevertheless, there are no simple natural laws here. Indeed, under-nourished people may be *less* attractive for some pathogens, such as the iron-hungry malaria parasite's presumed displeasure at finding itself in a host whose red blood cells are low on iron. After all, the pathogen's objective is merely to find an ample source of nutrients and energy in a host fit enough to then pass on the pathogen's offspring to another human host.

The relationship between starvation and epidemic disease depends, too, on context.[126] In England in the 1690s, when harvests failed and food prices soared, deaths from infectious disease remained fairly stable.[127] But a century earlier, epidemic outbreaks in the counties of Cumberland and Westmorland had followed soon after the great famine of bitterly cold 1597.[128] Many starving people from those counties traveled to Newcastle in adjoining Northumberland, where there was food—but also bubonic plague. The following year, plague broke out in Cumberland and Westmorland, presumably imported by those who had braved this dread disease in order to find food. A more recent example: among children conceived during the extreme "hunger winter" imposed on the Dutch by Nazi Germany during 1944–1945, severe undernutrition in early life was not related to subsequent adult immune system function, yet in villages in The Gambia, West Africa, children born during the annual hungry season are up to 10 times more likely to die, especially of infectious diseases.[129]

A focus on the *biology* of this relationship may obscure the roles of *social* conditions. Today, infectious diseases are overwhelmingly concentrated in the world's poor, marginalized, and crowded populations.[130] Their ready spread reflects low literacy, poor hygiene, ineffective sanitation, and, often, disordered social relations—all of which bear on future health prospects under climate change conditions. On the underside of modernity's ledger are the world's poorest 1–2 billion, many of whom still depend on local subsistence economies and live in conditions little different from those that prevailed several centuries ago. Many of those populations will, over the coming warmer decades, face similar types of climatic-environmental stresses and health risks as did premodern subsistence communities.

Niger in the West African Sahel offers an example. This landlocked country, chronically drought-prone, suffered severe food shortages in 2005, 2010, and 2012. These were compounded by regional conflict, weak government, and chronic undernutrition (the main cause of Niger's very high child mortality rate). The birth rate, averaging over seven per woman, is clearly unsustainable—but unlikely to decline rapidly. UN projections are that today's 17 million may triple to 55 million by 2050, surely boom-and-bust figures. While the number of mouths to feed escalates, Niger and West Africa as a whole are likely to experience significant heating and, in the northern Sahel, decreased rainfall and greater aridity. The West African summer monsoon, which originates in the Gulf of Guinea and usually reaches to 20°N (the northern tip of Niger), has been weakening and edging further south in recent decades. More hunger, unrest, and displacement loom as the world's climate changes.

Does the direction of a change in climate matter?

One final point is critical for interpreting historical food crises and the implications for future changes in the world's climate. The *direction* of temperature or rainfall change is not usually the main determinant of the size of the climate-related impact on food yields or infections.

The *magnitude* of the climatic change is more important; that is, the Goldilocks "just right" constraints apply. Crop and animal species are attuned, by both natural and managed biological selection, to their usual regional climate. Hence food yields are adversely affected by both excessive warming and cooling, and by either excessive rain or the drought conditions that often accompany temperature change.

Annual spring barley yields in the Czech Republic since the mid-twentieth century provide a good example (Figure 3.4). Barley is the most important spring cereal grown in Central and Western Europe, and when the spring rainfall falls outside the normal range to which that crop is attuned, both drier and wetter conditions can significantly impair yield. The graph for winter wheat yields is very similar.

There is, however, more to be learned, since most of the research on this topic has focused on the effects of climate change on cereal grains rather than on roots, tubers, horticultural crops, and feed crops. Hence

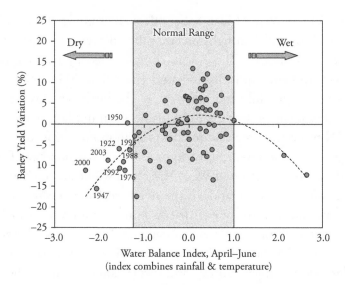

FIGURE 3.4 Relationship between late-spring normalized (soil) water balance (a function of temperature and rainfall) and barley yields in the Czech Republic during the past century. The dashed curve is the five-year running mean deviation in yield from the average value in the "normal range." Brázdil et al., "Variability of Droughts in the Czech Republic, 1881–2006."[131] Permission received from Springer.

much more is known about harvest response patterns in temperate climes than in the tropics.[132] In America, corn (maize) yields decline with unusually low temperatures and increase with warmer temperatures up to 29°C, and soybean yields increase up to 30°C. Temperatures above those optimal levels impair yield.[133] Applying these known relationships to future warming scenarios in Africa, and assuming sufficient rain, modeling indicates that two-thirds of Africa's present corn-growing areas would experience reduced yields for a 1°C warming.[134]

As for food yields, the risks of infectious disease outbreaks can be increased by both warmer and cooler conditions and by disturbances of either biological-ecological factors or social-demographic conditions. Other examples of this general pattern include the U-shaped daily death rate in response to daily temperature excursions beyond the thermal comfort zone, mosquito survival in relation to temperature extremes, and the health risks and benefits associated with intakes of most nutrients, ranging from too little to too much (known as Bertrand's Rule).[135]

Taking stock

These first chapters have outlined the context in which today's great environmental problems have arisen; the profile of those pressures at global scale, with discussion of threats to humans and other species from disruptions to Earth's operating system; the workings of the climate system and the process of human-driven climate change; and the many ways by which climatic changes affect human well-being, health, and survival.

The next chapter takes up the story of how changes in climate systems shaped early humans and their health, safety, and social stability.

4

From Cambrian Explosion to First Farmers

How Climate Made Us Human

DETAILS BLUR AS WE peer back through millions of years, but the outline of the story is clear enough. During the past 2–3 million years, our hominin forebears had to cope with an increasingly variable and cooling climate. Across those 100,000 *Homo* generations, survival and reproduction depended on maintaining biological and behavioral compatibility with constantly changing climatic and environmental conditions. Hence much of modern human biological versatility and adaptability, including several unique aspects of brain function, comes from evolution's selective winnowing within those ancient predecessor populations. The genes of the survivors, those best able to reproduce, are part of our genetic inheritance today.

That climate change has been a major source of natural selective pressure has long been known. Alfred Russel Wallace, the overshadowed younger contemporary of Charles Darwin and codiscoverer of evolution by natural selection, wrote that, among the variations occurring in every fresh generation, survival of the fittest occurred in response to the "changes of climate, of food, of enemies always in progress."[1] The corollary, of course, is that since biological evolution must focus on surviving the present, oblivious of the future, it provides no guarantee against extinction. Even so, a multivalent brain that enables cultural and behavioral adaptability and strategic forward thinking would surely help an animal species cope better with subsequent environmental changes. Indeed, it seems to have worked sufficiently well for our *Homo* genus ancestors during two million years of ever-changing climatic conditions for at least one *Homo* species to have carried the baton of survival into the present. In the next two centuries, our species faces a new challenge of greater, faster, and protracted climate change.

The last half billion years

Since the Cambrian Explosion of new life forms around 540 million years ago, there have been five great natural extinctions and many lesser ones.[2] The earliest extinction of multicellular life, though less destructive than its successors, occurred around 510 million years ago, apparently due to acute sulfurous shrouding, cooling, and oxygen deprivation caused by a massive volcanic eruption in northwest Australia.[3] Most of these catastrophic transitions were marked by climate extremes, volcanic activity, and altered ocean chemistry, especially rapid surface acidification of shallow coastal waters.[4,5] Three of the great extinctions appear to have occurred during cold episodes and two during hot episodes. The first hot episode, the "Great Dying" 250 million years ago at the end of the Permian, cleared the landscape for the proliferation of dinosaurs, and the second saw the demise of those "terrible lizards" 65 million years ago (Figure 4.1).

Temperature's role is complex; many extinction events followed a rapid pulse of cooling from the shrouding of the skies caused by volcanic eruption or asteroidal impact and then prolonged warming from voluminous release of carbon dioxide. In the catastrophic Great Dying, more than 90 percent of all species were wiped out over several million years of environmental shocks. The main cause was probably prolonged volcanism as continental plates collided to form Pangaea, causing a surge in atmospheric carbon dioxide levels and hence surface warming and acidification of the now oxygen-depleted

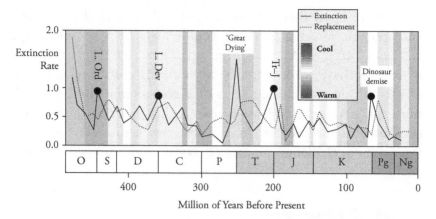

FIGURE 4.1 Relationship between temperature and the five major natural extinction events since the Cambrian "explosion" 540–510 million years ago, when small, complex, and then vertebrate marine animal species first evolved. The letters along the horizontal axis indicate geological periods: Ordovician, Silurian, Devonian, Carboniferous, Permian, Triassic, Jurassic, Cretaceous (K), Paleogene, Neogene. Temperature estimates are from oxygen isotope ratios in dated fossil shells, and rates of speciation and replacement are also shown. Blois et al., "Climate Change and Past, Present and Future Biotic Interactions."[6]
Permission received from AAAS.

oceans.[7] After the die-off, recovery and replacement took several million years.[8]

The great extinction of 65 million years ago that finished off the dinosaurs and pterosaurs was almost certainly caused by a massive asteroid strike in the Yucatan Peninsula region of Mexico and consequent extreme swings in temperature and atmospheric shrouding. Even so, that spectacular event was apparently a coup de grâce following a decline in dinosaur numbers during the preceding million years of climatic swings due to violent volcanism in the Indian subcontinent; the associated loss of herbivorous dinosaurs completed the demise of the "red in tooth and claw" dinosaurs.[9,10,11] During the intervening periods, lesser changes in climate have influenced the vitality, survival, and biological evolution of species everywhere.

Today's sixth extinction, in which the loss of species and ecosystems is occurring as fast or faster than during those earlier natural extinction events, is the culmination of a process that began quietly as humans spread around the world from around 70,000 years ago. This process gradually accelerated with the development of agriculture, and has been rapidly escalating with industrialized human pressures on the biosphere via overharvesting and habitat destruction.[12] Our actions are changing the profile of life on Earth, not just the conditions for it.

The past six million years: Pliocene and Pleistocene

Earth's average temperature has declined by around 8°C since the demise of the dinosaurs, which ended the Mesozoic era and ushered in the current Cenozoic era (the Age of Mammals, from 65 million years ago to the present).[13] After some slight warming in the first 10 million years, the Paleocene epoch, Earth has cooled during the five ensuing epochs: the Eocene, Oligocene, Miocene, and then, over the past six million years, first the Pliocene and from 2.6 million years ago the Pleistocene. For the past 11,000 years, we have been traversing the Holocene, a relatively warm interglacial period in the current ongoing succession of icy periods (or "ice ages") (Figure 4.2).

Aspects of the post-dinosaur cooling during the Cenozoic era are relevant to thinking about future human impacts on Earth's climate

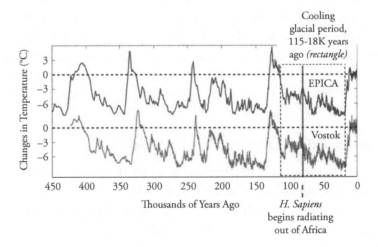

FIGURE 4.2 Time trends in global temperature (two Antarctica ice-core sources: EPICA and Vostok) and global ice volume during the past 450,000 years, showing changes during four glacial periods and four interglacial periods. Adapted from original by R. A. Rohde, "Ice Age Temperature Changes."[14]

system and the planet's future. At the junction of the Paleocene and Eocene epochs, when the background temperature was 6°C hotter than today, the global temperature suddenly surged upward by a further 5°C over a geologically brief period of several thousand years. This "Paleocene-Eocene Thermal Maximum" (PETM) was apparently caused by a massive geological "belch" of methane, perhaps from a rupture in the north Atlantic seabed; carbon dioxide release may also have been involved. The consequent extremes of warming and ocean acidification extinguished half of all species of tiny seabed zooplankton ("foraminifera").[15] The important, indeed alarming, difference between that PETM warming and today's dire prospect of a 3°C–4°C rise by 2100 is that the latter would occur 100 times faster. The future geological record of the "Holocene-Anthropocene Thermal Maximum" (HATM) will astound future paleoclimatologists. What on Earth did twentieth- and twenty-first-century fossil-fuel-burning people think they were doing? Over the past two million years of overall cooling, Earth's temperature has undergone increasingly great swings. These were primarily due to the slow changes in the relative strengths of the three Milankovitch cycles concurrently influencing the pattern of solar irradiation of Earth's surface.[16] During those two million years, there have been three recognizably different phases of global temperature variability, dominated early on by the 23,000-year axial wobble cycle, then the 42,000-year tilt/precession cycle, and, in the past million years, the 100,000-year orbital eccentricity cycle—hence the succession of approximately 100,000-year glacial periods ("ice ages").

Climate and the emergence of the hominins

Our family, the great apes, emerged as a branch of the primates during the Miocene around 15 million years ago, when global average temperature was around 3°C higher than today.[17] As cooling continued, Arctic ice-sheets began forming around six million years ago, and in eastern Africa, a significant evolutionary event occurred: the early hominin line was branching off from the ancestral chimpanzee line.[18]

There have been various theories about the environmental influences on that branching, many of them relating to either probable or identified shifts in climatic conditions.[19] The simplest early idea was that the long-term global cooling and drying trend had created, in eastern and southern Africa, a savannah landscape for which different attributes were needed if a great ape was to survive in the more exposed semi-open. That idea can now draw on specific evidence that as regional drying occurred 6–8 million years ago, the composition of plant communities in open terrain also changed.[20] The chemical isotope profile of fossil bones indicates that, over time, the early hominins were eating more of the later-evolving (C4) savannah-like grasses and sedges, along with papyrus and water chestnuts. The hominin emergence around six million years ago was also influenced by a period during which the 23,000-year Milankovitch wobble/precession cycle generated a fluctuating sequence of stronger and weaker tropical monsoons.[21]

Differences in local topography and climate may also have influenced selection pressures. One strong candidate is the climatic consequence of the formation of the Rift Valley, running north-south in Eastern Africa between Ethiopia and Mozambique. This great rift in Earth's surface is the result of the tectonic rupture of the African Plate into the Somali and Nubian Plates. As those two chunks of Earth's mantle were pulled apart, much of the intervening land surface sank to form a broad valley, while a mountainous ridge was thrust upward along the western wall of the valley. The rising slopes on the west side of this mountain range, much of it more than one kilometer higher than the valley, caused moisture-bearing winds to shed their rain, casting a rain shadow over the valley terrain. And so life for the forest-dwelling great apes west of the Rift continued as normal, while the drying of the eastern region with receding forest and more open grassland made a tree-dwelling life for apes impossible. More efficient and rapid upright mobility was now needed for traversing the savannah and finding ground-level plant foods. As legs became longer and landscape scanning ability increased, vulnerability to feline predators lessened. As tree-climbing declined, arms became shorter, further enhancing upright walking.

The long-term consequences of the appearance of this new branch of ground-dwelling apes were unforeseeable; Darwinian evolution was, as ever, about surviving the present, not planning for an unknowable future. Looking back from part of that "future," we see that these regional climatic changes around six million years ago had enormous consequences for subsequent life on earth. The now dominant modern human population looks set to reheat this planet to at least the late Miocene temperature by 2100. Of course, there can be no return to the Planet of the Apes—we cannot rewind evolution—but if increases in global temperature of 3°C or more were sustained for centuries, the environment and ecosystems inherited by our descendants would differ greatly, even unrecognizably, from today's world.[22]

As cooling continued during the Pliocene, changes in the Eastern African landscape influenced the evolution of early ape-like (*pithecine*[23]) hominins. Fossil finds have identified several species of Ardipithecines from 4–5 million years ago and their successors the Australopithecines, the "southern apes." The earliest Australopithecines were vegetarians who fed on plant foods typical of the open woodland and savannah environments, including fruits and tubers.[24] A later Australopithecine species, represented by the famous "Lucy" (*Australopithecus afarensis*) discovered in Ethiopia, existed around 3.5 million years ago. Over time, Lucy's descendants became omnivores and occasionally meat-scavengers who used stone tools to scrape and also to crack open the long bones of carcasses to get at the high-energy marrow.[25]

Around 2.4 million years ago, early in the Pleistocene, the *Homo* genus emerged (Figure 4.3). The transition from australopithecines was apparently led by changes in the jaw and its array of teeth; changes in brain size would follow. As the climate cooled and dried, forests receded and food sources continued to change. Antelope and other bovid species

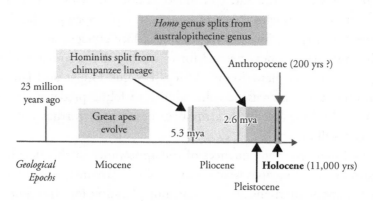

FIGURE 4.3 Evolution of *Homo* genus.
Source: Author.

had begun diversifying around 2.7 million years ago, meat-eating by scavenging hominins was becoming more common, and bushes and trees with different fruits and berries became more common.[26,27]

Climatic influences on human biological evolution

Every modification of climate, every disturbance of soil, every interference with the existing vegetation of an area, favours some species at the expense of others.

—JOSEPH HOOKER[28]

Temperature fluctuations during the prolonged Pleistocene cooling influenced many different aspects of human biological evolution. Climate change is in our bones, bowels, and brains, the result of both the *directed* and *plastic* forms of natural selection. The latter refers to a selective preference for flexibility of behavior and physiological functioning, adaptive features that would have enhanced survival, particularly during the very variable climate from 2.7 to 2.0 million years ago.[29] Although the *Homo* genus and its ancestral *Australopithecus* genus proceeded as two parallel branches for another million years, the Australopithecines could not keep up with the demands of an increasingly cold world nor with a more efficient and increasingly omnivorous food-acquiring competitor. Extinction soon beckoned. Selective pressures of a more directed and specific kind also influenced the early *Homo* line. Changes in food sources placed a survival premium on anatomical and metabolic features better suited to the new diet, including changes to the jaws and teeth and an apparent trade-off between two energy-intensive organs: the bowel and the brain.[30] Digesting meat needed much less energy than mulching and digesting a kilogram or two of fibrous leafy or tuberous foods. So, as the meat content of the early *Homo* diet increased, a shorter bowel was sufficient—and the redundant metabolic energy was available for an enlarging and energy-intensive brain. Such a brain enhanced hunting, tool-making abilities, and hence survival— and thus initiated a feedback loop that reinforced the evolution of the brain.

The subsequent controlled use of fire from well over a million years ago by *Homo erectus* enabled them to *cook* animal and vegetable foods, leading to a further reduction in tooth size. Indeed, via directed selection it probably further accelerated the size-for-energy swap between bowel and brain. In *Catching Fire: How Cooking Made Us Human*,[31] anthropologist Richard Wrangham proposes that the easier digestion of protein and increased availability of nutrients and energy from cooked foods accelerated the ongoing increase in brain size. Where apes and early hominins spent most of their waking hours gathering raw wild foods and then chewing them, as do herbivores today, human ancestors now had more spare time to make better tools, devise strategies, seek better food sources, create safer shelter, and indulge in rudimentary chatter—all of which, via feedback, enhanced the evolution of the brain.

The Pleistocene cooling culminated in an orderly succession of glacial periods over the past million years. These glacial fluctuations, each spanning 5°C–6°C of cooling followed by rapid rebound warming during a brief interglacial period, continued to shape the evolution of the *Homo* lineage. Our long-surviving older relative *Homo erectus* had ventured out of Africa from 1.8 million years ago, ultimately spreading throughout Eurasia and down into Southeast Asia. Subsequently several later-evolving *Homo* species drifted out of Africa, seeking new food sources, and spread mostly via the Near East into Asia Minor and southern Europe.[32] Meanwhile, conditions everywhere were becoming colder and drier as more of the world's water was locked up in great ice-sheets and glaciers.

Details about the types and times of branching within the *Homo* lineage are often tweaked and updated as new finds appear, but the story's outline is fairly well established. Sometime around half a million years ago, the Neanderthals, *Homo neanderthalensis*, emerged in partly ice-covered Europe, probably as an offshoot of *Homo heidelbergensis*, a direct descendant of *Homo erectus*. Subsequently, several hundred thousand years ago, the less well-defined species *(Archaic) Homo sapiens* evolved, another descendant of the *Homo erectus* line. In due course, as we will see, the archaics would provide the genetic stock for our species, the anatomically modern *Homo sapiens*.

The Neanderthals ranged widely over the cold and arid Middle East, the western regions of Central Asia, and the ice-free regions of Europe. As they evolved, they became attuned to surviving in that colder and glacier-laden northern climate. Their thick skulls, large noses, and stocky bodies helped to counter the cold environment and to warm the inhaled air.[33] During the middle stone-age period in Europe, the Neanderthals, equipped with their basic Mousterian blade-and-scraper stone technology and a diet of wild berries, fruits, and megafaunal prey, moved often as climatic conditions changed and great ice-sheets ebbed and flowed. Skeletal remains from Italy indicate that Neanderthal life-spans were short; two out of five died by age 15, and three-quarters of adults died by age 40.[34]

Far to the east, another surviving *Homo* branch was coping with a cooling and fluctuating climate change in eastern Eurasia. These Denisovans (*Homo denisova*) were first presumptively identified from fragmentary skeletal remains in the Denisova cave in southern Siberia. Although their species status is uncertain, DNA analyses indicate that they were closely related to the Neanderthals, with whom they shared a common ancestral lineage until around 500,000 years ago.[35] While the Neanderthals ended up mostly inhabiting Europe and West Asia on a predominantly east-west axis, the Denisovans occupied the east of Eurasia on a north-south axis that extended from Siberia to Southeast Asia.[36]

Throughout the Pleistocene, climatic changes exerted another type of selective pressure on human biological evolution, contributing to the rapid emergence of various *Homo* species over time. The changeable and cooling climate caused continuing movement of groups across the landscape, particularly during times of known climatic changes, resembling the better-documented evolution and migrations of particular mammal species during the same climatically unstable Pleistocene.[37] The choice of regional refuges sought out by different *Homo* species during periods of severe cold and food scarcity influenced their chance of survival and probably fine-tuned aspects of their evolutionary adaptation—just like Darwin's Galapagos Islands finches.

Emergence of modern Homo sapiens

Around 200,000 years ago, as Earth entered another 80,000-year glacial period, the anatomically modern human species *Homo sapiens* emerged from the "archaic" stock in Africa.[38] These early modern humans spent nearly all of their first 100,000 years in sub-Saharan Africa, coping for most of the time with a cool and dry world. Numbers were small, perhaps a total of 50,000 individuals dispersed in family groups. Stone tools were still primitive, and syntactical language had not yet evolved. Surviving each season was the focal task. Skeletal remains from caves in today's Israel from around 100,000 years ago suggest a curious episode when Neanderthals and modern humans alternated, perhaps overlapped, in their occupancy of the region.[39] Temperatures had been faltering and starting to fall since 115,000 years ago,[40] and both species ventured further afield for food sources.

All bets were off as to the eventual outcome for the *Homo* genus. Who would survive? Would it be the fleeter-footed African *Homo sapiens*, the heavy-browed European Neanderthals, their Denisovan cousins living further east, the diminutive "hobbit" *Homo floresiensis* whose bones have been found on the island of Flores in Indonesia, or the several regionally specialized subspecies of the older and smaller-brained *Homo erectus*? All had survived the previous Eemian interglacial, spanning 135,000 to 110,000 years ago. These *Homo* species now faced the next glacial period. After an initial rapid though jagged cooling episode, that next glaciation began in earnest around 80,000 years ago, reaching its coldest point, the Last Glacial Maximum, 18,000 years ago.

As the cold intensified, humans began wearing furs and hides, perhaps an early sign that anatomically modern humans were becoming more cognitively fluent and behaviorally modern.[41] This opened up a new ecological niche for lice, resulting in the emergence of the human-specific body louse, *Pediculus humanus*. DNA analysis shows that this human-adapted body louse split from the parent lineage of primate-infesting head lice (*Pediculus schaeffi*) around 70,000–100,000 years ago.[42] This newly evolved species of lice could thrive on the otherwise nearly hairless bodies of humans by living in their fur and hide clothing.[43] Much later, these body lice became the vector for the often deadly infectious disease typhus, known in recent times as "war fever" and "jail fever." In recent centuries, louse-borne typhus has caused countless millions of deaths, thriving in conditions of poverty, crowding, social disruption, and warfare.

Changes in climate have not only shaped many important aspects of human biology, anatomy, and behavioral scope—they have also shaped the evolution of various infectious agents and vectors that became scourges of human populations.

Out of Africa: The diaspora begins

The onset of cooling and drying in eastern Africa and the consequent changes in food sources was a stimulus for humans to range further afield. This was neither a planned migration to a known destiny nor a single mass movement. It occurred in punctuated fashion over several tens of thousands of years as various tribal groups ventured beyond their usual horizons in search of a better environment. The actual dates are uncertain, and estimates are often revised. More certainly, by around 70,000 years ago various groups were heading northeast toward the Middle East and West Asia, while others headed east around the Indian Ocean coastline.

Estimating the timing of these out-migrations depends largely on a heritable molecular clock that we carry in our body cells. The tiny power-generating organelles in every cell, called mitochondria, contain their own little package of genes—quite separate from the mainstream genes in the cell nucleus. Random mutations in mitochondrial genes occur at a steady "ticking" rate, and so the extent of difference in mitochondrial DNA between two populations indicates approximately how long ago they separated. Mitochondrial DNA analyses of blood-cell samples from diverse samples of today's regional groups of humans indicate that, following some earlier smaller migrations, larger numbers radiated out of Africa between 75,000 and 60,000 years ago, coinciding with a time when Earth's temperature was particularly cold around 65,000 years ago (Figure 4.2).[44,45,46]

There were earlier curtain-raiser excursions. Indeed, paleoclimatic and archaeological evidence indicates that, around 120,000 years ago, a few groups followed a well-watered corridor through the Sahara to North Africa's Mediterranean coast.[47] Further, a trail of African-style stone artifacts discovered in the southern Arabian Peninsula points to early eastward explorations around the coastal Indian Ocean from as early as 120,000 years ago—these were perhaps the trailblazers who first reached Papua New Guinea, Australia, and the Pacific coast of Southeast Asia.[48] That trans-Arabia crossing occurred during the Eemian interglacial, when warmer temperatures and wetter conditions greened the Saharan-Arabian desert zone, temporarily lowering the usual desert barrier to human dispersal.[49] Later, as glacial conditions settled in, other groups began probing eastward, aided by falling sea levels in both the Red Sea and Persian Gulf as expanding ice-sheets sequestered the ocean waters.

Some of the earlier east-migrating groups may have struggled and dwindled in numbers, and perhaps died out. Others may have been absorbed into much larger groups. Despite those unknowns, the mitochondrial DNA evidence indicates that one major successful out-migration best explains the similar genetic profile of all contemporary human populations outside Africa. Although unlikely, that surprising similarity could have resulted from some severe culling process among migrating populations fairly soon after leaving Africa, leaving a limited number of survivors with a smaller gene pool. There is one spectacular and plausible culling possibility, though the details are contentious.

Mount Toba eruption: A close call?

Around 74,000 years ago, the Toba *super*-volcanic eruption changed the global climate. This massive eruption in equatorial Sumatra is the largest known in the past two million years.[50,51] The volume of lava was twice the bulk of Mount Everest, and a vast amount of rock, sulfate particles, and ash was blasted into the skies. The cloud of ash and acidic sulfate aerosol spread globally, causing a darkened and cold "volcanic winter." Land temperatures at low latitude, in the northern hemisphere particularly, may have dropped acutely by 5°C–10°C, and chilly conditions persisted for half a decade.[52] The acidity of the atmospheric fallout was particularly lethal; the concentration of hydrogen ions rose in Greenland and Antarctic ice-cores 5- to 10-fold. Vegetation and wildlife in some regions died en masse. Geological ash layers up to a meter thick have been found in many parts of India and Malaysia, while some areas have layers several meters thick.[53] Clearly, the Indian subcontinent was the most exposed to the fallout.

The impact on already dispersed groups of archaic *Homo sapiens* and *Homo erectus* may have been rapid and deadly.[54,55] In particular, their populations in East and South Asia, the Middle East, and eastern Africa (see the affected zone in Figure 4.4) must have been at considerable risk.

The intriguing question is to what extent the dispersing populations of modern *Homo sapiens* and hence the future genetic profile of our species might have been affected.[56,57] Were groups of our species actually wiped out by cold and hunger? Any that were around the coastal regions of the Indian Ocean were at risk, and those in India would certainly have faced likely annihilation. Some recent paleogenetic research indicates that most of the groups of dispersing *Homo sapiens* left from northeastern Africa via the Arabian Peninsula a little later, around 65,000–60,000 years ago.[58]

Even so, the source *sapiens* population in eastern Africa may also have fared badly from the eruption, being directly under the westward-tracking volcanic plume. In the extreme, the number of survivors, largely confined to eastern Africa, may have been quite small;[59] maybe less than a quarter of a possible pre-Toba total of 100,000 survived. If so, the *sapiens* survivor groups in northeast Africa would have become the new source population for the main outward dispersals during the

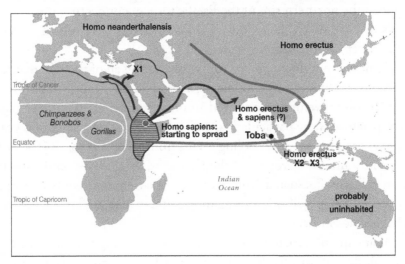

FIGURE 4.4 Map of the presumed deadly zone of the Toba eruption 74,000 years ago, within the outlined "U." Adapted from Weber "Toba–Aftermath: Climate and Environment."[60]

subsequent period 70,000–60,000 years ago. If so, our dominant base stock would be only around 74,000 years old. Meanwhile, the hominin populations of *Homo erectus soloensis* on Java and the diminutive *Homo floresiensis* on Flores may have survived, being upwind of Toba's ash, dust, and smoke.

Bands of humans continued spreading around the coastal region of South Asia, and eventually reached western, central, southern, and eastern Asia. By 60,000 years ago, the sea level had dropped sufficiently to enable groups to traverse land bridges and make water crossings. It became possible to walk across the entrance to the Red Sea between today's Eritrea and Yemen at the Bab el Mandeb straits. By 55,000 years ago, the first of several waves of ancestral Australian Aboriginal people may already have reached the north of that continent, while other branches of humans moved northward from Southeast Asia into East Asia. From around 40,000 years ago, some branches looped back into Europe from West Asia and the Asian steppes. At the other, eastern, end of Eurasia, sometime around the Last Glacial Maximum when the seas were 120 meters lower than today, several extended family groups had begun moving east across the still exposed and increasingly ice-free Beringian land-bridge. That bridge temporarily capped the northern rim of the Pacific Ocean, connecting the far corners of northeast Russia and northwest Alaska.[61] We return to their story later in the chapter.

Temperatures fluctuated continuously during all of these human excursions, including several major plunges and rebounds,[62] such as the rapid cooling events that occurred around 72,000 and 45,000 years ago. Temperatures were particularly unstable around 45,000 to 30,000 years ago, in part due to acute cooling events from increased volcanic activity around southern Europe and the Caucasus, and consequently vegetation patterns fluctuated from forest to open environments and back to forest. The distribution of mammals fluctuated similarly across this changeable landscape.

That climatic variability, particularly the recurring periods of intense rainfall 80,000 to 40,000 years ago, has been implicated in periodic bursts of human numbers and advances in basic cultural evolution. These changes in habitats, numbers, and social networks were evidently key factors in the technological innovations and the shaping of modern human behaviors in Africa.[63]

Encountering new infectious agents

The density of microbes is greater in warmer low-latitude regions. Hence hunter-gatherers in tropical regions were exposed to many different parasitic infections, such as leishmaniasis (spread by infected sand flies) and trypanosomiasis or "sleeping sickness" in equatorial Africa (spread by tsetse flies). The transmission of those infections needed local vector insects, and so their geographic range was restricted. In contrast, many person-to-person infections, including salmonella bacteria in food, amoebic dysentery, streptococcal arthritis, and various larger parasitic gut-infesting worms, could circulate within local tribal and family groups and travel with them without need for vectors. They became *heirloom* infections of the *Homo* genus,[64] mostly acquired in eastern Africa well before modern humans began dispersing. Many generations later, these infections were inherited by the early agrarians.

Throughout human existence, many new infectious agents have made a successful crossing from their natural animal host into the human animal. Today, these animal-source infections account for almost two-thirds of all infectious diseases in humans, and we are still acquiring more.[65] An increasing proportion of newly acquired animal-source infections in humans now come from "synanthropic" species that thrive and multiply in human-made environments.[66] In Paleolithic times, these "zoonotic" infections would have been picked up sporadically by hunter-gatherers from wild animals. Butchering and eating meat and using animal skins for rudimentary clothing and shelter brought contact with a widening range of bacteria and food-borne liver flukes, intestinal worms, and hydatid cysts.

Gut-infesting tapeworms are a very ancient group of parasites, so specialized by natural selection that they are incapable of independent existence. Instead, they passively absorb nutrients that are passing down their host's gut. Evolution also equipped them with a "stealth" capacity; they could therefore evade much of the new human host's immune-system surveillance and its defensive response. Early humans would have encountered these multicellular gut parasites often in undercooked or raw meat from animal prey.

Viewed from the pathogen's perspective, zoonotic infection was always a hit-or-miss affair. Human hunters and consumers of infected animals could usually only serve as incidental "dead-end" hosts, with no onward transmission to other humans. In those cases, either the humans were biologically unsuitable hosts for parasite reproduction, or their small nomadic bands lacked a sufficient replenishment of susceptible nonimmune persons to enable local circulation. This story of human acquisition of new infections from natural sources is continued in the next chapter.

Aboriginal Australia and the Last Glacial Maximum

The people who migrated from Southeast Asia to northern Australia at least 55,000 years ago completed humanity's first great sea crossing. Until 8,000 years ago, Australia was joined to Papua New Guinea by a landmass called Sahul. But even when sea levels were at their lowest, the traveling distance across the sea between Timor and Sahul was at least 90 kilometers. This pioneering group departed early enough to have been free from typical crowd diseases and animal-source infections, such as measles, smallpox, chickenpox, influenza, and cholera.[67] However, they faced an extremely hostile environment as the Earth cooled towards the glacial maximum of the late Pleistocene. In the northern hemisphere, ice-sheets reached their greatest extent about 26,500 years ago, but in the southern hemisphere this period was distinguished by intense aridity.[68] Protracted drought caused groups to gradually retreat to the foothills of mountain ranges and the riverine environments that still held reliable water supplies. Population densities decreased, and social networks became fragmented. The mortality rate in central Australia might have been as high as 10 to 25 percent.[69] Archaeologist Peter Thorley wrote, "The glacial maximum provides the first and most crucial test of human adjustment to a truly arid environment."[70] The Aboriginal people of Australia faced severe environmental and social stresses as the arid zone extended to almost the entire continent.

Skeletal remains suggest that during this cold and desiccated period, desert winds caused eye infections and diseases such as trachoma. "The very dry, dusty and wind-blown marginal areas of inland Australia that were occupied between 10 and 25,000 years ago would have provided an excellent environment for the transmission of this disease," wrote archaeologist Stephen Webb.[71] Cultural and socioeconomic adaptations such as increased mobility were essential means by which Aboriginal people managed to survive what was perhaps the toughest climate challenge humans have faced so far, albeit not without loss of life, increased disease, forced migration, and severe disruption to cultural and social systems.

Bad news for Neanderthals: Homo sapiens enters Europe

Groups of *Homo sapiens* began spreading into the European region from around 45,000 years ago, both from the Middle East and the Asian steppes. They had begun to make finely honed artifacts from stone, antler, and bone, foreshadowing an upward step into Aurignacian technology. In Europe, symbolic cave graffiti were most probably the work of the newly arrived *Homo sapiens* (Cro-Magnon Man). The very early symbolic art in Spanish caves, such as red ocher discs and stenciled hands, dates from around 40,000 to 35,000 years ago.[72] Those dates do not preclude early mural decorations by Neanderthals, but something else indicates that this advancing creativity was distinctively Cro-Magnon—their facial features were becoming less craggy as brow ridges receded, heads became rounder, and eye sockets less deep. These features, all affected by testosterone levels, suggest that aggressive behavior was making room for increased capacities for cooperation, compassion, and creativity.[73]

The Neanderthals, meanwhile, were adapting as best they could to growing environmental pressures from the continued cooling, the southward creep of ice and tundra, increased climatic variability, and dwindling prey species. They too were developing more refined and larger stone blade technology and improving their hunting tools. By 30,000 years ago, Neanderthals had mostly contracted into southwestern Europe, inhabiting the more open woodlands of today's southern Spain and Portugal.

Uncertainty remains about when, where, and why the Neanderthals died out. Some DNA evidence suggests that they may not have survived in their refuge in southern Spain beyond 30,000 years ago, although they may have hung on around the coast of Gibraltar for several thousand more years.[74,75,76] A major acute cooling event occurred at around that time, causing a severe drought in southwestern Europe that extinguished some mammal species and would thus have depleted food supplies for remnant Neanderthal groups. This may have been the final blow; their numbers had been gradually declining since around 50,000 years ago, as evident in the progressive reduction in their genetic diversity.[77]

We sometimes think of the Neanderthals as our rather slow-witted and ungainly ancestors, not well suited to the modern world. In fact, they stuck it out for more than twice as long as we modern *Homo sapiens* have so far managed to do, and they adapted their places and ways of living to a succession of four glacial periods. If climate change contributed to their final demise, as seems likely, it was at least not a climate change of *their* making. Will we, their cousins and the seeming success story of the hominin family, survive another 200,000 years if our slightly more advanced brains and behaviors create, rather than preempt, a global climate that may become unlike any that has existed for the past 10–15 million years?

The Neanderthals did not vanish without a trace; some modern humans may be carrying some of their genes. Some researchers believe interbreeding occurred between the *sapiens* and the Neanderthals during their shared occupancy of southern Europe and nearby West Asia. In non-Africans, around 2 percent of an average *sapiens* individual's genes are said to come from Neanderthals, and a small occurrence of interbreeding after the first migrations out of Africa is a plausible explanation.[78,79] But the actual genes acquired are not a standard package; they vary between individuals by "luck of the draw." In addition to Neanderthal genes for red hair, for skin type attuned to the prevailing chilly environment, and for an immunity-related gene enabling detection of "foreign-looking" amino acids,[80] other acquired genes include some that influence susceptibility to liver cirrhosis, Crohn's (inflammatory) disease of the bowel, and type II diabetes.

The Denisovans also may have interbred with *Homo sapiens*, particularly the branch that then spread through Oceania. Interbreeding would have been facilitated by contacts between these three "cousin" populations as they shifted habitation ranges in response to the fluctuating

Eurasian climate. Overall, an estimated 3–4 percent of the average individual human genome in populations outside Africa comes from interbreeding with those other two *Homo* species. In total, it appears that around one-third of all Neanderthal and Denisovan genes were passed on to *sapiens* humans.

Today's Tibetans, for example, have apparently acquired from the Denisovans a gene variant that enhanced oxygen uptake by red blood cell hemoglobin, facilitating human survival at high altitude.[81] The *sapiens* immune system was a particular beneficiary of this interbreeding, since both the Neanderthals and Denisovans had been adapting to Eurasian environments and their particular microbes for several hundred thousand years before coming into contact with the newly arrived *Homo sapiens*. The HLA genes of these two cousin-species of ours, crucial for launching antibodies against foreign microbial invaders, would have been fine-tuned by evolution long ago to counter the local infectious agents.[82]

New human vistas: The Pleistocene-Holocene transition and the rise of agriculture

The glacial period ended around 18,000 years ago, and the ensuing warming carried the world into the Holocene, by which time only one of the five later *Homo* species was left standing.[83] *Homo erectus* fossils have been found up to 70,000 years ago in Eurasia, and one subbranch survived to around 35,000 years ago in Java. The Neanderthals had apparently disappeared by 25,000 years ago.[84,85] The Denisovans probably died out around the same time. The diminutive *Homo floresiensis* had gone by around 17,000 years ago, along with the island's pygmy elephant, *Stegodon*.

During the post-glacial warming, massive ice-sheets several kilometers thick began to recede in northern Eurasia and North America, rainfall replaced snow at increasingly high latitudes, and new well-vegetated feeding grounds opened up for animals. As animal numbers increased, so did the numbers of hunter-gatherers, and as frozen rivers thawed and coastlines extended, fish stocks increased and fishing communities thrived. Prospects looked good for hunters, though less so for the megafauna. Was it the changing climate or *Homo sapiens* "*rampant*" that finished off the megafauna?

The postglacial change in climate evidently influenced megafauna extinctions,[86,87] partly via a substantial decline in the protein-rich forbs (flowering herbs such as today's clover) during and following the Last Glacial Maximum. This particularly hastened the extinction of large protein-dependent megafauna, including woolly mammoths.[88] Meanwhile, during this protracted thaw and warming, many large and medium-sized herd animals hunted by the Cro-Magnons in the European region, such as woolly rhinoceros and giant deer, went extinct, to be replaced by smaller and more dispersed game, including red deer, wild boar, and aurochs. Similar megafauna die-offs occurred in other continents. As the bonanza of megafauna disappeared, the geographic range of many other food sources also shifted, producing a gradual transformation in human diets and related behaviors. In some regions, the needs of hunter-gatherers probably exceeded the available yields of hunting and foraging.[89] New food sources and methods of acquisition were needed. Radical change was on the horizon.

The Natufians and the Younger Dryas

From around 14,000 years ago, small communities, called Natufian, had begun settling in the Levant, encompassing today's Israel, Palestine, Jordan, Lebanon, and much of Syria—part of the "Fertile Crescent" (Figure 4.5). Wild food supplies were now sufficient for a hunting-and-foraging subsistence economy based on settlement, not nomadism.[90] The earliest known Natufian settlement was at Abu Hureyra in today's northern Syria, upstream on the Euphrates River. Others were dotted along the eastern Mediterranean coastal region, through Lebanon down to southern Palestine. Some settlers noticed that gathering the seeds of high-yield patches of wild grasses caused incidental reseeding and regrowth of similar high-yield plants,[91] but there is no evidence of systematic "farming" by the Natufians.

Meanwhile, in nearby Egypt early settlements were forming along the fertile Nile Valley, which, with its abundant river-fish resources and well-nourished seed-bearing grasses on the river-plains, made hunting and gathering from a settled base easy. As the Late Pleistocene world warmed and became less arid, rains strengthened in the Nile's source

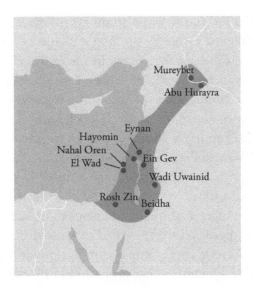

FIGURE 4.5 Map of Natufian sites in the Eastern Mediterranean. Wikipedia.[92]

regions of Ethiopia (the Blue Nile) and eastern Africa (the White Nile), and the annual flows and flooding of the river began to increase. Indeed, there was a tumultuous "Wild Nile" period around 13,000 years ago, with heavy rains, a northern expansion of the monsoon belt, and catastrophic flooding.[93] Food supplies faltered, and many settlements were abandoned. But more was to come.

Suddenly the global climate trend went into reverse. From around 12,800 years ago, average temperatures in the northern hemisphere plunged by several degrees Celsius in just several centuries. This recidivist burst of cold climate, the Younger Dryas episode,[94] persisted for over a thousand years. It coincided with the extinction of megafauna in both North America and Europe and with cultural changes in the Clovis Paleo-Indian population of southeastern North America. The episode may have resulted from the sudden release into the Atlantic Ocean of a huge pulse of fresh meltwater from vast inland lakes in Canada that arose from the melting Laurentide ice-sheet,[95] which would have also disrupted the release of Gulf Stream heat (Chapter 2) that warms the European region and Middle East. The abrupt cooling could also have been triggered by a massive comet explosion that apparently occurred over Canada around 12,900 years ago, causing shrouded skies, plummeting temperatures, and rising rates of megafauna extinction.[96,97]

Food supplies for human populations far and wide were affected. In Egypt's Nile Valley, the number of camp settlements declined dramatically during this time. Further, many of the excavated human skeletal remains have shattered skulls, evidence of violence. This presumably reflected intense competition for dwindling supplies of river fish and wild cereal grasses under the abruptly cooler and drier conditions.[98] The British Museum has a special display in its Ancient Egyptian section of several of these shattered and pierced skulls, dating from around 13 millennia ago. In Southern Africa, the number of stable settlements halved during this time,[99] while in the northern Fertile Crescent food shortages caused many Natufian settlements to disband. The Natufian settlements that persisted then resorted to more deliberate cultivation of cold-tolerant rye grass,[100] thus becoming forerunners of the more successful settled agriculture that evolved after the warming trend resumed around 11,500 years ago, ushering in the relatively stable climate of the Holocene.[101]

In North America, another part of the story was unfolding, with the arrival and spread of human groups and their stone technologies.[102] As noted above, several extended-family groups had drifted east across the temporarily exposed Beringian land bridge between Eurasia and North America around the end of the last glacial period. Subsequently, as north-south corridors opened up in the great melting ice-sheets that covered today's Canada, several of those human groups edged southward to warmer climes. Others pushed east and established an Amerindian way of life in the ice-bound Arctic.

Paleo-Indians first arrived in the southeast of the North American continent around 14,000 years ago after a north-south corridor had opened up between the two northern ice-sheets.[103] More migration followed, dispersing more widely across the open plains and giving rise to the dominant Clovis culture in the southeast of North America from around 13,500 years ago. Then suddenly there came the rapid Younger Dryas cooling, during which the megafauna in North America, including mastodons, camels, and saber-toothed cats, all vanished.[104] Some differentiation and fragmentation of the Clovis culture also occurred as environmental conditions changed and a hunter-gatherer way of life emerged based substantially on diets of smaller animals, roots, and berries.[105]

Then, from around 11,500 years ago, temperatures rose again, heading towards the warmer and less erratic Holocene climate. Agriculture was just around the corner.

Protoagriculture emerges

The tentative early phase of supplementary cereal cultivation, late in the Pleistocene, slowly evolved into grain-and-livestock farming systems. As new settlements spread around the Fertile Crescent, more systematic grain cultivation developed. But achieving food security was still not an easy off-the-shelf option; full subsistence agrarianism remained some way off. For a start, the range of edible plants underwent changes as the late Pleistocene gave way to the Holocene. In northern Syria, for example, rye grass and flood-plain plants were gradually replaced by barley, emmer wheat, pulses, lentils, peas, horse-beans, and vetches—many of which would support successful future agrarian life in that region.[106] In some places, ancestral local plants re-established themselves, although their yields in the now warmer climate were probably less than they had been a millennium or two earlier. The transition to agrarian living was not seamless; this was a trial-and-error process.

Why this radical transition in living occurred remains a complex and unresolved question. The fact that, within a span of around 2,000 years, agriculture appeared in half a dozen main centers around the world that apparently had no contact with one another implies something about the Holocene's new climatic-environmental conditions that favored harvesting and planting rather than digging and gathering.

Once rudimentary farming appeared, the die was cast; there would be no looking back. Over the preceding two million years, meat-eating and then the use of fire for cooking, heating, and hunting had been great steps forward in nutrition and survival. Agriculture, though, entailed a much more radical and comprehensive shift, entailing great changes in daily diets, work demands, social relations, political institutions, technologies, notions of property (and hence combat)—and health and survival. Gradual gains in harvest yields, food storage facilities, and shared knowledge about farming practices made early agrarian living an attractive option, especially where there were adequate wild plants and animals on hand for domestication.[107] This, as we shall see, was the case in various regions of the world.

Meanwhile, nature continued to create random genetic variants. During those early centuries of agriculture, several of the region's wheatgrasses fortuitously underwent spontaneous "polyploidy" multiplication of their total genetic DNA package. Of the five types of wheat species grown by early Egyptians, four were of polyploid type—emmer, durum, spelt, and bread-wheat.[108] These larger-sized grains were most likely to be harvested and later resown, and this increased the total yield of dietary staples and hence the daily food energy intake. Polyploid mutations of this kind occurred naturally in all the cereal grasses and other cultivars of early farming and, via selective cultivation, were destined to become the progenitors of the higher-yielding varieties that modernity has inherited.[109,110]

Living by bread alone

Yet agriculture, responding to new vistas opened up by the shift from the Late Pleistocene to the Holocene climate, would prove a mixed blessing. Beyond the greater amount of food that could be produced and the food security conferred by stored grains, there were significant deficits and risks. Early agrarian diets were restricted in their nutrient content, because they were based on a narrow range of cultivatable foods. Human skeletal remains from those early farmers show that the introduction of grains and other plants and the resulting contraction in dietary variety caused a decline in growth of bone and dentition and in general health and changes in patterns of daily activity.[111] In some regions, such as those with access to marine resources, the transition to agriculture had little effect on stature. However, the majority of studies show bodily growth and stature were compromised; adult heights declined by as much as 15 cm in populations that switched to agriculture-based diets and social systems.[112] In Australia, on the fertile River Murray floodplain, where the Holocene population began to increase in density and intensify its food production, diets switched from high-protein to high-carbohydrate foods. The change in nutrition was probably the cause of the anemia evident in bones from the region, and sticky carbohydrates caused plaque and caries in people's teeth.[113] Stephen Webb argued that crowding and sedentary lifestyles created a new "infectious environment," with "the highest frequency of non-specific infection in Australia ... found in the central Murray area."[114] Non-specific infection (types of bone infection) occurred in many populations around the world with the transition to agriculture.[115]

The climate-influenced disappearance of the megafaunal meat bounty, previously available to hunter-gatherers from mammoths, woolly rhinoceroses, giant deer, and other prey animals, further jeopardized dietary quality and hence agrarian health and vitality during the first two millennia of farming. This led eventually to moves to ensure readier access to animal foods. The domestication of herd animals emerged around 10,500 years ago probably in the Zagros Mountains around the northeastern reaches of the Tigris River.[116] Sheep and goats, ungulates that moved in more concentrated herds, especially where surface water was sparse, were the first to be domesticated; they were followed later by wild boar, hence pigs. From around 10,000 years ago, cattle (from wild aurochs) were herded and then domesticated. Once herded and penned, the selective culling of animals favored smaller size, plumper products, and docility. Milk, wool, and hides were also important. The addition of meat and milk to the agrarian diet gradually spread throughout the Fertile Crescent and into northern Africa,[117] providing a more assured and high-quality protein supply.

The domestication of plants and then land animals for food is now many thousands of years behind us. Agriculture created new forms of living, property ownership, and social hierarchies, which produced new risks for the well-being and safety of communities as diseases emerged and predatory raids by outsiders increased. Agricultural activities also resulted in increasingly damaging pressures on soils, watersheds, and much else in the natural environment. We are now on track to complete the takeover of the world's food resources, as we deplete the ocean's wild fisheries and become reliant on aquaculture. Overall, humans have usurped more than one-quarter of the planet's primary photosynthetic product to feed itself.[118] The Faustian bargain's offer of opportunity and resources to achieve abundance is being pushed to new limits as the environmental base of the world's food system increasingly shows the strain.

Concluding remarks

As early farming emerged in the Levant from around 11,000 years ago, the world's climate was settling in to a global average temperature about 6°C warmer than at the end of the last glacial period.[119] From around 18,000 years ago, the temperature rose by an average of 1°C per thousand years—fast by geological standards, but a mere one-thirtieth of the likely warming rate of, say, 3°C during this century. The Holocene's global average temperature has varied modestly around an otherwise fairly steady figure of 15°C. On a century-to-century scale, regional temperatures have varied within a range of plus or minus 1°C, with many larger, though often local, transient fluctuations.[120] This relative stability in climatic conditions has facilitated the rise and spread of agriculture and much that followed from it: the growth of towns, cities, and civilizations.

The Fertile Crescent has long been regarded as the clear front-runner in the origin of agriculture. In fact, studies of the genetic ancestry of regional food-plant staples, plus archaeological findings, show that agriculture emerged in disparate parts of the world much closer in time than previously thought.[121,122,123] Indeed, the Younger Dryas cooling, evident in the Middle East, impinged elsewhere and spurred communities in China, the Papua New Guinea highlands, Central America, and the Andean-Amazonian region to seek more secure supplies of staple plant foods.[124] For those five particular regions, the earliest evidence of domesticated staple foods (wheat, rice, taro and other tubers, corn, and cucurbits, respectively) lies within the range of 10,500 to 8,000 years ago.[125]

In northern and western Africa, from around 7,000 years ago, parts of the then moist, well-vegetated, and heavily populated Sahara enabled domestication of millet, sorghum, and cowpeas. But poor soil and limited rain throughout much of the region, along with the relatively low calorie value of these crops, precluded their being intensively farmed, especially during a prolonged drying period that set in for five centuries about 6,300 years ago. The cultivation of millet, sorghum, and peas spread in and beyond the Sahelian region in West Africa, but the limited harvest surpluses constrained population growth. Indeed, substantial land-clearing for agriculture did not occur in the more southern, monsoon-dependent regions until around 4,000 years ago.

Meanwhile, farming in Egypt's fertile, well-watered Nile Valley was gathering pace, with cereal grains, fruits, and, later, some vegetables. Settlements in Egypt and neighboring southern Mesopotamia would soon set the pace for the evolution of social integration and the features of what we call civilization.

Holocene climatic stability facilitated the development of agriculture by allowing adequate time for learning how to domesticate a particular plant in a particular region and then for its selective breeding to improve the size of seed, height of stalk, and retention of seeds on the stalk, contrary to nature's need for plants to disperse their seeds. The domestication of wild herds of animals was facilitated similarly. Indeed, the overall cultural transition from hunter-gathering to settled agrarian living benefited from climatic constancy, allowing trial-and-error approaches to building dwellings, laying out settlements, corralling livestock, improving farming tools, developing a barter economy, and reshaping cultural values and social relationships.

Overall, this process typically spanned several thousand years— longer in the Americas than in Eurasia.[126] This accords with Jared Diamond's conclusion that not only were the natural biological resources for agriculture, including animal husbandry, more abundant in Eurasia, but ideas, methods, and food types were more easily exchanged across Eurasia's similar east-west climatic zones than via north-south exchanges over differing climate zones and through forested terrain in the Americas and eastern Africa.[127]

The planting of wild grasses and other crops, initially low in yield and dependent on rain, temperature, and sun, was labor-intensive and risky. Reliance on just a few select species was an inherently less resilient strategy than having access in the wild, during most climatic periods, to a wide and seasonal range of plant and animal food species. The relative stability of the Holocene climate may well have made the critical difference between success and failure of this novel system of food production. Over subsequent millennia, the vagaries of weather and changes in prevailing climate would take a recurring, sometimes disastrous and lethal, toll. But farming was here to stay.

5

Spread of Farming, New Diseases, and Rising Civilizations

Mid-Holocene Optimum

AS THE EARTH WARMED after the last glacial maximum, temperatures fluctuated. About 9700 B.C.E., temperatures rose again suddenly and began to stabilize, marking the beginning of a new geological epoch, the Holocene. The landscape continued to change, but not so fast that a single generation of humans would have noticed. Ice-sheets and tundra were receding in Eurasia, and over time human groups, both hunter-gatherers and then early farmer-pastoralist communities, adjusted their ways of living to warmer conditions and different rainfall patterns.

Small-scale farming and herding emerged on all nonpolar continents during the period 8500 to 6000 B.C.E., predominantly in the northern hemisphere, while human numbers were creeping up. These great changes in environmental conditions and subsequent cultural practices had a profound influence on the foundations of human health and survival: food sufficiency and quality, water supplies, contacts with infectious agents, modes of settlement, and social relations. A new era in human ecology was looming. Farming increased food production, but the switch to dependency on a few staples decreased diversity of diets and created an annual agricultural regime more susceptible to climate shifts. Close contact with animals, standing water in irrigated environments, and denser settlements provided opportunities for microbes, pathogens, viruses, and parasites to cross species barriers and infect and spread among human populations.

During the Early Holocene, from about 9700 B.C.E. to 6000 B.C.E., the earth was subjected to the competing stresses of high solar influence and still massive melting ice-sheets. From around 6000 B.C.E., the majority of ice-sheet melting had abated, allowing the stabilization of the Earth's climate into what can be called the Mid-Holocene Climatic Optimum (approx. 6000 to 3000 B.C.E.). This was a change in climate that spanned 3,000 to 4,000 years. Warming was most evident in the northern hemisphere, influenced by the peaking of solar radiation at higher northern latitudes as the 23,000-year Milankovitch "wobble" cycle maximized northern sun exposure for several millennia. The Milankovitch cycle also drew the rain-bearing Inter-Tropical Convergence Zone (ITCZ) further north.[1] Warming in the north influenced human populations in many ways, including changes in food sources and nutrition and later, as settlements expanded, by exposure to new infections from domesticated and urban-pest animal sources. In contrast, this phase of the cycle diminished the warming of the equatorial-southern Pacific and thus curbed El Niño activity and its drying influence in Eurasia.[2]

Settled agricultural living enabled birth rates to rise. Young children could be weaned earlier onto cereal foods and milk, enabling women to conceive again, and families were no longer subject to size constraint by the logistics of nomadic journeys. Lives and property were now a little more secure. However, the natural vicissitudes of annual weather were an ever-present hazard to agrarians: crops needed rain.

Human adaptation to "unnatural" agrarian diets

During the first phase of the Mid-Holocene Climatic Optimum much of Europe experienced temperatures that were periodically 1°C–2°C warmer than the Holocene average temperature, though mostly in the summer months.[3] Weather patterns were relatively stable, as was the "positive" North Atlantic Oscillation, and stronger westerly winds were bringing more rain from the Atlantic.

These improved climatic conditions assisted the early spread of farming from the eastern Mediterranean region into the Balkans. Later, a brief period of unfavorable climatic conditions beginning around 4500 B.C.E., marked by drying on both sides of the Mediterranean, may have stimulated farming communities in southeastern Europe to seek moister climes by spreading further northwest into the forests and woodlands of Europe, adding further momentum to the diffusion of the farming idea.[4] Spreading west and northwest from Fertile Crescent origins, farming settlements and their cleared land gradually replaced the sparsely populated forested world of European hunters and gatherers. This was not an invasion in which farmers overwhelmed hunter-gatherers. Rather, the radical idea of *farming* itself was spreading, not hordes of *farmers*.[5] The agrarian idea and its apparent benefits gradually caught on, leading to hybridization of ancient and modern ways of living as farming spread, kilometer by kilometer, across Europe's natural landscape.

As farming spread in Europe, West Asia, and northern Africa, the reliance on wild foods was supplemented and then replaced by cereal grains and ready supplies of meat and milk. Diets were becoming increasingly agrarian; in particular, there were now a greater reliance on staples and a reduced diversity of foods. This posed physiological and metabolic demands on a Pleistocene hunter-gatherer biology that had evolved in a setting of seasonally varying and diverse wild food sources. So, once again, natural selection was called into play. Individuals whose genetic profile enabled more efficient digestion of these dietary novelties, including cereal-grain with its gluten protein and the milk sugar, lactose, survived best on this "unnatural" agrarian diet.

Lactose digestion is an interesting example: human infants naturally digest lactose in maternal milk, thanks to the digestive enzyme lactase. After weaning, the gene controlling lactase production switches off, since, in nature, there is no further need in humans and other mammals on milk-free adult diets to use metabolic energy to run a now superfluous genetic app. However, the advent of dairying changed all that. There was now potential benefit in retaining the capacity to digest lactose; hence natural selection duly favored those genetically variant individuals whose lactase remained active, following the random occurrence of a particular genetic mutation soon after dairying emerged.[6] They are the "lactose tolerant" individuals. (Here the word "tolerant" is not a compliment; it refers to those who have forfeited an ancient mammalian genetic switching mechanism.)

Today, there is a geographic gradient in the proportion of lactose tolerant persons across regions of Europe; likewise, for tolerance of gluten from wheat. These gradients are echoes from the past; they reflect differences in the length of time since each particular region first encountered these new dietary exposures as farming spread north and west through Europe.[7,8] The shorter or less intense the population's exposure, the higher the residual proportion of so-called "intolerant" persons. Hence gluten intolerance (or "celiac disease") is more prevalent today in the British Isles than in the Balkans or Turkey.

Scandinavians are the most lactose tolerant regional population in Europe. They were not the first to drink milk—farming reached that region well after its peri-European origins in the Fertile Crescent—but their colder climate restricted the availability of cereal grains and enforced greater reliance on animal foods. However, the story goes further; another advantage to biological health presumably reinforced selection favoring lactose tolerance. Sun exposure stimulates the skin to synthesize vitamin D, which then enhances the uptake of dietary calcium, without which bone weakness occurs. But because of the lower exposure to sunlight at higher latitudes, and since dairy foods are rich in both calcium and the vitamin D precursor molecule, lactose tolerance would also benefit bone growth.[9] Weakened limb bones are bad enough, but the lack of calcium and vitamin D can even distort the female pelvis via daily inward pressure from the hip sockets. And a misshapen pelvis jeopardizes birthing, because the large fetal head may be obstructed by a deformed narrow pelvic opening.[10,11] It might seem paradoxical that the Amerindian populations living above the Arctic Circle have retained their Stone Age ancestors' dark skins. However, a diet of fish and marine mammals provides plenty of vitamin D, and, besides, the intense ultraviolet glare of an icebound landscape damages lightly pigmented skin. Natural selection, ever pragmatic, acts to balance the two risks.

The retention of lactase activity was not confined to European populations. The early domestication of cattle in the Middle East also led to the migration and spread of cattle herding into northern Africa.[12] During the Early Holocene warming period, the stretch of land spanning the northern Arabian Peninsula and the Sinai desert was green and suited to pastoralism. Migrating herders, now able to move with their flocks from Palestine into Egypt and northern Africa, may have carried lactose tolerant genes into northern Africa, or spontaneous lactase-retaining mutations may also have occurred among the several African pastoralist tribes—eastern Africa's Masai, the Sahelian Fulani, the Tuareg of Mali, and others.[13] Animal bones show that soon after 8,000 years ago, cattle, sheep, and goats were widespread across the Saharan savannahs.[14] Early dairying in North Africa is plausible, given the rapid advance from herding to milking that had occurred in the Middle East. Corroboration comes from pictorial evidence of Saharan cattle herding, in the region's extensive rock paintings—including depicting the actual milking of a cow (Figure 5.1). Further, chemical isotope analyses show traces of dairy fats in food residues on unglazed pottery in Libya, dating from 7,000 years ago.

Dairy-cattle herding broadened the food security base for the Saharan pastoralists, but there were no certainties about long-term climatic security, and before long the Saharan environment began to cool and dry as regional climatic zones shifted.

FIGURE 5.1 Rock art from Teshuinat II rock shelter, southwest Libya. This tracing shows Saharan pastoralists with pots and cattle. Reproduced in Dunne et al., "First Dairying in Green Saharan Africa in the Fifth Millennium BC."[15]

Permission received from Nature Publishing.

The Sahara dries: Food and water shortages

Ten thousand years ago, Saharan Africa was much wetter and greener than the desert we know today. A temporary northern shift of climatic zones meant that the West African Monsoon, which blows in from the Gulf of Guinea, carried its rain further inland.[16,17,18] The resulting abundance of water, cereal grasses, and game animals enabled Saharan hunter-gatherers to live a semi-sedentary life in the Early Holocene, as is evident from their pottery-making.

A drier intermission occurred in the Sahara during the several centuries immediately before and after 6000 B.C.E. Vegetation and wildlife receded, the newly introduced domesticated cattle had less pasture, and various hunter-gatherer and pastoralist groups withdrew south towards the moister Saharan fringes. When wetter climes returned, refurbishing the savannah-like vegetation and revitalizing rivers and waterholes and many large animal species, the Sahara was again repopulated by human groups.

By 6,500 years ago, a strong cooling and drying phase that lasted for half a millennium had emerged in the northern hemisphere, interrupting the Mid-Holocene Climatic Optimum. This coincided with a weakening of the 2,300-year Hallstatt solar cycle, the same cycle that influenced subsequent cooling periods, especially in the northern hemisphere, centered at approximately 2500 B.C.E., 350 B.C.E., and 1600 C.E. (the nadir of the Little Ice Age).[19] There was now a gradual southward shift in the rain-bearing equatorial ITCZ and the drier subtropical ridge in the Mediterranean southern coastal region, causing widespread drying that encompassed the recently verdant Sahara. Well-established nomadic pastoralist communities and their herds suffered from increasing food and water shortages and migrated toward the Sahara's wetter fringes.

As desertification of the Sahara intensified, it extended later in the seventh millennium B.C.E. to severe drying in the eastern Sahara's Chad Basin, rendering much of the land uninhabitable. Displaced groups from that basin migrated further east towards the middle reaches of the Nile Valley, where they and their successors developed communities of mixed farming. One particular displaced group, from Nabta in the semi-desert region of southwest Egypt, took their more advanced cosmology, megalith technologies, and social complexity with them when they resettled in the Nile Valley. These and other in-migrating groups added cultural and technological depth to the early agricultural ventures in the fertile Nile Valley.

The Tenereans

The history of the Tenerean people during the drying in the eastern Sahara provides insights into complex patterns of biosocial evolution and cultural adaptation in response to climate fluctuations.[20] The re-vegetation and drying of the Saraha was one of the most extreme climate shifts during the Holocene. Excavations at Gobero, a paleolake in present-day Niger, revealed much about the people who occupied this area. Archaeologists uncovered scores of occupation artifacts, including human burials spanning multiple generations. The first inhabitants abandoned the lake as it dried about 6200 B.C.E. About 1,000 years later the humid climate returned, the lake refilled, and people believed to be the Tenereans occupied the site. Researchers have gleaned information about their food sources, their skeletal size and health, their artifacts, and their burial practices. Identifiable remains from 54 animal species, 20 kinds of trees, and 30 kinds of shrubs, grasses, and algae were found at the Gobero site.

Food leftovers, mostly shells, bones, and teeth found in middens, reveal a diversity of animals that were eaten, including clams, catfish, and tilapia, along with lesser intake of antelopes, hippos, crocodiles, and turtles. Drought did not always result in famine. Excavations at Gobero in the Chad Basin show how the Tenerean people adapted to the wetter period that lasted from around 5500 B.C.E. to 3000 B.C.E.,[21] followed by dessication of the region. Domesticated cattle were eaten less often, which suggests that the main diet was that of a subsistence economy based on fishing, hunting savannah vertebrates, and gathering grain from selected wild grasses. The subsequent onset of drying is mirrored in the changing content of middens at Gobero, showing that dietary sources had become more opportunistic and diverse as aridity increased. Similar findings from that time, including some skeletal evidence of impaired nutritional status, have been made further west in the Sahara as the climate changed and drier conditions prevailed.[22]

This Saharan climatic experience, in contrast to many other important historical droughts, did not result in a clear sequence of drought and famine for three reasons. First, the Saharan drying progressed over 500 years or more. Second, those affected were mobile nomads able to relocate gradually without forfeiting property and with access to adjoining land. Third, the migrants remained flexible in their dietary sources and preferences, free of the settled agrarians' prime dependence on rain and sun for harvests of specific staple foods—a fateful dependence that became increasingly clear later around 3000 B.C.E. in Mesopotamia, Egypt, and the Indus Valley.

Egypt and Sumer: Rivers, rainfall, and early civilizations

Early settlements in Egypt and Mesopotamia benefited from major river systems and climates that produced good regional monsoons. Egypt's population was concentrated along the Nile Valley; early Sumer's population settled between the Tigris and Euphrates Rivers.

Early writing appeared in both societies around 3000 B.C.E., mainly for administrative purposes in Egypt and for accounting and trading in Sumer. Figure 5.2 shows the timeline for the central span of the Holocene from 6000 B.C.E to year zero B.C.E.

The Nile Valley: Early stirrings

Settlements in the verdant Nile Valley were already forming by the beginning of the Mid-Holocene. The in-migration from the drying eastern Sahara provided an extra boost. Daily life and politics were becoming more complex. Competition between communities emerged, and before long, local kingdoms were vying for territory and supremacy. By around 3300 B.C.E., the foundations of the mighty civilization of Egypt were forming.[23] Nile Valley settlers were supported by the unusually fertile riverine environment. Their fields were rejuvenated annually when, in late summer and early spring, this great river overflowed its banks and spread a thin layer of rich silt over the flood plains. This nourishing annual cycle, fed by the headwaters of the Blue Nile and the White Nile, would underpin Egyptian civilization during its subsequent growth over the next three millennia (and, indeed, to the present day). The Greek historian Herodotus

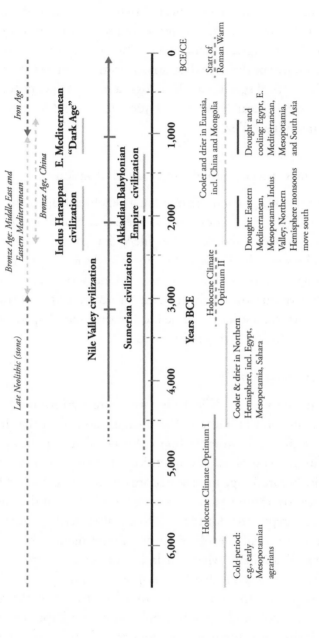

FIGURE 5.2 A timeline during 6000 to 0 B.C.E., focused on climate change (in Eurasia and Egypt), the growth of those early civilizations, and the successive technological ages.

Source: Author.

wrote of Egypt's "gift of the Nile," the object of great envy around the Mediterranean world.

But the Nile was also unpredictable, and its variable flow brought episodes of famine and death for Egypt. A recurring climatic influence on this annual rhythm was the impact of the El Niño Southern Oscillation on rainfall in highland and northern Ethiopia, the Blue Nile's origin. During El Niño events, the range and strength of the South Asian summer monsoon (and the related East African monsoon) were affected, resulting in less rain falling in those highlands.

The Nile flow faltered briefly around 3400 B.C.E., and then, several centuries later, it declined markedly. The resulting food crisis and hunger fueled tensions, causing conflict between local kingdoms. The eventual resolution led to the political unification of Upper Egypt, extending north from Aswan to the southern fringe of the Nile Delta. The new and powerful rulers imposed a centrally organized management of the river water flow to reduce the risks to farm yields in the Nile Delta. At about that time, simple pictograms on clay tablets came into use, primarily to record and to instruct, including prescribing the requirements of water system management. The vagaries of the Nile and the need for centralized control of the hydraulics of food production were a major stimulus for a politically unified and more socially stratified Egypt from around 3200 B.C.E. The First Dynasty emerged, and by the middle of the next millennium, Egypt's Old Kingdom was established.

Mesopotamia: Beginnings of a city-state civilization

The small agrarian communities that had formed soon after the Early Holocene Climatic Optimum in southern Mesopotamia were growing. The cooler and drier phase during the Mid-Holocene prompted the harnessing of river water for irrigation. The heartland of southern Mesopotamia comprised the floodplains of the lower Tigris and Euphrates Rivers before they joined and flowed into the Persian Gulf (Figure 5.3). Farmers in southern Mesopotamia took advantage of the rich alluvial soils along the southern reaches of this dual river system, fed from the north by the winter rains blown in from the Atlantic.

Those soils had been further enriched by the temporary inland extension of the Persian Gulf when the post-glaciation rise in sea level

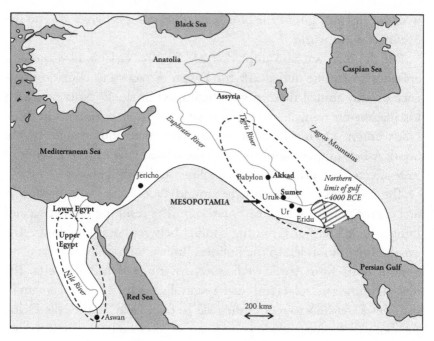

FIGURE 5.3 The Fertile Crescent region, indicating the civilizations of early Egypt and Mesopotamia inside dashed ellipses. Adapted from Free Maps.[24]

peaked around 6,000 years ago, approximately a meter higher than today.[25] The refilled gulf extended several hundred kilometers further inland, to the margins of the elevated site of the future Sumerian city of Ur.[26] The low-lying marshlands of today's southern Iraq, inhabited by the Shia Muslim marsh-dwellers, were temporarily part of a shallow seabed. The 1,000-year cooling intermission in the middle of the Holocene Climatic Optimum caused the top of the gulf to retreat,[27] leaving behind a layer of nutrient-rich river silt spread across the erstwhile river delta, now coastal land.

Mesopotamia's climate is the product of a complex set of seasonally varying weather systems: the Atlantic circulation (west winds, warmth, and seasonal rain) driven by the North Atlantic Oscillation, latitude variations in the arid high-pressure subtropical zone, the Southwest Asian summer monsoon, and periodic cold dry air from the Siberian High.[28,29] For almost 2,000 years, the positive Atlantic weather pattern of the Holocene Climatic Optimum persisted. This,

along with the continuing development of methods of river-based irrigation, ensured the gradual and successful spread of agriculture in
both the southern deltaic and northern regions of Mesopotamia. By
around 5000 B.C.E., the landscape of southern Mesopotamia, Sumer,
was dotted with small towns, each comprising several thousand people
and surrounded by a network of smaller rural communities. Sumer
was emerging as the world's first example of region-wide agriculture
based on a trading network of connected villages and towns.

Here, too, was a pointer to the growing tide of human numbers
that would follow later in the Holocene. The *more food = more people*
equation was about to gather momentum. As the population grew
and the boundaries of Sumer's farmland pushed outward, irrigation
became increasingly necessary. Even so, and despite reduced rainfall
during the Mid-Holocene, this alluvial plain with nearby marshlands
and steppes as sources of wild birds and animals for hunting remained
attractive real estate. Hence many farming groups and villagers from
further upriver moved southward, and, as larger towns such as Ur and
Uruk formed, the foundations of the Mesopotamian civilization were
being laid.[30]

Settlement and agriculture: Diets, microbes, and human health

In this new, evolving agrarian environment, regional climatic trends and
fluctuations continually influenced patterns of daily and seasonal work
and the composition and nutritional adequacy of diets. The move from
foraging to farming involved a partial replacement of (hunted) meat
with cereal grains; hence diets now contained less protein and more carbohydrates. The decreased diversity of food sources reduced the range
and intake of micronutrients (vitamins, iron and other trace elements,
and some important amino acids for protein-building). Studies of skeletal remains from many sites around the world indicate that the move
to agrarianism resulted in various deficiency disorders that affected the
strength and micro-architecture of bones (causing osteoporosis) and
had detrimental effects on dental growth and oral health, eyesight,
and various metabolic and hormonal activities. Other health deficits
included higher rates of iron deficiency anemia and increased susceptibility to infection.[31] These nutritional consequences affected the growth

of children and teenagers in particular, and the overall health, strength, and longevity of adults.

Princes, priests, and peasants: Status and stature

In Mesopotamia, Egypt, and other centers of agricultural development, early settlements and villages grew and slowly evolved into larger towns, increasingly connected by trading networks; higher-order city-states and civilizations would follow. Skeletal remains reveal information about stature and bone health in different segments of these societies, indicating that social stratification intensified as settlements became more centralized and political networks became stronger. Rulers, priests, warriors, merchants, laborers, and peasant-farmers: these groups formed the basis of political structures, power, and privilege. The higher up the political pyramid you were, the greater your access to food. Adult height in the urban wealthy exceeded that in the rural workforce. Some of the skeletal evidence indicates that infectious diseases took their greatest toll on the poor and the peasantry. Bone inflammation (osteitis), tuberculosis, and dental disease, for example, are more prevalent in the bones of the underprivileged classes. Much of this difference in health and growth between classes reflected nutritional deprivation, while some of it was likely the result of chronic and repeated infections.

Differences in stature and health are evident on another fundamental axis of comparison: agrarians versus hunter-gatherers. In early agrarians, there are signs of early-life growth stunting, of arrested growth (presumably famine-related), of weakened bone micro-architecture and shorter adult stature by comparison with near-contemporary hunter-gatherers. In the wider eastern Mediterranean region during the post-glacial warming period from around 16,000 to 12,000 years ago, hunter-gatherers were approximately five centimeters taller on average than their counterparts in the early farming communities (which formed around 10,000 years ago).[32] Interestingly, the hunter-gatherers' stature exceeded that observed in skeletal remains dating to the Last Glacial Maximum 20,000–18,000 years ago, when life in southeastern Europe was at its coldest, hardest, and most food-insecure. There may have been a further contributory influence to the approximately 10-centimeter height deficit in both men and women at the Last Glacial

Maximum. Many prey animals and other food sources were shifting geographically and declining in number, and humans might have undergone a "micro-evolutionary adaptive process" that matched body size to the available amount of food and energy.[33]

Farming boosted the food yield per unit of land in favorable climatic conditions. This enabled larger families, and so the population grew, and farmers pressed more land into agricultural service. But all was not rosy; agrarian living required long hours of toiling in the fields, and it brought a new and crucial dependence on seasonal rainfall and, as irrigation evolved, on annual river flows. Given the natural year-to-year climatic fluctuations, food shortages would henceforth become frequent, and for the next 10 millennia, periodic food crises would follow.

Here was further grist for the mill of human evolution. The vagaries of climate posed recurring risks to the health and reproductive capacity of agrarian populations. Could biological evolution chance upon some genetically variant physical or metabolic adaptation that would help maintain fertility during times of undernutrition? The polycystic ovary syndrome (PCOS) suggests a fascinating example. It has long been a puzzle that the prevalence of the variant form (the PCO allele) of the relevant gene for this ovarian abnormality is so high in many adult female populations: typically 30–45 percent in a diversity of regional populations. Why would such an apparently detrimental gene, causing multiple cysts of the ovaries, have been conserved over many millennia? After all, in modern and increasingly overweight populations those individual women with the PCO genotype are much more likely to develop increased insulin resistance, leading to type 2 diabetes and ovarian infertility. But—and this is the key to the proposed explanation—in underweight women that same PCO genotype confers a *higher-than-average* level of fertility.

A small research group (to which I belonged) explored the published epidemiological and experimental research and concluded that in past times of serious food shortage and weight loss, often the result of climatic adversity, agrarian women with the PCO genotype would have been better able to conceive and bear children.[34] Since evolution sets a particularly high premium on fertility and child-bearing, this reproductive advantage would have been strongly selected for. Our "fertility first" hypothesis, with deep historical roots in evolution's selective responses to agrarian food shortages, thus offers an explanation for the retention of the PCO genetic allele. It was once an asset to the tribe or community, but in the relatively recent circumstances of modernity it is becoming a liability.

Pioneer microbes: Crossing the species barrier

Remarkably, we owe the origin of most serious infectious diseases to the conditions which led to our cultural heritage, the city states made possible by the planting of crops in the flood plains of Mesopotamia, Egypt and the Indus Valley.

—THOMAS MCKEOWN, *The Origins of Human Disease,* 1988[35]

As farming, land-clearing, animal husbandry, denser settlements, and trading connections spread, patterns of contact between humans and microbes began to change. The advent of the Holocene climate and environment fostered a way of living in which people were more intimately exposed to novel infectious agents, both from domesticated animal sources and from locally proliferating pest species such as rodents.[36,37] This radical change in human ecology began around 7,000 years ago. The acquisition of new infectious diseases during the agrarian era is at odds with a long-standing assumption that all such diseases derived from early Stone Age hunter-gatherer times[38]—when, in Thomas Hobbes' view four centuries ago, lives were "solitary, poor, nasty, brutish and short."[39] Certainly, early hunter-gatherers in Africa acquired many animal-source parasitic infections from butchering, meat-eating, and hides, but these tapeworms, liver flukes, and sleeping-sickness trypanosomes were not directly transferable to other fellow humans. But beyond a few infectious organisms such as pinworms, salmonella, and staphylococcus bacteria, hunter-gatherers had very few infectious diseases that were transmissible person-to-person and across generations to Holocene agrarians.

A brief ecological comment about the phenomenon of infection may be useful. For many, though not all, members of the microbial world, infecting other living organisms is simply their way of life. To survive and replicate, they too must find a way of dining at nature's table. A microbe that gains entry to another living organism then has access to nutrients; in the case of viruses, with no metabolic needs, infection enables the virus to connect up with the host's cellular genetic machinery in order to replicate. Infection usually does not cause actual infectious disease; most infecting microbes are more discreet lodgers. The occurrence of actual infectious disease in the host signifies either that the particular microbe-human relationship is a relatively recent one evolutionarily or that it is an evolved way of gaining assisted passage to another person by influencing the host's behavior and functioning by, for example, causing sneezing, coughing, and diarrhea.

The spread of farming-based communities, the increases in their population size and density, and the subsequent development of cities during the fourth and third millennia B.C.E. created conditions in which animal-source microbes (often aided by fortuitous mutations) could jump the species barrier and gain a foothold in a new host, *Homo sapiens*.[40] A minority achieved greater security of tenure by being able to circulate continuously within large human populations. Only a tiny minority of would-be pioneer-microbes succeeded, and they became the progenitors of now familiar infections in Late Holocene human populations.

During the Mid-Holocene, humans acquired new infections such as measles, smallpox (now extinct), chickenpox, mumps, the common cold, influenza, typhoid, malaria, and many others.[41] The measles virus may have been one of the earliest to make a successful species crossing and become an endemic human-crowd disease. A 6,200-year-old grave has revealed evidence of schistosomiasis, transmitted from bovine hosts via water snails, in an irrigated region in the Euphrates river basin.[42] The origin of one of the great and persistent killers, tuberculosis, remains uncertain. In the Americas, New World tuberculosis may have been transmitted from seals and sea lions, which were hunted around the coasts of Peru and northern Chile.[43] Cattle have a closely related mycobacterial disease (*Mycobacterium bovis*) long thought to be the likely source of the human disease via agrarian-pastoralist contact with herded cattle during the Early Holocene. But tuberculosis lesions have been found in a *Homo erectus* skeleton from half a million years ago, and it is even possible that early humans initially transmitted the disease to cattle.[44,45]

Over the past quarter century, the advent of virulent new strains of influenza and, in 2003, of SARS (severe acute respiratory syndrome, a life-threatening viral disease originating from forest-source animals captured in southern China for the live-animal food trade) are merely a continuation of this long-running story. The most calamitous modern example is that of HIV/AIDS caused by the human immunodeficiency virus (HIV), a successful go-it-alone genetic mutant of the simian immunodeficiency virus (SIV), apparently acquired sometime early in the twentieth century, quite probably via the bush meat trade in Central Africa.[46]

Of course, the mammals and birds from which we have acquired most of our infectious diseases are themselves late-stage hosts in longer-running evolutionary narratives. Bacteria, viruses, and their coterie of plasmids and phages predate the evolution of multicellular life by more than two billion years. They can directly exchange packages of genetic material as a form of fast-track, sexless, genetic randomization and supplementation, a process that virologists call "viral chatter." This, along with random mutations, facilitates rapid adaptation to nature's antibiotic chemicals—and, now, to modern medicine's synthetic antibiotics.

Not surprisingly, there is no specific historical evidence of climatic influence on the emergence of new human-adapted infectious diseases from animal sources. That may change during this century as more rapid and substantial climate change affects, for example, the number, geographic range, and migration patterns of various naturally infected animal (including bird) host species and hence their contact with livestock and with humans.[47]

Concluding remarks

The Mid-Holocene Climatic Optimum, extending over several millennia, opened up timely opportunities for a successful switch to settled agrarian living in much of the northern hemisphere. Its acceptance and then its rapid spread in parts of Eurasia represented the implicit acceptance of the Faustian bargain—here was a chance to increase food supplies, to enjoy some basic comforts of settled living, to control territory, and, for the more ambitious or scheming, to sequester new wealth and power. But it soon became clear that this was not all unalloyed gain. Food crops and livestock were at the mercy of inclement weather and, longer term, of fluctuations in climate. Food shortages and famines struck often and hard. In later years, new feverish, often fatal, diseases emerged, particularly in increasingly crowded urban settings, but were not recognized as infections spread by microbes for many millennia. The agrarian advance nurtured by the Mid-Holocene Climatic Optimum would also lead to heightened conflict, warfare, and conquest . . . as we will see in the next chapter.

6

Eurasian Bronze Age
Unsettled Climatic Times

THE STORY NOW MOVES beyond the mid-Holocene. By around 4000
B.C.E., viable agrarian settlements had appeared in many parts of the
world. Not only could larger populations be supported, but surplus
food produced by toiling farmers enabled the differentiation of labor
and social status. Settlements expanded, made trading connections, and
formed larger collective polities. Hierarchical authority and power began
to replace horizontal flows of local information and decision-making.

The vagaries of climate, however, lurked on the horizon. Agrarian
societies, with their increasing dependence on harvest staples, were
painting themselves into a corner. Also, as populations grew and settlements coalesced, mutant strains of animal-hosted microbes that made
a successful crossing from livestock or urban pests to humans took advantage of larger, intermingling host populations. A few of these adven-
turers, such as the measles virus, not only initiated new epidemics but
continued circulating, between outbreaks, as endemic "crowd diseases."
Measles, a microbial success story, is still with us today.

The advent of property, food stores, and occupied land in nearby
populations stimulated both war and conquest, each having diverse,
debilitating, and often bloody consequences for health and survival.

Mesopotamia dries: The rise and decline of the Akkadian civilization

Climatic conditions in Sumer, sitting at the meteorological crossroads
of the Middle East,[1] began changing about 3600 B.C.E., one-third of

the way into the fourth millennium B.C.E.[2] (Figure 6.1). There was a general cooling and drying in the northern hemisphere as the first phase of the Holocene Climatic Optimum waned and as the Icelandic Low and Siberian (Asiatic) High circulations intensified, funneling colder air southwards.[3] Rainfall declined in southern Mesopotamia, compounded by a southerly drift of the rain-bearing Inter-Tropical Convergence Zone (Figure 6.2) and the regional monsoon.

Further west, the Sahara was changing from green to brown, and Egyptian agriculture was faltering. As rainfall declined and arrived later in the year, farming became more difficult; farmers now needed to make a year-round effort, with double-cropping and shorter fallow periods. By extending their irrigation systems, the Sumerians compounded another problem: several centuries of overirrigation and deforestation had already begun to turn the soil saline. More irrigation meant more salinity, which further reduced harvests, and so the Sumerians began to replace wheat with salt-tolerant barley. Nevertheless, food shortages persisted, causing widespread hunger, undernutrition, and, for some, starvation. Local communities began building defenses to protect their crops from pillaging, while some outlying settlements were abandoned.

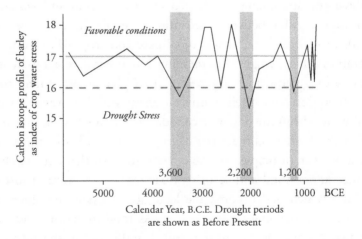

FIGURE 6.1 Periods of peak drought stress in early Near Eastern agricultural settlements, estimated from the isotopic composition of extant barley grains. For values below the dashed line, severe drought conditions are likely. Adapted from Riehl et al., "Drought Stress Variability."[4]
Permission received from PNAS.

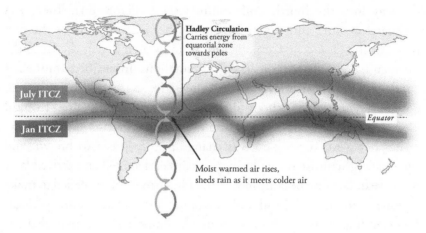

FIGURE 6.2 Northern hemisphere summer-winter oscillation of the Inter-Tropical Convergence Zone (ITCZ) as it follows the interseasonal north-south excursion of the sun.

(top shaded line—northern summer location; bottom shaded line—northern winter.) Over longer time periods, the actual latitude range of the ITCZ oscillation also shifts either northward or southward. The Hadley Circulation (three cells per hemisphere) relocates with it, causing latitude shifts of other climatic zones.[5]

Source: Author.

These climatic changes unfolded in slow motion and would have been imperceptible between generations. People did not have written records to serve as benchmarks of climate and crop yields from earlier generations. How might an emerging but preliterate agrarian society adapt to long, slow, unrecognized climate change other than to have each generation treat "their" climate as the norm? This was a very different situation from the one we face today, where, first, we have a reasonable sense of where climate trends might lead, and, second, we know that, at least in principle, we could avert much of that climate change.

These agricultural and security anxieties caused political instability in Uruk, which had become the leading Sumerian city since eclipsing Ur as the regional trading center. As food insecurity and hunger increased toward the end of that fourth millennium, the authority of rulers was eroded, and hard-pressed farming communities resorted to raiding others. Skirmishes broke out between neighboring cities, and general social unrest grew as impoverished and hungry farmers and pastoralists crowded into urban settlements or migrated upriver to the as yet less climatically stressed northern Mesopotamian region. These were some of history's first large groups of climate refugees, a category we can expect to hear more about during this century.

Cities evolve: Infectious diseases, writing, and bronze

From around 3300 B.C.E., conditions became warmer, moister, and better suited for farming and the progress of city-state civilization. This second phase of the Holocene Climatic Optimum spanned the next thousand years, a period during which momentous changes occurred in Mesopotamian society. In particular, the Neolithic (Late Stone) Age gave way to a future dominated by smelted and alloyed metals: tin, lead, and then the alloy bronze.

We have already noted how several of the human-adapted infections acquired from increasingly close contact with domesticated and pest animals, such as measles, established themselves and became part of human life. In later centuries, as trading networks and military campaigns extended more widely, the chances of both transmitting and receiving novel human-adapted infectious agents increased further. The region's climate changes were often subtle, if not overt, contributors to these increased risks of acquisition and transmission of disease-inducing bacteria, viruses, and other microorganisms. For example, the aridity that beset Mesopotamian agriculture during the mid-fourth millennium B.C.E. assisted some of these flickering infections to graduate into established endemic "crowd" infections, as farms and pastures were abandoned and people sought security and jobs in increasingly crowded city settings.

Around 3000 B.C.E., written records emerged in the Eastern Mediterranean region. Clay tablet and stylus, and later, stone and chisel, were used to record information and transactions.

Cuneiform script on clay tablets appears to have originated in Mesopotamia, while early Egyptian hieroglyphic records on both clay and stone date from around the same time. In China, writing evolved along the banks of the Yellow River during the Shang dynasty in the second millennium B.C.E. While these pictographic records were developing, fortuitous discoveries with fires and metallic ores led to the smelting of tin and copper, and thence to the production of bronze, a harder and stronger metal alloy comprising both tin and copper.

At much the same time, the Bronze Age emerged in the Middle East, transforming farming equipment (plowshares, axes, and wagon wheels), housing construction (saws, hammers, etc.), garment-making and leatherware (needles and blades)—and, inevitably, military weapons. These commodities, along with metal-fortified ship hulls, began to reshape commerce and trade. As seafaring, trading, and population size increased around the eastern Mediterranean coastline, the early Bronze Age civilizations of Palestine, Greece, and Crete emerged to bear comparison with those of Mesopotamia and Egypt.

As bronze implements and inscribed tablets became more widespread, so did the early symptoms of a change in climate in southern Mesopotamia and the wider Mediterranean region. A negative shift in the North Atlantic Oscillation weakened the incoming moist west winds, resulting in reduced winter rainfall and hence in river flows and crop yields.[6] These cooler and drier conditions, in combination with overworked and damaged soils, excessive irrigation, a hungry populace, and weakened government, left the enfeebled Sumerian confederation easy prey to the imperial ambitions of their northern neighbors, the Akkadians. Around 2300 B.C.E., the Akkadians conquered the central Mesopotamian region and, subsequently, Sumer in the south. Their ambitious and militarily able ruler, Sargon, now controlled a much-enlarged empire stretching from the Persian Gulf to the northern reaches of the Euphrates River. The Akkadian Empire, however, would soon be fatally afflicted by the northward spread of the same aridifying change in climate, bringing a severe drought that lasted in central Mesopotamia for nearly two centuries.

As the Akkadian Empire became drier, harvests declined, hunger spread, and social and political unrest increased.[7,8] In the more marginal farming areas, the change in climate heightened preexisting environmental, cultural, and social stresses.[9] The unknown author of the Curse of Akkad wrote:

> For the first time since cities were built and founded,
> The great agricultural tracts produced no grain . . .
> The gathered clouds did not rain, the masgurum did not grow. . .
>   He who slept on the roof, died on the roof,
> He who slept in the house, had no burial, People were flailing at
>   themselves from hunger.

Archaeological evidence points to a continuing rapid decline in cereal grain production and a contracting population as people migrated, dispersed, or reverted to subsistence pastoralism. In their search for water and pasture, many northern Amorite pastoralists moved down the Euphrates River valley with their goats and sheep. Conflict with southern Mesopotamian farmers resulted when the herders' free-range herbivores found new fields of cultivated food to eat. Despite the walls built by the Akkadian imperial rulers and threats by local authorities, the numbers of displaced nomads and drought refugees were too great to resist. Invaders overran the militarily weakened Akkadians, especially the Guttians from the Zagros foothills to the east.[10,11] The Akkadians and their agricultural base were largely undone by this regional drought, and the empire disintegrated soon after 2000 B.C.E.

Mesopotamia had landscape diversity, internal trading strengths, and the flexibility of multiple local governments. As Akkad declined, the Babylonian settlements in eastern Mesopotamia acquired greater cohesion, helped by lying further upstream on the Euphrates than Sumer, with good river-water flows and more reliable rainfall. After 2000 B.C.E., the riverside city of Babylon led the restoration of confidence and power in Mesopotamia. In 1792 B.C.E., Hammurabi, long-time ruler of the Babylonian Empire and celebrated codifier of laws, came to power (Figure 6.3).

Mesopotamia as a whole, however, continued to suffer from a cooler, drier climate and harvest losses. The population halved in size.[12] Clay tablets and stone carvings attest to misery, conflict, starvation, and epidemic outbreaks. Letters and other texts refer to droughts and their devastating effects.[13] One fragmentary letter discovered in Nippur in central southern Mesopotamia pleads that "young men (or servants) should not die through drought." A record of an Old Babylonian incantation, addressed to the god Enki, recounts various disasters: "he has cast over me famine, thirst, drought, cold, and misery."

During this prolonged cooler and drier period, the northern hemisphere's monsoon systems receded southward as El Niño conditions gathered strength. This shift in climatic regime reduced the rainfall in the Atlas Mountains in Northwest Africa, the Gulf of Oman at the entry to the Persian Gulf, the Indus Valley (home of the Harappan civilization) in present-day western Pakistan, Tibet, and northern China.

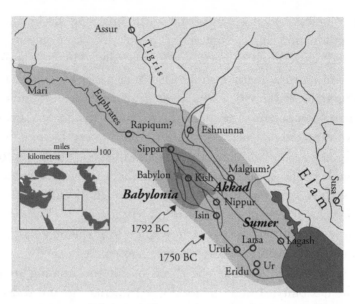

FIGURE 6.3 Hammurabi's Babylonia, showing the imperial territory at his ascension (1792 B.C.E. dark shading) and death (1750 B.C.E. lighter shading). River courses and coastline are as at the time and differ from those of today. Wikimedia Commons, "Hammurabi's Babylonia."[14]

Many semisedentary dry-field farmer-pastoralists around China's now sluggish Yellow River were displaced, causing conflict with older settlements to the south.[15] Hunger and impoverishment increased death rates, and population numbers declined. In contrast, further south still, the Yangtze region wet-rice farmers, with their long-established water conservation and management systems, coped better with the drying.

In northern and eastern Europe, this widespread northern hemisphere cooling brought wetter and colder summers. Village-based farming was still in the early stages of development in north central Europe (today's northern Germany and southern Scandinavia) and the eastern Ukraine region north of the Black Sea. These agrarian communities, typically aligned with local megalith kingdoms, were precariously located at the eastern inland margin of the Atlantic-influenced temperate-subtropical climate regime of western Europe, where they were vulnerable to shifts in climate. The advent of increasingly inclement weather, particularly cooler and wetter summers, prevented the production of sufficient hay for cattle, a major staple of their farming. As climatic conditions became increasingly unfavorable for crops and livestock and hunger worsened, the peoples in this region began to emigrate in waves, heading to the east and south. This, as it turned out more than 2,000 years later, was a forerunner of the similar climatic pressures on barbarian tribes that would overrun the Western Roman Empire.

Some scholars believe this wave of migrations from late in the third millennium B.C.E. may have included the original dispersal of the Proto-Indo-European peoples.[16] One theory suggests their origin might have been in the eastern Ukraine steppe region. From there they radiated outwards to southern Russia, Attica, Anatolia in southeast Turkey, Iran, and India. Indo-European people may also have reached Xinjiang in northwest China, according to anthropologists and archaeologists who have studied contemporary facial features and historical information about clothing, hair, and cultural habits.

Climates and life in the Indus Valley

Further afield in the upper Indus Valley, today's Thar Desert between western Pakistan and Rajasthan, India, the Harappan civilization evolved and flourished during the period 2500–1700 B.C.E. (Figure 6.4). This society had formed as Neolithic village communities along the rivers became interconnected. The villages enjoyed good soil and abundant monsoonal rain, with river flows secured by the year-round melting of Himalayan snow. Today that region is arid, but during the warm third millennium B.C.E., the Indian summer monsoon reached further inland. At their peak, the Harappans farmed more land than the Mesopotamian and Egyptian agricultural systems combined. Wheat, barley, dates, and melons were staple foods; elephants, water buffalo, and rhinoceroses were abundant. The Harappans grew cotton successfully, and their weaving finesse dominated the production and trading of textiles in the region.[17]

The three great civilizations in Mesopotamia, Egypt, and the upper Indian subcontinent all lay predominantly within the latitude band 28°N–33°N, where their climatic regimes were influenced by the northern hemisphere's dry subtropical ridge (typically centered on 30°N). Hence they were sensitive to north-south shifts in that high-pressure drier zone. During the first and second phases of the warmer Holocene Optimum periods (ca. 6000–4500 B.C.E. and 3300–2300 B.C.E.), that

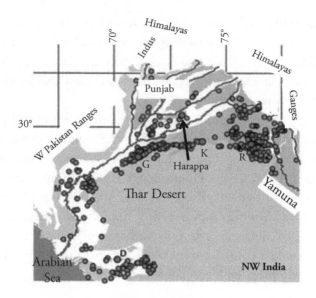

FIGURE 6.4 Early and mature Harappan sites (3200–1900 B.C.E). Adapted from Giosan et al., "Fluvial Landscapes of the Harappan Civilization."[18]

Permission received from PNAS.

drier zone had shifted northward, drawing stronger summer monsoon rains behind it that replenished the life-giving rivers and agricultural systems of Mesopotamia and Harappa. Meanwhile, conversely, much of Central Asia became drier as the subtropical ridge moved northward.[19] Then from around 2100 B.C.E., in a global climate now increasingly influenced by El Niño conditions, the summer monsoons contracted south, causing water shortages and drought in Egypt's Old Kingdom, the Akkadian Empire, the civilizations of Mycenae and Crete, and the Indus Valley.

Agriculture in much of the Indus region has always depended on the annual river flooding in addition to direct rainfall. As cooling and weakened monsoon rains occurred and the moist Atlantic west winds weakened late in the third millennium B.C.E., the Harappans' harvests declined.[20,21] As in Mesopotamia, attempts to compensate by extending farmland would have deforested much of the area and reduced soil moisture, while intensified irrigation would have increased soil salinity. By improving the management of river flows and floods, areas that had previously been prone to heavy flooding could now be brought into production. But eventually the annual rejuvenating river flows fell too far, and serious food shortages occurred. This prompted regrouping of urban settlements at locations of major reliable flooding, especially in the southeast Punjab, and a number of these cities became larger and grander as rural communities abandoned their farms. That urban concentration spawned a further set of troubles.

The reasons for the Harappan decline during the next several centuries were complex. Regional drying and reduced river flows, environmental deterioration, urban crowding, rises in infectious diseases (including leprosy and tuberculosis), social tensions, and increased skull-shattering violence all contributed.[22] The civilization contracted in size, cohesion, and food security, and from around 1800 B.C.E. it began to fade from the history books of the future. Several centuries later, this largely abandoned region was resettled by Indo-European nomadic herdsmen during the same extensive regional cooling and drying period.

Eastern Mediterranean: Late Bronze Age collapse and the Sea Peoples

Late in the Bronze Age, the Eastern Mediterranean economy was booming, buoyed up in particular by trade in copper, gold, ivory, incense, jewelry, textiles, clothing, and grain. Then, late in the thirteenth century B.C.E., a severe region-wide climatic change emerged that lasted for four centuries, spanning 1200–800 B.C.E. This disruptive cooling and drying event helped bring to a close the rich and innovative Bronze Age culture and its network of Aegean, Egyptian, Syro-Palestinian, and Hittite civilizations,[23,24] ushering in a temporary "Dark Age" in the Eastern Mediterranean region.

Colder and wetter conditions had emerged further north in Europe a century or two earlier, as the North Atlantic cooled and the westerly rain-bearing winds weakened. This, capped off by a temporary southward intrusion of the cold conditions of the Siberian High system, meant that warmed moist air no longer reached the Eastern Mediterranean. The region's annual rainfall halved. The ensuing widespread drought caused crop failures, extreme food shortages, region-wide warfare, and a mix of migrations, including the more threatening in-migration of the mysterious "Sea Peoples" from further afield. Far to the east, as invasions and economic and political turbulence engulfed the eastern Mediterranean, South Asia was also affected by this climatic change. As summer monsoon rains declined, crop yields fell in north-west India (today's Rajasthan), accelerating the long-term desertification of the Punjab's Thar Desert, the erstwhile Harappan home base.

The Sea Peoples

The severe drought conditions around the eastern Mediterranean caused widespread disruption and displacement. Greek historian Herodotus described emigrating masses of Lydians leaving Anatolia (southeastern Turkey) because the severe drought had caused "a great scarcity through the whole land of Lydia." They headed for the Mediterranean coast and may have become part of the so-called Sea Peoples, the large bands of displaced, sometimes piratical and marauding, peoples who roamed those eastern coasts on foot and by boat, as the great regional drought took its toll on the hinterland (Figure 6.5).

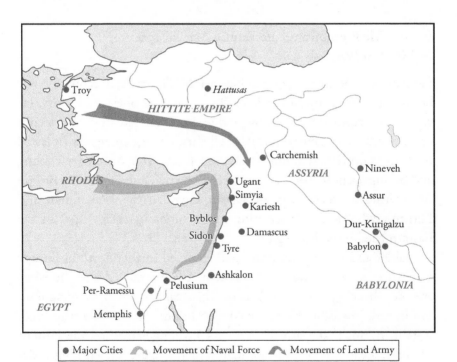

FIGURE 6.5 The Sea Peoples migrations into and around the Eastern Mediterranean during the second millennium B.C.E. was one of the largest and most important in ancient history. Adapted from Salimbeti, "The Greek Age of Bronze: Sea Peoples."[25]

There is much speculation on the identities of the Sea Peoples. Some scholars believe they may have been descendant groups of the Indo-European speakers who had migrated into southern Europe, including the Hellenic peninsula, and the west-central region of Turkey during the later years of the second millennium B.C.E.

The Sea Peoples' names appear differently in the written records from several of the region's societies. The Egyptian renditions were vowel-free, leaving historians to deduce the missing letters. The major groups of Sea Peoples had names like Shekelesh, Tjekker, Ekwesh, Peleshet, and Tursha.[26]

The Philistines are an interesting example of one of the groups thought to comprise the Sea Peoples. The Israelites, recently resettled in Canaan, called them the Peleshtim, a word that appears ancestral to the words "Philistine" and "Palestine." Archaeologists have assembled pottery fragments, inscriptions in temples, and diagnostic evidence of patterns of burning cities to the ground that point to the Philistines as one of the marauding and invading Sea Peoples.[27] Their archaeological footprints appeared first on the Turkish coast of the Sea of Marmara connecting the Black Sea to the Mediterranean as they made their way by sea and overland in vast numbers, the ships manned by warrior-sailors. Terrifying advance reports heralded their arrival in the Mediterranean.

Meanwhile, the severe Eastern Mediterranean drought ground on relentlessly. In northern Anatolia, the powerful Hittite kingdom's subjects were starving, despite the kingdom's great prosperity and diplomatic and military success during the preceding two centuries. Ancient letters from the Hittite kingdom begged neighboring powers for grain shipments to avert famine.[28] One Hittite king wrote pleadingly, "It is a matter of life or death!" Another from the south of the kingdom sent an SOS stating, "If you do not quickly arrive here, we ourselves will die of hunger." Stele carvings and clay tablets describe grain shipments to the Hittites from Syrian and southern Anatolian sources, and even from their erstwhile foe Egypt. The Egyptian pharaoh, Merenptah, recorded that he sent grain "to keep alive the land of Hatti."

The Hittite kingdom, however, was by now seriously enfeebled by starvation and social disorder and would soon suffer terminally at the hands of groups of Sea Peoples as they fought and plundered their land-route way through Asia Minor.

During the middle years of the second millennium B.C.E., the Mycenaean civilization on the Peloponnesian Mediterranean coast, the forerunner of Classical Greece, had been serially stressed by fallout from the massive Thera volcanic eruption in mid-millennium and then by protracted regional warfare. Several centuries later, Mycenae and Crete suffered further from the dramatic succession of earthquakes that beset the eastern Mediterranean during several decades around 1200 B.C.E. and then by the regional drought.[29] Weakened and diminished, Mycenae was easy prey for the Sea Peoples later in the twelfth century. This confluence of calamities would speed the collapse of the Bronze Age in the eastern Mediterranean around the end of the second millennium B.C.E.

Intense drought pressed heavily on all the major urban sites associated with the Late Bronze Age collapse.[30,31] The resulting hunger, starvation, displacement, and social and political disorder led to a Dark Age, a period of several centuries of widespread cultural decline, conflict, and warfare. This regional upheaval and the striving for stronger weaponry probably accelerated the development of the Iron Age. Out of this cultural and political darkness would eventually emerge new civilizations and power relationships: initially the mighty Assyrians, the Phoenicians, and the Philistines on the Mediterranean coast (each of those two with probable origins in Sea Peoples settlers), and later the Greeks and Romans.

However, there was still more mayhem to come. Seeking places to settle, the Sea Peoples caused great upheavals, pillaging, destruction of cities and city-states, and a general transformation of the political landscape around the coastal region. Jealous of the fertile Nile Delta lands, early in the twelfth century B.C.E. they joined with the Libyans to attack Egypt. The army and navy of Rameses III managed to repel and defeat this desperate and determined foe, seizing its stocks of cattle and killing several thousand of the invaders. The severed hands and uncircumcised penises of the slain were duly tallied for the royal Pharaonic records.

Concluding remarks

Agriculture had allowed civilizations to flourish during the Eurasian Bronze Age, but it also left many of these societies more vulnerable to climatic shifts. Urban centers with permanent, sedentary populations had emerged during benevolent climatic conditions. With their changed diet, now dominated by a few staple foods, burgeoning civilizations became dependent on continued satisfactory surpluses from the harvesting of cereal grains. Advances in metallurgy created new wealth and property to protect and covet, and it also facilitated the creation of new metal weaponry. When climatic conditions changed, such as increasing aridity, some people adapted by leaving the great civilizations and returning to subsistence pastoralism, while others migrated in search of a chance to make a living elsewhere.

Sometimes the changes were too rapid for successful adaption, and famine, disease, and conflict were the result.

Temperatures continued to decline during the early centuries of the first millennium B.C.E., both in Europe and China.[32] This was also the time of the Homeric Minimum in solar activity. Central and northern Europe cooled by around 1°C. On the European plains, crop yields fell, and food shortages and starvation occurred. Skeletal remains indicate a rise in death rates, especially in children. In the Alps, increases in rain and snow caused glaciers to advance and the levels of lakes to rise. Around 800 B.C.E., the lakeside dwellers of Zurich had to abandon their settlements.[33] In the Nordic region, as temperatures fell and rainfall increased from the seventh century B.C.E., conditions deteriorated and much of the population reverted to simpler village-based farming. Persistent rain leached the nutrients from their soil, the area of bogland increased, and farmland productivity fell. This, along with the disruption of its southerly trade routes by the now widely radiating Celtic migrations and by Scythian raiders, heightened the isolation of the Nordic peoples from continental Europe. The economic setback and distancing from mainstream European culture and trade lingered until the revitalizing influence of the Renaissance years in Europe.[34]

In the final two centuries B.C.E., warming conditions in central Europe allowed the La Tène Iron Age civilization in today's southern Germany, generally assumed to have been Celtic, to expand widely through central and western Europe during the next five centuries. Further north, there were further pulses of Germanic tribes extending their farming northward to the shores of the Baltic Sea. Their numbers increased throughout this warm period. Indeed, it was more than a passing local intermission; the Mediterranean climate regime was beginning a slow extension northward which continued for about five centuries. The Classical Optimum period in Europe would last through the first three centuries of the Common Era (c.e.). This warming, during which the more reliable "Mediterranean" summer warmth and winter rains now extended further north in Europe, would not reach the civilization-nurturing heights of the earlier and longer Holocene Climatic Optimum. But it would tip various historically momentous scales along the way.

7

Romans, Mayans, and Anasazi

The Classical Optimum to Droughts in the Americas

THE STORIES OF THE Roman Empire and the Mayans are well known and have fascinated generations of scholars, artists, storytellers, and history enthusiasts. Less familiar are the ways in which the changing climate contributed to the rise and fall of these civilizations, and of the Anasazi, among others in North America. This chapter examines the fates of different societies in three climatic periods: the warm Classical Optimum (300 B.C.E. to 350 C.E.), cooler conditions in the Dark Ages (500 C.E. to 800 C.E.), and drought in the Americas (950 C.E. to 1250 C.E.). Recent gains in the reach and resolution of paleoclimatology have enabled more detailed reconstruction of climate and health relationships. Beginning around 300 B.C.E., Europe and the Mediterranean experienced a prolonged period of warm and stable climate—often termed the Roman Warm. Historian John L. Brooke has labeled the ensuing "remarkable" 600 to 800 years of benevolent climate conditions the Classical Optimum, and he suggests that the effects were global.[1] A positive North Atlantic Oscillation (NAO) pushed warm winds west towards Scandinavia, glaciers retreated, and the Mediterranean settled into its characteristic pattern of dry summers and winter rainfall. In the wake of the spread of farming and rising fertility rates, the estimated global population was approaching 200 million. Cities were becoming larger and grander, trade routes were extending, and armies and their iron weaponry were ranging further afield. So too were various infectious agents, many of them beneficiaries of the new and intensifying transcontinental contacts among China, Rome, South Asia, the Middle East, and North and East Africa.

During this period, the Mediterranean sustained "the deepest landscape transformation in antiquity."[2] Scattered populations increased and coalesced into forts and cities, supported by thousands of new farms.[3] By around 300 C.E., however, the Classical Optimum began to wane. Ice-melt events cooled northern Europe, and by 500 C.E. the strong NAO reversed, bringing a deep cold. The shifting climatic conditions placed enormous pressure on the civilizations that had transformed their socio-ecological systems during conditions more favorable to agricultural productivity and human health.

The Classical Optimum

This period followed the cooler conditions that had prevailed in the northern hemisphere for much of the final millennium B.C.E.[4] In continental Europe, the climate is influenced mainly by three regional climate regimes: the North Atlantic regime in northwest Europe (moist westerlies from the ocean, especially during the NAO "Iceland high"), the temperate continental regime in northeast Europe (drier cooler air from Siberia), and in the south, the subtropical-temperate Mediterranean regime (drier and warmer westerly winds, enhanced during a positive NAO "Azores high").

One distinctive feature of the Classical Optimum influenced the Roman Empire's fortunes (hence the Roman Warm label). For several centuries, the Mediterranean climate, well suited to Roman open-field farming practices, extended much further north in Europe.[5,6] Paleoanthropologist Carole Crumley wrote:

> This pivotal period in the history of the West offers a powerful illustration of the relation between climate and society, organizational structure and long-term (*longue durée*) history . . . Throughout Western Europe, from ca. 300 B.C. to A.D. 250, the climate was Mediterranean—warm, dry, and unusually stable. The usually volatile climate of the continent was less variable than at any time since the middle Holocene.[7]

This northern extension and then retreat of the Mediterranean climate system has recently attracted wider attention in studies on the fate of the Western Roman Empire.[8,9] Information about changes in the regional climate has long been fragmentary and devoid of modern scientific understanding. In 1789, English historian Edward Gibbon completed his monumental six-volume *The History of the Decline and Fall of the Roman Empire*.[10] His was primarily a moral tale in which he attributed the decline primarily to the loss of civic virtue and wise government, the use of co-opted barbarians as mercenary soldiers as Roman military strength and commitment weakened, and the spread of the pacifying distractions of Christianity with its promise of a better life hereafter.

We now know much more from diverse proxy indicators about climatic and environmental changes in Europe during the early centuries C.E.[11,12] In particular, as the Mediterranean climate system extended north by many hundreds of kilometers from around 250 B.C.E., it brought warmer, drier summers and mild, wet winters to much of Western Europe, especially Gaul, an important area of the Roman Empire at its height.[13,14] This coincided with an unusually long positive phase of the North Atlantic Oscillation (NAO) which brought stronger and warmer west winds to the Mediterranean basin and southern and central Europe, producing drier summers and wetter winters.

Feeding a growing empire

Agricultural production at home and abroad was critical to the expansion and military might of Rome. The main dietary staple was cereal grain, much of it from North Africa and Egypt. The northward excursion of the Mediterranean climate facilitated the spread of Roman agricultural practices from peninsular Italy to the alpine foothills and then into southern and central Gaul.[15] Roman grain and wine agriculture, well suited to the Italian plains, used a large-farm monoculture system based on the North African, Spanish, and southern Italian latifundia model that the preimperial Romans had evolved centuries earlier. That system, powered by a low-paid or enslaved workforce, sought maximum output from just a few crops, mainly wheat and millet. As the empire expanded, particularly to the northwest, this system displaced the region's traditional Celtic mixed farming based on a mixed-crop and cattle combination that was more resilient in the colder and more variable northern climate.

Egypt's Nile Valley, Rome's major offshore breadbasket, also enjoyed unusually good conditions for grain-growing during the core period of the Roman Warm, from around 30 B.C.E. to 150 C.E. The annual Nile floods were frequently of good volume, fed by good rainfall in the Ethiopian highlands. Pliny the Elder wrote, in the first century C.E., that when the height of the Nile reached 14 cubits (about seven meters), a good harvest would follow, while a height of 12 cubits or less meant poor harvests, food shortages, and possible famine. For the moment, food supplies for the Roman Empire in those first two centuries C.E. were secure. Well-fed troops were able to extend the Empire's boundaries further, encompassing all of Celtic Gaul, southern Britain, parts of today's Germany, and richer territories in the East such as Dacia (Romania) and Mesopotamia.

As imperial power and fortunes grew, Roman agricultural practices were introduced further afield. Wine growing was extended into these northern climes, including southern England.[16] The recently resuscitated white wine Viognier comes from a modest-yield grape introduced into central Gaul during the Roman Warm. Meanwhile, beyond the northeastern imperial borders, in the adjoining northern region of Central Europe and extending to the coastal lands of the Baltic Sea and North Sea, the warmer climate had enabled several Germanic tribal groups to extend their farming territory and to grow in numbers. A little to their east, Slavic tribes had done likewise. However, as temperatures and rainfall began a slow decline in northeastern Europe from early in the third century C.E., tribal populations began to probe southwestward across the imperial border, seeking better farming territories.

Cooling began around 300 C.E. associated with a prolonged period of strong El Niño influence and a cooling of the North Atlantic sea surface. Cooler and wetter conditions appeared first in northeastern and Central Europe, extending later into Western Europe. This shift, and the southern contraction of the itinerant Mediterranean climate, brought 1°C cooler conditions to northern Gaul, and drier winters than two centuries earlier. As rainfall decreased and weather became more variable, harvest failures became more common. The resulting food insecurity compounded the Empire's third-century crisis of economic, social, and military difficulties, capped off by political instability as rival generals and strongmen fought for power and high office.

By 350 C.E., the climate had steadied, and a half century of moderate warming suffused the northwestern provinces. But the social, political, and agricultural foundations of the Empire's earlier successes had been seriously shaken. Historian Peter Heather, in *The Fall of the Roman Empire*, notes that agricultural output, population numbers, and local economies began to contract during the fourth century, first in Rome's northern European provinces of Gallia Belgica and Germania Inferior,[17] while agricultural declines emerged later in the southerly provinces. The early northern contractions, suggests Heather, were influenced by the disproportionate regional impost of Rome's increased military-funding taxes on farmers, exacerbated by more frequent barbarian incursions along that northeast border. Even so, the broad north-before-south geographic sequence of provincial decline also fits with what one might expect as the Mediterranean climate retracted southward. Meanwhile, food yields in Britain and the Middle East remained high, as did rainfall and climatic stability, and the local economies prospered during the late fourth and fifth centuries.

Much of Western Europe reverted to its historically more usual temperate-zone climate, and the task of feeding the empire's growing urban populations and its border garrisons via conventional Roman agricultural practice became more demanding. Food shortages bore down increasingly on the local colonized Western Europe populace, particularly in poorer rural regions. The resultant nutritional deficits and lessened strength for toiling in the fields would presumably have further eroded farm productivity. Overall, regional climate changes, increased taxes, and the commandeering of local food harvests were all contributing to hunger and undernutrition, particularly in continental Western Europe. Serious regional food crises sporadically occurred, around three-quarters of which have been attributed to episodes of climatic extremes and associated droughts.[18]

The leaders in Rome now faced the politically critical challenge of feeding the swelling urban citizenry as well as the far-flung military. This problem was heightened by periodic agricultural downturns in Egypt as the annual rains in the Ethiopian highlands and the Nile floods became more erratic during the cooler fourth century. Yet with parts of the Empire's provinces in western North Africa coming under pressure from early waves of barbarians, especially the Vandals, Egypt was now the one substantial and secure source of cereal grain imports able to satisfy the great urban appetite of Rome. Meanwhile, the mounting toll from infectious diseases was further depleting the manpower, morale, and management of the Empire. A crisis was looming.

Infectious diseases: Local epidemics, major urban plagues

During the cooling and climatically more variable phase in the late third century and much of the fourth century C.E., the populace was exposed to epidemic infections now circulating more widely. Climatic influences may often have played a role, but the prime cause lay with the introduction and reintroduction of microbes picked up via military campaigns in distant corners of the Empire and by intensified patterns of long-distance trade via overland and sea routes. During those two centuries, there was a marked decline in the population living around the Mediterranean coastal regions and a generalized increase in rates of infectious diseases, infant deaths, and mortality in the elderly.[19] The frequent combination of food shortages, undernutrition, impoverishment, and social instability during the cooling climate, as well as the Empire's troubles later in the third century, would have predisposed many citizens to contracting infections.

A comprehensive analysis of famine and pestilence in the Roman Empire during 284–750 C.E. by Dionysios Stathakopoulos reveals many connections between climatic changes, food crises, and epidemics.[20] Among the more than 200 documented epidemics were several regional outbreaks of smallpox, one of which occurred in the eastern Roman Empire in 312–313 "in the midst of famine apparently caused by the lack of winter precipitation." Stathakopoulos writes: "In urban locations the epidemic and the famine resulted in a high death toll, which was even greater in the country. Those who escaped starvation were apparently particularly consumed by the infection." Beyond the enfeebling effect of hunger, rural vulnerability was heightened by lack of immunity among those spared many of the earlier urban epidemics. These constantly flickering epidemic outbreaks sapped the energy and economic productivity of the rural sector and heightened the empire's woes.

During 166–190 C.E., in the heyday of the Roman Empire, the first great epidemic, the Antonine Plague, swept around the Empire for several decades, seriously eroding population numbers, military strength, and social stability. The disease first broke out in the Middle East and was brought by the army to Rome, where at its height it caused up to 2,000 deaths a day.

Several emperors died from this plague, including Marcus Aurelius Antoninus (hence "Antonine"). Papyrus-inscribed records from Roman

Egypt indicate that around one-fifth of that population also died from this disease. As the epidemic spread through much of the Empire, it caused 7–10 million deaths, including more than a quarter of the population in some parts and one-sixth of the army.[21]

Galen, the most celebrated physician of the Roman era and an eyewitness to the event, described the epidemic in his treatise *Methodus Medendi*.[22] The plague was great and prolonged, he wrote, and those afflicted typically had fever, diarrhea, an inflamed throat, and a pustular skin eruption that appeared around the ninth day of the illness. Although Galen's description is not definitive, the disease, while plausibly due to measles, is generally attributed to smallpox.[23] Indeed it was probably the European debut of this highly contagious disease of South Asian origin. Since the germ-transmitted nature of such disease was not then understood, it is little wonder that chroniclers struggled to understand how differing sets of disparate symptoms might each be caused by some particular external agent or miasmatic influence.

Although variations in climate apparently had little to do with the actual introduction of smallpox into the Empire, the disease is known to spread more readily in cool and dry conditions.[24] Such conditions were returning to parts of western Europe as the Roman Warm receded and winter rains decreased, and may have facilitated its spread. That aside, the Antonine Plague contributed significantly to the incipient weakening of the Roman Empire, a weakening that followed the second-century imperial triumphs of Emperors Trajan, Hadrian, and Marcus Aurelius Antoninus.

Half a century later, the Plague of Cyprian occurred between 250 and 270. Bishop Cyprian of Carthage provided an eyewitness account (which earned him the naming rights). He thundered from the pulpit: "How suitable, how necessary, is this plague and pestilence . . . the just are called to refreshment, the unjust are carried off to torture." The disease resembled the Antonine Plague and may have been either smallpox or measles.[25] At its height, the epidemic caused several thousand deaths daily in Rome. Bishop Cyprian's biographer later wrote:

> . . . a hateful disease invaded every house in succession of the trembling populace, carrying off day by day with abrupt attack numberless people . . . All were shuddering, fleeing, shunning the contagion, impiously exposing their own friends, as if with the exclusion of the person who was sure to die of the plague, one could exclude death itself also.[26]

This second hammer blow, compounding the earlier losses of population and military capacity caused by the Antonine Plague, exacerbated shortages in both the agricultural workforce and the Roman army, further weakening the empire's administrative cohesion, economic reserves, and border defenses. Entering the fourth century, Rome's morale and population strength were under siege from the food crises and epidemics associated with the variable and cooling climate during the third and fourth centuries and from increased barbarian incursions across the empire's northeastern boundaries, which disrupted regional administration. The empire, already straining at managing its complex military campaigns and defenses, laws, taxes, and layers of government and bureaucracy,[27] was now contracting in territory and military might. Historian William McNeill concludes that the hundred years that spanned the Antonine and Cyprian plagues, plus many other regional epidemics, so depleted the empire's population and economy that there was an insufficient social base to support the huge state apparatus and the imperial military machine.[28] The Western Empire's ensuing decline led to an abrupt end late in the fifth century C.E.

The unheralded advent of these major epidemics, including new microbes coming from far afield, underscores the fact that the Roman Empire was no longer a lone and insulated power in Eurasia. No longer did all roads led to Rome; many now led to Central Asia and China. As long-distance east-west trade intensified across the Central Asian steppes, and as other trading and political contacts were made with a widening circle of polities, the spread and transmission of infectious disease moved up to a transcontinental scale.

The fall of Rome—and decline of the Western Empire

The context and contributory causes of the fall of the Western Empire are famously complex—and beyond the reach of this text and author. Summarizing the outlines of the above account, the unusual Roman Warm climate had weakened in the third century and largely receded by the fourth century C.E., replaced in much of western Europe by its default cooler and wetter climate. The empire's dominions in western, central, and northern Europe had been under serious external attack since that same time as Rome's border defenses weakened, army recruitment faltered, and food production declined.

Now the restless populations of Germanic and Slavic tribal peoples, beset by increasingly farm-unfriendly northern climates and diminished food yields, were being pressured even more, this time from the east by incoming Hunnic (Asiatic) peoples. A particularly severe and prolonged half-century drought had occurred in those Eurasian Steppe grasslands in the mid-fourth century, especially in the eastern region,[29] and the displaced nomadic tribes were seeking habitable territory and sustenance, along with some bonus plunder (see Figure 7.1). Like a great population pump, the steppe has long drawn in nomadic peoples and their horses during favorable climatic times and then, during arid times as pastures thinned, has forced them out to seek sustenance further afield.[30] These westward-migrating Asiatic nomads put further

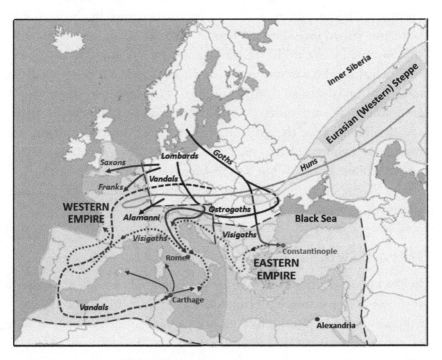

FIGURE 7.1 Map of main invasion routes of major barbarian tribes during the third to sixth centuries. The approximate borders of the Western and Eastern Empires are shown with long-dashed lines. The Western Empire collapsed around 500 C.E. Adapted from Andreas Kunze, "Distribution map of Europe," *Wikimedia Commons*.[31]

pressure on the borders of the Western Empire, as well as on the barbarian populations that stood between the Huns and those borders.

Skirmishes with Roman border garrisons had increased during the third century as the Alemanni, Vandals, and other Germanic-Suebi tribes began to invade Roman territory on a broad northeastern front. The Romans, now seriously short of military manpower, resorted to recruiting some of the invaders to serve in border garrisons—a risky deal that would sow the seeds of future political leadership turmoil within the Empire.[32] But a demographic and political sea change was under way. As regional cooling continued and farm yields faltered, this "great migration" period, the *Völkerwanderung*, gathered momentum.

During the fourth century, waves of barbarian invasions flowed west and south across the Western Empire. The Western Empire and its northwest African colonies were overrun and occupied by the incoming Germanic tribes, including the Franks, Vandals, Alemanni, Lombards, and Saxons, by the mid-fifth century. In 476, finally forced to concede defeat on the western front, Emperor Theodosius designated the then imperial co-capital, Constantinople, as the empire's official seat of government. From Constantinople a succession of emperors sought, with limited success, to regain western territory and political control. Justinian I (who ruled 527–565 C.E.), supported by two able generals, made the most headway, but his campaign was undermined by the (bubonic) Plague of Justinian and its dire economic, social, and military aftermath.

The Pandemic of Justinian: European debut of bubonic plague

In 542 C.E., the city of Constantinople (today's Istanbul), the eastern and now dominant capital of the Roman Empire, experienced a catastrophic epidemic. Within four months, a third of the city's population of half a million was wiped out: rich and poor, young and old.[33] The estimated 150,000–200,000 deaths were agonizing and often grotesque. Pain and delirium drove many to suicide, jumping from city walls or from cliffs into the Golden Horn estuary. The emperor, Justinian I, was also infected but survived. Over the next two centuries, this deadly disease caused an estimated 50–100 million deaths within the eastern Mediterranean region, West Asia, the Arabian Peninsula, and Europe.[34]

Justinian, ruler of a shrunken empire, had ambitions to regain lost territory and power. He had some early success—until this violent epidemic depleted the military ranks and the government coffers. For urban citizens, however, funding and feeding imperial military expeditions had meant more taxation and longer bread lines. Constantinople needed grain, and while some shipments could still be commandeered from the few remaining North African vassal states, much had to be purchased via Egypt from the powerful Christian kingdom of Aksum (approximating today's Ethiopia).[35] To hedge against possible siege by the emboldened Ostrogoths and the expansionist Sassanians (Persians), extra granaries were built in Constantinople.

The epidemic in Constantinople appears to have been the first major incursion of bubonic plague into Europe, occurring eight centuries before the much better known "Black Death" pandemic.[36] This dread disease is spread by blood-feeding fleas from infected black rats (*Rattus rattus*) (Box 7.1). The rat populations initially acquire the infection via occasional contact with the natural hosts of the plague bacterium (*Yersinia pestis*), the ground-dwelling rodents that live in vast underground many-chambered burrows located in climatically suitable regions, including Central Asia, southern China, Mongolia, northwest India, the Arabian Peninsula, and the Great Lakes region in eastern Africa.

Black rats frequently cohabit with humans, and their prodigious reproductive capacity can sustain the infection by providing a continuing source of blood-meals for their fleas. However, many of the infected rats die, and if rat numbers dwindle, hungry fleas often jump ship to reside in the clothing of humans, close to an acceptable blood-meal substitute. Human infection follows, and can then circulate in the population via the rat-human interface or, less often, via person-to-person pneumonic (respiratory) spread. Archaeological skeletal trails show that black rats spread widely in the Red Sea and Macedonian regions during the last millennium B.C.E.,[37,38] having migrated widely out of their ancestral habitat in India as regional trade and military contacts increased. These rats would have enjoyed easy passage throughout Central and Western Asia, northeast Africa, and the Mediterranean region as fellow-travelers with trading caravans, ships, and armies.

BOX 7.1 Profile of bubonic plague

Over time, the relationship between microbe and infected host often evolves towards being less virulent—thus improving the microbe's opportunity to replicate and launch its progeny. The *Yersinia pestis* bacterium has coevolved to survive in its natural wild rodent hosts without causing severe sickness. This infectious agent, however, has had insufficient contact with human hosts for a biological détente to have yet evolved. Hence its dread effects.

Within days, the victim experiences swelling of lymph nodes in armpits and groin, blackened by internal hemorrhages. The word *buboes* come from Greek *boubôn*, a swollen gland in the groin. The bacterial invasion also causes bleeding in internal organs and blood clots that lead to gangrene of the feet and hands. Liver damage acidifies the blood, causing spasm of the throat muscles and a strangled gutteral sound. Victims became delirious, recorded contemporary historian Procopius, and "suffered from insomnia and were victims of distorted imagination, for they suspected men were coming to destroy them, and they would rush off in flight, crying at the top of their voices . . . In some cases death came immediately; in others, after many days . . . but all succumbed."[39]

The sights, sounds, and smells of Constantinople in 542 C.E. must have been appalling. If a fiendish deity sought to devise a cruel affliction for miscreant subjects, this was surely it. When the city ran out of room to bury the dead, new burial pits were hastily dug outside the city walls and bodies were compressed tightly into the limited burial space by trampling them "as in a wine-press."[40]

The initial outbreak in Constantinople burned itself out after four months. The rapidly declining numbers of still-susceptible persons became a limiting factor, and the rat population may have declined after depleting the city's granaries.[41]

The pandemic's cause and origin

Whether *Yersinia pestis* is the true cause of bubonic plague has long excited argument among historians.[42] However, reconstruction of the entire genome of *Y. pestis* based on bacterial DNA samples from bodies buried in communal graves in Bavaria late in the Justinian pandemic indicates that the particular bacterial strain involved was a newly emerged *antiqua* variant of *Y. pestis* that subsequently died out.[43]

The geographic origin of the Plague of Justinian is also debated. Was it from northeast Africa, the Central Asian steppes, or India?[44,45,46,47] Did it travel west with Silk Road traders from Central Asia, the ancient Eurasian heartland of the plague, or by overland caravan or sea trade from India via the Red Sea? The sequence and timing of the plague's spread around the Mediterranean coast argues against the Silk Road and India-trade possibilities. Besides, since 540 c.e. the powerful Sassanian army had closed the Silk Road route across northern Mesopotamia.[48]

Much of the evidence points to a northeast African origin. Indeed, local outbreaks of bubonic plague may have occurred in northern Africa many centuries earlier.[49] Rufus of Ephesus, a medical writer from the first century c.e., wrote that a plague-like epidemic had apparently occurred in Egypt and Libya three centuries earlier, with clearly described buboes (groin tumors) recorded by the physicians of the time.[50] Even earlier evidence comes from paleobiological research on material dating to 1500–1000 b.c.e. in the Nile Valley.[51] Other historical documents from the region describe epidemics with swollen buboes, including the Hittite archives (northern Turkey) and Egypt's Ebers papyrus, dating from around 1500 b.c.e.[52]

The sudden outbreak of bubonic plague that occurred in July 541 in the Egyptian trading port of Pelusium (Figure 7.2), as well as its timing, accords with a northeastern African origin of the Plague of Justinian. The contemporary Roman historian Procopius, a leading military advisor and chronicler, described the course of events.[53] Nile river cargo boats unloaded and loaded at Pelusium, as did trading ships from eastern African ports, India, and the Red Sea coastal ports. A manmade canal connected the Red Sea port of Clysma (today's Suez) with the Nile as it entered the delta. Plague-infected black rats from various southern sources could therefore have reached Pelusium, where grain-storage granaries provided a deluxe fast-food facility. Later in 541, the infection would have traveled from Pelusium to Egypt's major seaport Alexandria, where the plague was therefore

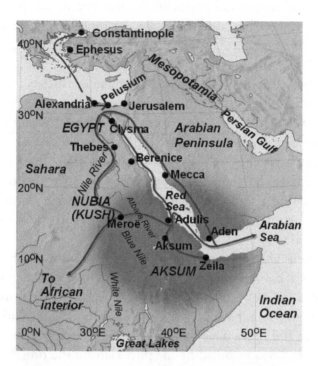

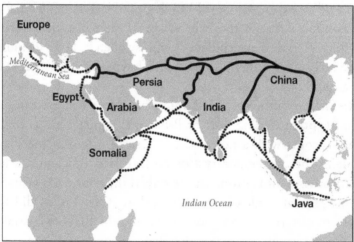

FIGURE 7.2 The northern Silk Road and southern spice (Eastern) trade routes. The sea routes around the horn of Arabia and the Indian subcontinent were Aksum's specialty for nearly a millennium.

Source: Aksum, Classwell Images; Silk Road, Wikimedia Commons.[54]

festering when the grain ships set off for Constantinople in March–April of 542.

How might the plague have reached Pelusium? Grain shipments from Aksum via Pelusium to Constantinople are a likely vehicle. The grain shipments would have reached Pelusium by River Nile grain boats or merchant ships plying the busy Red Sea coastal trade, or possibly by cumbersome overland caravan. Besides Procopius, there were several contemporary reports of plague activity in Aksum. Evagrius Scholasticus, a Roman scholar, wrote that the plague "took its rise in Aethiopia, as it is reported, and made a circuit of the whole world in succession, leaving, as I suppose, no part of the human race unvisited by this disease."[55] John of Ephesus reportedly stated that the plague began "in the people of the inland regions of ... Kush, the Himyarites, and others."[56] Was this separate or recycled information? Aksum, with its capital at an altitude of around 2,000 meters, thrived economically on local agricultural production (especially wheat), gold and iron reserves, artisanal activities, and commercial profits from its bustling Red Sea port of Adulis. The population fed itself from the fertile grain-growing region in Aksum's northern uplands. And where there are grain storage and transport facilities, black rats and their fleas will not be far away.

Family-tree (phylogenetic) analysis based on the bacterial DNA samples from exhumed plague victims also points to a northeast African origin.[57] The DNA of the *antiqua* variant of *Yersinia pestis* that apparently caused the pandemic is a close genetic relative of the *antiqua* variant found in the Great Lakes region today.[58] The infection may have entered the rat-human cycle from its wild-rodent base around the Great Lakes in eastern Africa[59] and then spread by land, river, or coastal trade to Aksum. Bubonic plague was probably smoldering in several eastern seaboard ports further south, from where the busy coastal sea-trade could spread the disease northward to Pelusium.[60,61]

But these local routes, by land and sea, faced a serious obstacle: the high temperatures and low humidity in the latitude zone 15°N–25°N would have made survival and reproduction very difficult for rats and, especially, fleas (Box 7.1). So, without some quite unusual relief from extremes of heat and dryness, the infection was always unlikely to reach the Mediterranean coast. Some unusual relief occurred in the late 530s.

An unusual half decade of extreme cooling

Contention about the pandemic's origin persists, mostly based on phylogenetic branching and time lapses.[62] But much less attention has been paid to explaining the timing: why 541–542 C.E.? Well, there is indeed an unusual candidate explanation.[63] An abrupt fluctuation in the regional climate occurred just before the plague outbreak in Pelusium. The global and regional temperature had plummeted by around 2°C–3°C during 536–538 C.E. and remained low for the next five years before easing back to its original level by the late 540s, as is well illustrated in Figure 7.3. During the cooling, rainfall increased markedly in the Mediterranean region, including very unusual flooding in the Arabian Peninsula, sufficiently extreme to destroy the infrastructure of the ancient Kingdom of Saba in Yemen. That increase in rainfall would have caused a temporary increase in regional humidity.

Procopius wrote that in the year 537 "a most dread portent took place. For the sun gave forth its light without brightness … and it seemed exceedingly like the sun in eclipse, for the beams it shed were

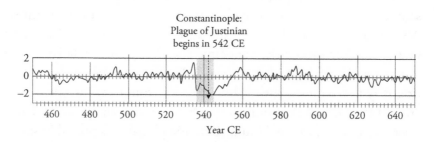

FIGURE 7.3 Fluctuations in annual temperature in western Europe and Scandinavia during 450 to 650 C.E. (based on tree ring data from seven regional sites). The graph shows unusual cooling during 536–545 C.E. The temperature data have been "de-trended" over the 200-year period to display only the short-term variations in temperature. Larsen et al., "New Ice Core Evidence for a Volcanic Cause of the A.D. 536 Dust Veil."[64]

Permission received from John Wiley and Sons.

not clear."[65] John of Ephesus described an 18-month period during the years 536–537 when the days were greatly darkened. During that time, he wrote, crops failed, the wine from unripened grapes was sour, and unusually heavy snow fell in Mesopotamia.

The cause of this rapid cooling event, with its great atmospheric dust-veil and darkening of the skies, is not known for certain. The chemical evidence from Greenland ice-core layers points strongly to a massive volcanic eruption dated to approximately 540 C.E.,[66] but the impact of either a meteorite or a comet shower is also plausible.[67] The eruption, probably in present-day Rabaul (in Papua New Guinea)[68] is estimated to have been twice as massive as that of Tambora in 1815 (see Chapter 9).

This 536 event had extraordinarily far-reaching climatic impacts. Oak trees in the British Isles and throughout Europe displayed extremely low growth rates during the ensuing decade, as did trees in Scandinavia, northern Russia, Mongolia, southwest United States, Chile, and Argentina. In America, near today's Mexico City, the mysterious and monumental civilization of Teotihuacan was plunged into conflict and chaos during the 540s and collapsed within a decade. In China, the Northern Wei Dynasty fell, as freezing weather, crop failures, and famine enveloped the population, three-quarters of whom died. And in Central Sweden, archaeological research has identified a social and demographic crisis seemingly triggered in the exceptionally cold year 536, the "Fimbulvinter."[69]

This crisis in Sweden led rapidly to replacement of the simpler egalitarian Early Iron Age culture with a more hierarchical Late Iron Age society, reflected in a distinct change in the type of graveyard monuments favored by the nouveau-riche elite.[70]

But back to the travelling fleas—in ancient Upper and Middle Egypt and the adjoining Red Sea, at approximately 15°N–25°N, the usual high temperatures of 33°C–40°C on boats traveling north on the Nile or Red Sea coastal routes in the post-harvest season would have been too hot for flea survival and reproduction, and perhaps also for rat survival. The tolerable temperature range for critical aspects of flea biology is approximately 20°C–30°C,[71,72] which fits with the fact that most of the dozens of recent outbreaks of the plague have occurred in regions with mean annual temperatures of 24°C–27°C.[73] These grain-shipment journeys of around 2,000 kilometers took several weeks, depending on the route. To sustain the cycle of rat-flea infection, a sufficient number of black rats was required. This was easily achieved by reproduction or by replenishment in ports visited along the route—ports where river cargo boats could also pick up the related native Nile rat, the region's longstanding natural host of the plague bacterium. The flea population, however, was less easily replenished. The usual mid-journey temperatures, well above 30°C, would have impaired flea reproduction, and would have been too hot for multiplication of the bacteria in the flea gut before regurgitation into the bloodstream of the bitten host.[74,75] The Red Sea coastal temperatures are some of the hottest in the world. However, the unusual transient cooling during 536–542 C.E., along with wetter and more humid conditions, would have facilitated the infection's successful northward journey to Pelusium from Aksum or environs.

The pandemic spreads

During the second half of the sixth century, the plague spread east into Asia Minor (Turkey and the southern Caucasus), killing up to a quarter of the human population there. For several decades it was rampant in Persia. It subsequently reached China as flea-infested rats hitched rides with trade caravans along the Silk Road. Ancient records show that during 636–655 a sequence of plague-like epidemics spread eastward along the Silk Road. The Chinese authorities recorded serious "plague epidemics" in today's provinces of Shansi, Gansu, Ningxia, and Shensi.[76]

The pandemic also extended west through (today's) Greece, Italy, and France and into Spain. Via France, it reached the British Isles in the seventh century. The Venerable Bede, a scholarly monk and writer secluded in a monastery on Scotland's Lindisfarne Island, recorded that in 664–666 C.E. the plague "ranged far and wide and caused fierce destruction. It ravaged Britain and Ireland with cruel devastation." The plague also returned several times to Constantinople and ultimately caused the death of an estimated half of its urban population.[77] By the early seventh century, population numbers within the Eastern Roman Empire had been dramatically culled by this plague and secondarily by the resultant social disorder, crop failures, and starvation. Many of the region's ancient towns and cities were abandoned and disappeared.

This great pandemic lasted for over two centuries and may have killed 50–100 million people across Eurasia, especially around the eastern Mediterranean region. Procopius, on the spot as contemporary historian, wrote: "During these times, there was a pestilence, by which the whole of the human race came near to being annihilated." This assessment reflects just how dire the state of the world must have looked from the Roman citadel.

Six centuries later, bubonic plague again entered Europe. This time it originated in Central Asia and entered Europe through the Black Sea trading ports. The ensuing Black Death spread through Europe over the next five years, killing around one-third of the population and persisting with flickering outbreaks in Europe, the eastern Mediterranean, and North Africa for several centuries. A further six centuries on, during the second half of the nineteenth century, the third plague pandemic began in China, gathered momentum over several decades, and then spread around much of the world via shipping from the great seaport of Hong Kong. The infection then established footholds in far-flung native rodent populations in the Middle East and North, Central, and South America.

Today, outbreaks in local human populations flare up from time to time in locations close to either the newer or older rodent-reservoir locales. Several thousand new cases are reported to the World Health Organization every year.

All three pandemics occurred in the wake of different configurations of climatic changes, as we will see. Several commentators have noted that not only did the three pandemics all arise at times of climatic fluctuations but each subsided when stable climatic conditions reappeared.[78] What might be the prospects for bubonic plague in a future climate-change world?

Drought in 8th- and 9th-century Central America: The Maya

The Classic Maya civilization in Mesoamerica, with its mix of grandeur, cruelty, astronomy and mathematics, pictographic script, mighty stone edifices, and the romance of a "rain-forest civilization,"[79] is an enduring popular source of mystery and fascination.

The Maya had deep historical roots; there are records of a preceding, simpler civilization of villages and primitive agriculture from around 1500 B.C.E. The Classic Maya period extended from roughly 300 to 900 C.E. Urban centers were spread throughout the lowlands of Mexico's Yucatán Peninsula and in part of Central America, including today's Belize, Guatemala, El Salvador, and Honduras (see Figure 7.4).

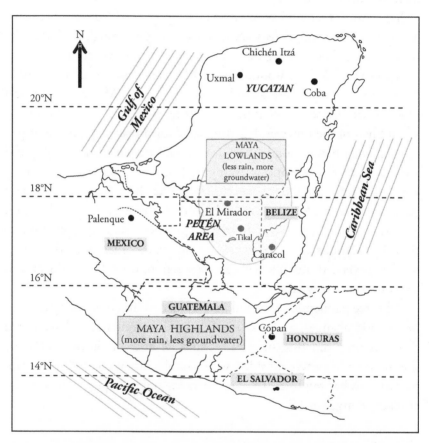

FIGURE 7.4 The main lowland and highland Classic Maya centers. The civilization spanned all or part of today's Mexico, Belize, Guatemala, Honduras, and El Salvador. Adapted from Demarest, *Ancient Maya: The Rise and Fall of a Civilisation*; Turner II and Sabloff, "Classic Period Collapse of the Central Maya Lowlands: Insights about Human-Environment Relationships for Sustainability."[80]

There have been many theories as to why, after half a millennium of success, the Classic Maya civilization declined: shifts in political power, changing local fortunes of river-based trade, overpopulation, exhaustion of arable land, resentment of ever-increasing appropriation of wealth by insatiable rulers, territorial conflict, and so on. Meanwhile, later studies point to a significant influence of regional climate change.

First, though, some background information. Visitors to the ruins of Mayan cities can be shocked by accounts of the bloodthirsty goings-on. The Classic Maya may have used human sacrifice less extensively than did the rulers of the contemporaneous city of Teotihuacán, further north in the Mexican highlands, or, later, the Aztecs.[81] Even so, ritual and retributive human sacrifice by the Maya was common. During a visit to Uxmal, a major Maya center in the northern lowlands of Yucatán, I learned about three settings that involved human sacrifice.

First, the rain god Chaac required regular supplication, and, in this region of cavitated porous limestone, a common ritual was to sacrifice selected citizens and throw their bodies into cenotes (deep natural limestone sinkholes, from which fresh water was available). Second, the Mesoamerican ball game *öllamaliztli*, played in an open court with a heavy rubber ball to be thrown through a stone hoop, was often used for ceremonial purposes. In some games, the final glorious moment for the winner was to die by ritual sacrifice in a state of drug-assisted ecstasy. The contest also served as a winner-take-all method of resolving serious disputes between neighboring cities. Under the diplomatic protocol of the day, the defeated ruler was then ceremonially sacrificed.

Third, and more brutal, prisoners captured in warfare were treated mercilessly. In Uxmal, as elsewhere, one unpleasant practice was to forcibly bend and bind individual prisoners into a ball shape and then roll them down the 72 steep stone steps of the Grand Pyramid. (I counted the steps on my way down, taking special care not to stumble and tempt historical fate.)

This was a civilization without wheels, metal tools, sails, or beasts of burden, living in an environment with rain-forest soils of mediocre fertility, few navigable rivers, and limited freshwater supplies. Annual monsoonal rainfall was very seasonal: nearly all the rain fell in a six- to seven-month period. This seasonal rainfall, plus water from re-engineered cenotes in the porous limestone bedrock, was crucial to Maya survival and social stability, and the ever-present threat of drought (as monsoons varied in strength and range) explains why the Maya lavished so much religious fervor and worship on the fearsome rain god Chaac. Centuries later, following the Spanish conquest, Bishop Diego de Landa of Yucatán wrote: "Nature worked so differently in this country in the matter of rivers and springs, which in all the rest of the world run on top of the land, that here in this country all [rivers] run and flow through secret passages under it."[82]

Yet not only did the Maya forge a successful way of life, despite dense tropical forest, mediocre soils, and a paucity of surface water, but the highly advanced Classic Maya civilization flourished for more than half a millennium. They developed methods of drought-resistant food production, essential for feeding ever-larger urban populations— most notably permanently raised fields, ponds, and forest gardens. This in turn led to the invention and construction of micro-watersheds.[83] The plant-based Mayan diet was dominated by corn (maize) and often supplemented with wild plant foods. The domesticated turkey was the main source of animal protein. Since the Maya had no large domesticated mammalian livestock, unlike the better-endowed early European agrarians, any additional meat came from wild rabbits, armadillos, monkeys, deer, and macaws.

Overall, their successful methods of gardening, field agriculture, and water management, along with good rains in the seventh and first half of the eighth centuries, especially in the Maya lowlands, enabled the total population to peak at around 10–15 million by 700 C.E.[84] Around that time, Maya society built many of its most imposing ceremonial and civic edifices; Temple IV in Tikal, built in 747 C.E., soared about 65 meters skyward, the height of a 20-story modern building. Trade and connectivity between centers had also been on the rise, accelerating the accrual of wealth—though also intensifying pressures on arable land and water supply.

But the threat posed by a serious shortage of rainfall to food yields and drinking water supply could not be avoided. After a flourishing half millennium of demographic, cultural, architectural, and technological advance came a period of declining rainfall and droughts during 750– 950 C.E.[85] Food shortages, conflicts, and internal migrations followed in the wake of this change in climate. It appears that the change in climate played a significant role in the serial decline and subsequent collapse of many Mayan centers.[86] More specifically, was climate change a central actor, part of an interactive and diverse cast, a walk-on extra, or merely loitering in the wings?

The Classic Maya decline

The decline and then collapse of the Classic Maya civilization was neither a single-cause event nor a synchronized process.[87,88,89] The cities in the central lowlands were the first to be abandoned, but some centers further north in Yucatán survived into the early eleventh century c.e. The early stage of the decline was marked by an upsurge in competition for trade and power, warfare between centers, social unrest, and political instability. In a further attempt by rulers to bolster their faltering power and prestige, there was also a brief flurry of construction of ever-larger stone edifices—analogous perhaps to the final, ever-larger, stone statues carved by the Easter Islanders as their trees (and rollers) ran out.

But the wall of social stability and political authority had been irreparably breached. Morale declined; hunger, disorder and violence flared; power structures weakened and dissolved; population numbers shrank; and urban dwellers dispersed into simpler decentralized settlements. In due course, the rain forest gradually began to reclaim its alienated territory, eventually engulfing much of the farmland, urban gardens, and great stone edifices.

Skeletal remains from Mayan burial sites dating from the late eighth century show an increased proportion of infant, child, and adult female deaths.[90] Skeletons recovered from Copán, in the eastern Maya highlands (in today's Honduras), show the telltale signs of undernutrition: weakened porous bones; Harris lines in long bones, indicating arrested growth; and micronutrient-deficiency stress lines in teeth. These several hundred sets of bones from Copán date predominantly from the period 650–850 c.e., and indicate that the nutrition-related health of both the elite and the lowly was deteriorating.[91] There is no specific evidence of infectious disease in the skeletal remains (other than dental caries). Unlike the early farmers of the Fertile Crescent and southeast Europe, societies throughout the Americas had no large domesticated livestock other than the Incas' llamas as a ready source of zoonotic infections. Hence until the Spanish conquistadores and their microbes arrived, there is no evidence of infectious disease outbreaks.

Monsoonal retreat, droughts, and disorder

Under increasing population pressure in the late seventh century, agricultural production faltered in many Mayan centers due to combinations of soil degradation, increased reliance on marginal land, the silting up of rivers and irrigation channels, and declining rainfall. By the mid-eighth century, some peripheral Mayan centers in the southeast, straining for sufficient food and water for their enlarged urban populations, became politically unstable and were among the early collapses. The first of a series of debilitating droughts had struck. The comprehensive archaeological studies of Richardson Gill led him to conclude, "Large Maya cities collapsed in four phases of abandonment, spaced about 50 years apart, around A.D. 760, 810, 860, and 910"[92] (Figure 7.5). Maya society depended critically on the annual monsoonal rains. The inland reach and strength of the summer monsoon varied as the average position of the Inter-Tropical Convergence Zone meandered north or south of the equator. The further south it went, the more the monsoon rain fell offshore in the Pacific waters. Building on earlier lake-bed pollen studies in the Yucatán Peninsula, the reconstruction of regional drying trends and major droughts during the period of Maya civic decline was subsequently enhanced by laboratory techniques that track temperature and rainfall variations. First, high-resolution chemical analyses of coastal sediment layers have estimated changes in onshore rainfall and in oxygen isotopes in seabed sedimentary layers of shell-remnant carbonate, and second, finely slicing a half-meter-long stalagmite from the peninsular southeast into several thousand wafer-thin slivers has yielded a detailed picture of annual variation in rainfall (Figure 7.5).[93,94,95]

These findings present a fairly coherent picture. Around 700 c.e., rainfall had begun a downtrend, lasting until the early 900s, during which three droughts each of several decades duration occurred, centered at around 770, 825, and 910 c.e. (Figure 7.5). Rainfall then steadied until the end of the tenth century, before plummeting dramatically into a century-long drought that coincided with the final

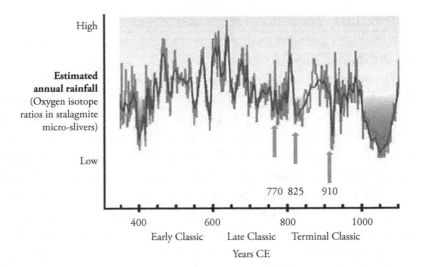

FIGURE 7.5 Annual rainfall during the Classic Maya civilization, estimated from serial micro-slices of a stalagmite from a cave on the eastern periphery of the Central Maya Lowlands (near Caracol). Adapted from Kennett et al., "Development and Disintegration of Maya Political Systems in Response to Climate Change."[96]

collapse of several surviving centers in the more water-secure northern Yucatán.

These findings suggest that during that 200-year period of societal decline, an increase in southerly excursions of the summer monsoon system reduced the northward reach of annual rains into the Mayan lands. The droughts may also have been exacerbated by the microclimate impact of extended land-clearing for crops, for urban expansion, and for making lime and lime plaster for building.[97] The sequence of droughts, each causing around a one-third reduction in annual rainfall,[98] resulted in a water-supply crisis, falls in food production, and great stress on the social and political fabric. Archaeologists have found evidence in many Maya centers of social stress, architectural decline, and violent conflict clustered in time around these droughts.[99,100] Depending on local circumstances and levels of resilience, Maya centers both great and small suffered, contracted in size, and often succumbed to violence and destruction—some more rapidly and turbulently than others. Among the cities abandoned early was the flourishing Late Classic city of Tikal in the southern-lowlands Petén area.[101] Some studies have found episodic increases in the prevalence of nutritional deficiencies and in the proportion of child-age skeletons during the drying period, along with apparent instances of survival cannibalism.[102]

The uneven timing of regional collapses of Maya centers seems paradoxical. The northern lowland centers of the Yucatán Peninsula, although in the most rain-deprived region when the summer monsoon shifted south, generally survived longer than centers in the Central Lowlands and the highlands of Central America. But the northern region had better groundwater sources, and trading routes and commodities such as obsidian may have shifted in their favor as river traffic became less easy.[103] Overall, though, it is reasonably clear that the advent of drier conditions and serious droughts was a major, perhaps critical, terminal stress on the Classic Maya civilization. From his extensive archaeological studies, Richardson Gill concluded: "Among all the potential causes of the deaths of millions of Maya, famine and thirst is the only explanation that accounts for the pattern of their survival around long-term, stable sources of drinking water and their death elsewhere."[104]

Droughts in the Americas: The American Southwest (800–1250 C.E.)

In the 1130s, a major drought emerged in the southwestern region of today's United States, the greater San Juan Basin. This was the second of three droughts that occurred during the eleventh to thirteenth centuries (Figure 7.6). This region, clustered around the "Four Corners" junction between (clockwise from top right quadrant) Colorado, New Mexico, Arizona, and Utah, was home to several preliterate corn-farming peoples, the pueblo-dwelling Anasazi, Hohokam, and others.

These cultures each combined a mix of hunting, gathering, and, increasingly, corn farming—and hence were particularly dependent on adequate rainfall for their crops.

Early in the ninth century, the global climate was influenced by a significant shift from El Niño to La Niña dominance, resulting in a warming of the temperate-zone northern hemisphere that lasted several centuries known as the Medieval Warm Period.[105] The northern

Atlantic and Iceland-Arctic region warmed first, from around 700 c.e., followed soon by Scandinavia and then, a century or two later, by Europe.

The first of the three droughts, during 990–1060 c.e., was a northern "relative" of the severe eleventh-century drought in southern Mexico that had terminated the Yucatán remnants of the urban Classic Mayan civilization. In the background was a long-term trend of sustained cooling and drying with fluctuating extremes, extending into late in the fifteenth century and qualifying as the region's longest-lasting extreme climatic event in the past thousand years.[106]

The second and third droughts, beginning in 1135 and 1276, respectively, extended eastward sufficiently to influence the "Cahokian" culture in what is now southwestern Illinois, alongside the Mississippi River (Figure 7.7).

The megadrought that began in the mid-twelfth century in the North American southwest would shape the fate of those several neighboring civilizations and cultures over the next two centuries. It may have been the result of an unusually dominant and prolonged period

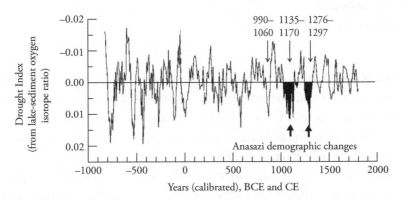

FIGURE 7.6 Variations in drought conditions over 2.5 millennia, based on lake-floor sediment isotope profile, Pyramid Lake, Nevada, United States southwest. The three droughts centered on the San Juan Basin region are shown. Based on Figure 9.b of Benson et al., "Possible Impacts of Early-11th-, Middle-12th-, and Late-13th-Century Droughts on Western Native Americans and the Mississippian Cahokians.[107]
Permission received from Elsevier.

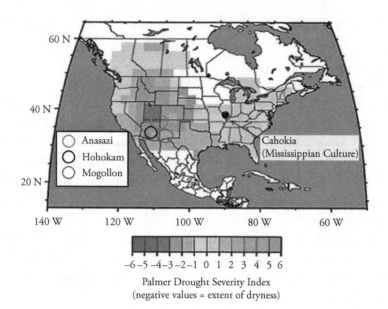

FIGURE 7.7 Range and severity of the mid-12th-century drought in southwest America. Approximate locations of the three main pueblo peoples are shown, as is the Cahokian-Mississippian settlement. Adapted from Cook et al., "North American Drought: Reconstructions, Causes, and Consequences.[108]

Permission received from Elsevier.

of La Niña conditions, conditions that characteristically caused drying in the western and coastal region.[109] This twelfth-century drought also extended south into Central Mexico, where it presumably contributed to the final dissolution of the Toltecs—the Aztecs' mysterious predecessors—by curtailing harvests of the staple crop, corn (maize).

Back in the North American southwest, the widespread regional drought took a heavy toll. Most is known about the reversal of fortune of the Anasazi ("the ancient ones"), especially the pueblo-dwelling communities who lived in Chaco Canyon and the surrounding region at the Four Corners junction. The Anasazi and Hohokam people, as subsistence farmers, had become expert over many centuries in managing the limited and variable water supplies and in knowing where crops, especially corn, grew best and yielded most, all the while maintaining knowledge of wild plant foods as backup options.[110]

Anasazi food yields were, however, always hostage to climate, and the foods were deficient in vital micronutrients. Most importantly, we know now that corn lacks three important protein-building amino acids: lysine, isoleucine, and tryptophan. In modern cultures, the risk of significant deficiencies is minimized by consuming a wide variety of other foods, or by laborious food preparation methods that maximize micronutrient absorption. However, carbon-dated bones from that period show an unusually high prevalence of weakened and porous bone structure in the Chaco Canyon communities.[111] This condition of the long bones, often evident in the eye socket, is caused by chronic deficiency of micronutrients, particularly iron.

Nevertheless, having established a fairly reliable supply of food, local populations had grown substantially during the sixth to tenth centuries. This, of course, increased the risks they faced as their numbers encroached increasingly on the region's carrying capacity.[112] In due course, as drought conditions emerged, widespread conflicts broke out, especially during the twelfth and thirteenth centuries. Many farmers and their families fled from their fields.

Before that period of drought and conflict, however, the Pueblo peoples had built some of the largest towns in North America, with characteristic communal housing. On the plains, the pueblos were ground-level dwellings constructed with tightly packed rock and earth walls, their roofs supported by wooden posts and beams. In some canyon-dwelling communities, the Anasazi built extraordinary dwellings within the sheltered cliff faces on massive canyon walls. At Chaco Canyon, in response to faltering rainfall in the ninth and tenth centuries, they built several multistoried "great houses" near major water drainages. The much-photographed Pueblo Bonito, replete with hundreds of rooms, could house around 1,000 people.

Then, in the twelfth century, came the prolonged droughts and water shortage that would cause crises and violence during the thirteenth century. As rainfall on the Mesa Verde plateau decreased, the water flowing down through Chaco Canyon dwindled and corn harvests shrank. Corn needs winter rain to germinate and start growing, and summer monsoonal rain for its continued growth. As with the Sumerians and Harappans three millennia earlier, the remaining Anasazi farmers tried to compensate by pushing their irrigation system to new limits, but with the accompanying topsoil erosion, deforestation, a dwindling supply of native piñon nuts, and a now excessive population size, the local environment's carrying capacity was clearly exceeded. Here was yet another example in which climate change pushed an already stressed system over its viable threshold.

The Anasazi of Chaco Canyon and the Mesa Verde began to lose cohesion and to contract. Further south, the Hohokam and Mogollon Pueblo cultures were less seriously affected.[113] The increasing levels of malnutrition, starvation, and thirst exacted an increasing toll on the Anasazi people. Further stress and conflict appear to have arisen from new arrivals moving onto the adjoining Colorado Plateau. Remains of fortifications from around 1250 C.E., along with signs of burned villages, arrowheads in skeletal body cavities, and bony evidence of scalping, all suggest violent conflict, and the skeletal remains of seven dismembered bodies indicate ritual or conflictual cannibalism.[114] Under these deteriorating conditions, many of the Anasazi dispersed, migrating along preexisting lines of trading and cultural affinity and mostly heading south, where they resettled.[115]

An echo of that protracted desiccating past is sounding louder today in that same region. The great drought of 2012 in the American West set uncomfortable new records, and there have been several other droughts in that region since the extreme drought of 2000–2004. Fast-forwarding, climate modeling suggests that extreme droughts in the American West, once rare, may become the "new normal" during this century as warming proceeds.[116]

The Cahokian-Mississippian Culture

Cahokia, located in present-day southwestern Illinois, was the main center of a Middle Mississippian culture that flourished from the mid-tenth to mid-fourteenth centuries. This was a period of stable and favorable climatic conditions. The city of Cahokia and the surrounding villages and farms became the most extensive urban center in prehistoric North America, but the agricultural networks extended widely within the American Midwest and Southeast. The Cahokian diet, replete with corn and squash, included deer, fish, waterfowl, nuts, and seeds—all abundant in the Mississippi Valley and the surrounding uplands.

Urban Cahokia was dominated by a vast human-made earthen mound, the lofty base for a grand ceremonial stone edifice that overlooked a huge plaza used for communal feasts, religious festivals, and sporting events. The Cahokian culture peaked with a regional population of around 40,000 in the later decades of the thirteenth century—just as Europe's Medieval Warm was fading. By the time Christopher Columbus landed in the New World, late in the fifteenth century, Cahokian society had collapsed. The city of Cahokia itself had been abandoned, leaving a sparsely populated Mississippian hinterland in which successor generations led simpler lives of local subsistence.

Was it infectious disease, food shortage, climatic change, societal disorder, or warfare that influenced this demise? The large population had presumably overexploited wood resources for fires and construction and had cleared forests for building the city and its surrounding village communities. The denuded watershed would have become prone to runoff, erosion, and unseasonable summer floods that damaged pre-harvest crops. Social unrest, hunger, loss of faith in rulers, political upheaval, and violent conflict were the likely, familiar consequences.

The second and third droughts shown in Figure 7.6 each coincided with cool-phase (negative) "minima" of the Pacific Decadal Oscillation (PDO), which has an approximate 50-year cycle-reversal time. During negative phases of PDO, as for the El Niño phase of the El Niño Southern Oscillation, the southwest becomes drier, whereas the positive PDO phase brings increased rainfall and wetter-than-average conditions. The drought conditions of the mid-twelfth century extended east to Cahokia; they did likewise during the third drought, late in the thirteenth century.[117]

A further clue comes from the fact that the rise and fall of the late-Mississippian Cahokia complex coincided with the rise and fall of the Anasazi society. Indeed, both the Four Corners region and Cahokia had undergone rapid growth in agriculture, construction, and population between the mid-eleventh and mid-twelfth centuries, before the droughts struck. This pair of protracted droughts, separated by a century, almost certainly played a major role in the later-stage weakening and demise of the Cahokian-Mississippian culture. The unusually prolonged dry conditions presumably curtailed crop growth and harvests, causing stress, hunger, and potential social disorder. The priests and shamans had failed the people. The social and political structures were weakened. The good days were in the past.

Concluding remarks

Three climatic features distinguish the period spanning the origins of the Roman Warm Period around 250 B.C.E. through to the end of the great period of drought in the American southwest and Central America around 1600 C.E. They mostly relate to prolonged changes in regional climatic conditions, particularly drying trends and severe droughts.

First, the experience of the Roman Empire illustrates the subtle but pervasive influence of long-term shifts in regional climate systems, particularly on agriculture, food sufficiency, nutrition, susceptibility to infection, and social stability. Adversity applied on both sides of the empire's border. Within, the empire's farmers, townsfolk and military legions became hungry, and infectious disease outbreaks were more severe. Outside the borders, the swelling populations of Germanic, Slavic, and, later, Hunnic tribes were pushing for new lands; a deteriorating climate impaired northern and eastern European and peri-Baltic farm yields; and aridity affected the Asian steppes. Invasions of the Western Empire intensified, and Rome fell.

Second, the experience of the Classic Maya civilization underscores the power of sustained drying and severe droughts to weaken populations and undermine polities. Whatever the full complement of factors contributing to the Maya decline, a protracted shift in seasonal rainfall systems over several centuries, punctuated by sustained drought episodes, caused widespread hunger, nutritional disorders, and deaths. This was an extra stress that an already overpopulated and environmentally compromised civilization of multiple city-states could not absorb.

Third, the experience of the pueblo-dwelling societies of the American southwest (Anasazi and others) during the twelfth and thirteenth centuries attests to the power of a megadrought to overwhelm the resources of populations already living in marginal climatic and agricultural conditions. The central problem was a lack of water as aridity encroached and a prolonged megadrought emerged. In the next chapter, we will encounter increases in hunger and starvation, infectious disease outbreaks, and violent conflicts, all due in part to the climatic changes, environmental stresses, and social turbulence that characterized the colder periods of the Little Ice Age in Eurasia.

8

Little Ice Age
Europe, China, and Beyond

PICTURES OF WINTER ICE-FAIRS on frozen rivers in seventeenth-century central and northern Europe are familiar. Less well known are the many other consequences—social, environmental, military, political—of the unusually cold period throughout Eurasia that lasted from around 1300 to 1850 C.E. For many it was a time of social instability, food shortages, epidemic outbreaks, impoverishment, and miserable or violent deaths. Climate-related crises helped foment persecutions, armed conflicts, and the overthrow of dynastic rulers.

This was the "Little Ice Age," the name coined by American glaciologist François Matthes in 1939. The cooler conditions that emerged in western Europe from late in the thirteenth century followed a dip of around 1°C in global temperature after a massive volcanic eruption in Indonesia in 1257 that disrupted harvests and caused an increase in epidemic outbreaks throughout Europe.[1,2] Both the earlier Medieval Warm Period (around 950 to 1250 C.E.) and the Little Ice Age were influenced by multicentury changes in solar activity, reflecting the 2,300-year solar Hallstatt Cycle.[3,4] During the first 200 years of the Little Ice Age, temperatures fluctuated, with a warm respite of six decades around 1400 C.E., after which cooling prevailed until 1500. The cooler conditions then receded until 1560, after which the climate plunged into a longer, colder second phase (Figure 8.1).

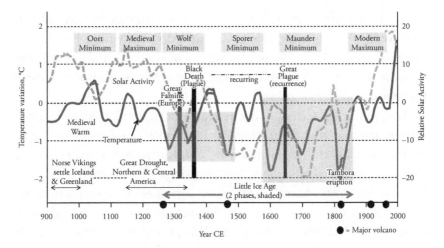

FIGURE 8.1 Variations in European temperature (relative to twentieth-century average) and in solar activity during 900–2000 C.E. Several major historical events are shown, as are the two phases of the cooling. Temperature graph adapted from Büntgen et al., "2500 years of European Climate Variability and Human Susceptibility," Solar activity graph: "Solar cycle," Wikipedia, http://en.wikipedia.org/wiki/Solar_variation.[5] Permission received from AAAS.

There is no consensus on when the Little Ice Age began or ended, but most scholars place it between the thirteenth and nineteenth centuries. The cooling was not confined to Europe; it occurred throughout much of middle- and higher-latitude Eurasia and North America, while in eastern China the temperature fell by around 2°C during 1250–1350. The climatic changes associated with the Little Ice Age in the northern hemisphere may have been global in scope. In the seventeenth century, for example, the northern cold was mirrored by substantial weakening of the South Asian summer monsoon, resulting in droughts that killed millions in northwest India.

Europe's average temperature during 1300–1850 was around 1.5°C lower than in the preceding Medieval Warm period, and 2°C–3°C lower than later in the half-century around 1575–1625. Many shorter-term fluctuations of several degrees were interspersed within and between decades. There were periods of intensely cold winters with easterly winds, and other periods with milder winters, heavy spring rains, and early summer storms. There were local droughts, sometimes with cold conditions, sometimes hot. Conditions varied within Europe. For example, in the Tatra Mountains region in southern Poland the temperature actually rose a little around 1600 while western and most of Central Europe was cooling.[6]

As the Little Ice Age deepened, Europe's hierarchical feudal system and patchwork of kingdoms, baronies, and fiefdoms began to fray during the latter years of the fourteenth century. A less rigid and oppressive society was evolving, though still riven with inequalities. Class-based codes even pertained to men's shoes: the higher one's class, the longer and more decoratively curled could be the extended "toe" of the shoe. This period was characterized by the growth of towns, commerce, trade, and cathedral building and the flourishing of the arts. Even so, daily life remained harsh for the rural majority, and average life expectancy at birth was no more than 30 to 35 years.

Throughout this period, harvest failures, famines, and starvation occurred frequently, until there was some relief in the eighteenth century. The Little Ice Age was also a time of recurrent outbreaks of bubonic plague. The second pandemic, beginning eight centuries after the outbreak of the Plague of Justinian, first emerged as an epidemic in central-western China in the 1330s, followed by the notorious "Black Death" in Europe. Other epidemic scourges, including typhus and smallpox, brought suffering and death to many. So did dysentery, associated with unsanitary living conditions and compounded by urban crowding, poverty, flooding, and disorder following extremes of rainfall.

The Great Famine, northern Europe (1315–1322)

The transition from the Medieval Warm Period to the onset of Europe's Little Ice Age late in the thirteenth century was a troubled time. Climatically, the El Niño signal was strengthening, causing cooling in northern latitudes and warmer and drier conditions with weakened monsoons in lower-latitude Asia. The gradual change in climate during the thirteenth century saw a fall in grain yields and in protein production from cattle and sheep.[7] Europe's northern farmland harvests failed as temperatures fell, and local cattle-and-fisheries economies were hit especially hard. In England, hunger and malnutrition increased and life expectancy declined by nearly 10 years between the late thirteenth and late fourteenth centuries.

As economies contracted, poverty and insecurity became widespread. By the early fourteenth century, the feudal system began to weaken, and hostility towards local Jewish communities increased throughout Europe after two centuries of tolerant coexistence. The long-simmering religious dispute about usury boiled over as cash-strapped kings increased taxes on Jewish moneylenders, who then were forced to charge higher interest rates on loans. In 1290, Edward I of England issued the *Edict of Expulsion* that expelled all Jews, and continental rulers took similar actions, forcing fractured Jewish communities to head east to Poland and beyond.

The fourteenth century, according to climate historian Hubert Lamb, brought "wild, and rather long-lasting, variations of weather in western and central Europe."[8] An extreme low pressure system formed in the mid-latitude North Atlantic Ocean[9] as the North Atlantic Oscillation moved into negative phase, which boosted the moisture content of westerly winds and brought heavier rainfall to continental Europe.[10] Even so, the opening few years of the fourteenth century were benign, and many farmers took renewed heart and expanded their sowing; fresh optimism pervaded southern England's grain-growing region and vineyards. Gradually, though, cooler weather settled in, with wetter summers and earlier autumns. From 1312, the weather began to fluctuate, rains became heavier, and by 1315 the countries of northern and central Europe faced the onset of a severe and prolonged famine—the Great Famine of 1315–1322, perhaps the single worst subsistence crisis in Europe's recorded history.[11]

During much of this climatically extreme period, most of Europe's grain crops failed, and famines resulted. In southern England, yields declined precipitously during the years 1315–1317, and there was a threefold surge in the grain price (Figure 8.2); the death rate doubled, as did land sales, and crime rates escalated fivefold.[12] Meanwhile,

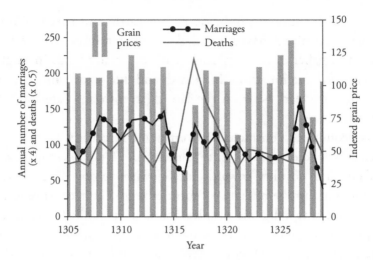

FIGURE 8.2 Annual variations in the grain price and estimated rates of marriage and death in southern England during the Great European Famine, 1315–1322. Note the unusual occurrence of two consecutive back-to-back harvest failures in 1315, 1316, and 1317. Single-year harvest failures and price rises, as in 1321, had little effect on marriages and deaths; people coped. Adapted from Campbell, "Nature as Historical Protagonist: Environment and Society in Pre-Industrial England."[13]
Permission received from John Wiley and Sons.

the corollary of this negative phase of the North Atlantic Oscillation (NAO) was that northern Scandinavia reaped temporary agricultural benefit, courtesy of a strong high-pressure system emanating from the Icelandic region and bringing warmer winds and less drenching weather.

In that first dreadful year, 1315, the relentless heavy rain damaged crops and rotted much of the seed grain even before it germinated. Many families desperately sought out edible substitute foods, such as nuts, roots, leaves, nettles, and bark. Malnutrition was widespread. In the following year, the cold and wet weather persisted, food reserves further diminished, and the death rate began to rise, irrespective of status: both peasants and noblemen suffered. As desperation grew, draft animals that normally helped plow the lands were killed and eaten, as were any unspoiled seed grains.

Food shortages were not limited to reductions in grain yields; meat was also in short supply. Sheep, cows, and oxen died in large numbers, mostly because of epidemic outbreaks in herds and flocks. The highly contagious measles-like viral infection rinderpest is thought to have swept through Europe's herds of cattle, appearing for the first time around 1318 and then spreading to England, where by 1320–1321 herd sizes had halved.[14] This disease had serious knock-on effects, since there was less manure to replenish the fertility of the soil and fewer oxen to plow the fields, leading to bankruptcies and the abandonment of some farm holdings.

Meanwhile, the relentless rain leeched nitrates and other nutrients from the soil, rendering plants more susceptible to disease and to infestation by rusts, mildews, and molds. Blighted plants would usually have been discarded uneaten. However, given the severity of the famine, many people were impelled by hunger to eat blighted plants, often suffering illness or poisoning. Some mycotoxins that would have contaminated infested plants are known to disrupt the body's immune system; they can also cause neurological and behavioral disturbances. One, rarely seen today, is ergot poisoning, which causes "St. Anthony's Fire," notable for its psychotic effects and dancing mania. Ergot, a fungus that affects rye and other grains in weakened plants and damp conditions, contains the psychedelic compound lysergic acid, of LSD fame. In medieval times, outbreaks of St. Anthony's Fire were common in populations subsisting on rye bread. Ergot poisoning caused gangrene and burning pain in the extremities, convulsions, hallucinations, and severe psychosis. Death often followed. Sometimes entire village populations were afflicted with visions, and pregnant women often miscarried. During the Great Famine, outbreaks of ergot poisoning were common.

The worst of the winters occurred in 1317–1318, described by one Englishman as the harshest in a thousand years: "a thusent winter ther bifore com nevere non so strong." During most of this half decade, dreadful weather prevailed throughout central and northern Europe, with incessant and often torrential rain, floods, mud, and cold.

Harrowing accounts described widespread starvation, epidemics, escalating death rates, class conflicts, sodden rain-drenched warfare, unspeakable human behavior, and widespread violence and theft. The horrors of this catastrophic time reverberated in folklore memory and children's stories for several centuries.

In conditions of extreme starvation, social disorder, and poverty, infectious disease epidemics are usually not far away. Wolfgang Behringer writes that in 1315, in addition to hunger and starvation, a "grim pestilence" was spreading in much of Europe.[15] Records do not identify the exact nature of this pestilence; it may well have been a mix of different infectious-disease scourges. The disease or diseases killed up to one in every three persons in the most afflicted localities in the Netherlands, France, England, and Scandinavia. In many communities, the local cemeteries were overwhelmed and emergency burial pits were dug outside the town walls.

Systematic parish registers of deaths had not yet come into wide use at that time, and the best estimate of the overall mortality toll in northern Europe is that 5 to 10 percent of the population perished. Death rates in towns and cities tended to be higher than in the countryside, sometimes 10 to 20 percent.[16] Of course, those statistics do not capture the suffering of many thousands of starving and penniless people. According to one contemporary account: "Their movements become slow, their voices still. Their skin grows pale. Lacking their usual food, they give themselves over to strange diets. They graze like cattle. And, eventually, they are buried in mass graves without benefit of the rites of the Church."[17]

There were reports from England, Poland, and the Baltic regions of desperate parents killing or abandoning young children in order to reduce the family's number of unfed mouths, and of famished people resorting to eating the dead.[18] (However, the reports of cannibalism that have often surfaced during great famines in history are nearly always anecdotal and difficult to verify.) Grimm's fairy tale *Hansel and Gretel* is thought to reflect the widespread accounts of the abandoning of children and of cannibalism during this Great Famine. In *The Name of the Rose*, a novel set in 1327, novelist-philosopher Umberto Eco imagines what it must have been like during that dreadful famished time.[19] The narrator recalls the horrors recounted by the mysterious monk Salvatore:

> And the worst among the worst accosted boys, offering an egg or an apple, and then devoured them ... He [Salvatore] told of a man who came to the village selling cooked meat for a few pence, and nobody could understand this great stroke of luck, but then the priest said it was human flesh, and the man was torn to pieces by the infuriated crowd.[20]

This Great Famine was a time of roaming preachers and eccentric religious sects, and a time of grim and determined maneuverings by the Catholic Church to seek out and eliminate heresies. As the masses realized that the church's leaders could not persuade God to dispel the famine, faith in the mainstream church and in prayer receded. Disbelievers and heretics grew in number, as did retributive trials and burnings at the stake.

After 1321–1322, the climate steadied and food shortages receded. The food supply recovered to relatively normal conditions by 1325, by which time the depleted population had also begun to increase again. Even so, people had been so enfeebled by hunger, malnutrition, and infectious diseases such as pneumonia and tuberculosis that the recovery of morale, social order, and productive work took many years.

The causes of the famine

The causes of the Great Famine include climatic, demographic, environmental, social, and economic factors. In his comprehensive study, historian William Jordan describes a complex interplay of social, cultural, and environmental factors, including recent changes in agricultural practices.[21] Jordan concludes that the famine promoted a period of social instability in Europe. It coincided with the beginnings of the grinding Hundred Years War in Europe, when the drenching rains of 1315 and 1316 helped bog down the armies of Louis X of France, preventing them from crushing the rebellious Flemings. Other historians have seen the famine as either symptomatic of or contributory to the incipient decline of the feudal system in Europe via its disruptions of work patterns and of class relations.

Further massive disruption of life, religious authority, and social structures in Europe lay ahead. The Black Death would erupt in just three decades' time, its lethal impact heightened by the enfeebling and destabilizing impact of the Great Famine.

The Black Death: Bubonic plague strikes again

Reams have been written about the human, social, and economic devastation caused by the Black Death in mid-fourteenth-century Europe and the adjoining Middle Eastern and North African regions. In a half decade of blitzkrieg-like spread, beginning in 1346, the bubonic plague wiped out one-third to one-half of the European population.[22] Florence, in Tuscany, was an early victim: half the population died, while the other half survived by either closeting themselves away in countryside villas or, as Giovanni Boccaccio wrote, choosing to live for the day in a defiant, drunken, and debauched manner. Many, he wrote, "ate lunch with their friends, and ate dinner with their ancestors in paradise."[23] The causes of the disease are complex, but climatic influences may have facilitated its rapid spread.

The Black Death probably originated from naturally infected wild rodents in Central Asia. From there the infection spread into central-western China.[24] An epidemic of apparent bubonic plague raged in Hubei province in 1334, killing over three-quarters of the population. Severe outbreaks occurred in several more southern provinces. During the 1340s, with the east-to-west fuse now ignited, the disease spread west along trade, travel, and military routes via the Central Asian steppes, the shores of the Caspian Sea, and the Black Sea Genoese trading port of Kaffa (now Feodosia). After entering southern Europe by way of Italian seaports, in the course of six years the plague fanned out through continental Europe, the British Isles, Scandinavia, and the Baltic States (Figure 8.3).[25] The pandemic, though, was much more than a European disaster; while the disease persisted in China, it also spread widely in the eastern Mediterranean, the Middle East, North Africa, and the Arabian Peninsula. Altogether, around 75 million deaths may have occurred during that first great regional mosaic of lethal outbreaks during the late 1340s and 1350s.

FIGURE 8.3 The spread of the Black Death in Europe during 1346–1353, including western Russia, the Middle East, and North Africa. In the year preceding the major Black Death plague outbreak, urban centers were contaminated by ships returning to port at the end of the sailing season. A small outbreak occurred but was suppressed by cold autumnal and wintry weather; it lay dormant in the rat colonies until re-emerging in the spring with warmer weather. Benedictow, *The Black Death, 1346–1353: The Complete History*.[26]
Permission to reproduce received from the author Ole Benedictow.

The plague flared up frequently throughout Europe in subsequent centuries. Accounts of outbreaks also abound from North Africa and the Middle East.[27] Major outbreaks occurred in Venice in 1555 and in London's Great Plague of 1665–1666. The train on the London Underground's Piccadilly Line follows an unusually sinuous course between Knightsbridge and West Kensington as it passes the respectfully undisturbed burial pits of that great 1660s epidemic. Milan's experience in northern Italy during the fifteenth and early sixteenth centuries illustrates the typically heavy death toll from the plague. Milanese doctors were required to examine all deaths and write clinical reports of each postmortem examination. Milan's archives contain the details of approximately 115,000 deaths that occurred from 1452 to 1522.[28] Approximately one-fifth of these deaths were said to be due to "plague," while another third were attributed to (unspecified) "fever."

During the fifteenth and sixteenth centuries, the yearly frequency of recorded outbreaks in Europe increased threefold, followed by a lull as increasingly cold conditions set in from 1600, only to be hugely exacerbated by the chaos, displacement, and poverty of the Thirty Years War (1618–1648) (Figure 8.4). Following that war and the extreme epidemic in England during the mid-1660s, including the Great Plague in London, the bubonic plague gradually receded in Europe, disappearing by around 1800. Several different aspects of climate are thought likely

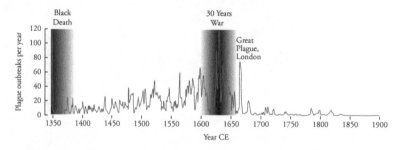

FIGURE 8.4 Yearly distribution of 6,929 documented bubonic plague outbreaks in the greater European region during 1347–1900. Local frequency ranged from one outbreak in 544 locations to 146 outbreaks in London. Büntgen et al., "Digitizing Historical Plague."[29]

Permission received from Oxford University Press.

to have helped shaped that half-millennium rise and fall of bubonic plague intensity—but more about that below.

Was it really bubonic plague?

Was the Black Death actually due to bubonic plague? That question has persisted for many decades, spawning many countertheories.[30,31] The rapid spread of the Black Death, its extremely high death rate in infected persons, the unusual seasonality (summertime outbreaks in cool temperate countries), and the paucity of dead rats in recorded accounts, have raised questions.[32,33] Might this dread disease have actually been caused by anthrax, the Ebola virus, or a lethal influenza? Since, like all forms of life, microbes evolve biologically over time, and bacteria naturally mutate at a high rate, a fourteenth-century strain might have caused an epidemic considerably different in profile from those caused by descendant strains centuries later.

There are several other interesting considerations. First, the apparent lack of dead rats may reflect the fact that human body lice can act as person-to-person blood-feeding disseminators of the plague bacterium, with no need for rats as the bacterial reservoir.[34] Body lice were rife in and beyond Europe during much of the Little Ice Age, particularly as lice-friendly unwashed woolen clothing became more affordable and widely used.[35] Second, the new molecular biology is illuminating: the DNA of *Yersinia pestis* has been identified in skeletal remains recovered from the plague burial pits in London and elsewhere in Europe.[36,37] This implicates two particular bacterial strains in the Black Death, apparent ancestors of today's existing *orientalis* and *medievalis* variants.[38,39]

And third, the plague's seasonal pattern, unusual for an infectious disease, has a likely two-phased explanation. During winters, spread would often have occurred via person-to-person transmission, including the "pneumonic" (exhaled) form of plague, nocturnal huddling among the poor, and via the ubiquitous lice infestations. It is quite likely that much of the transmission was of the pneumonic kind. In above-average warm summers, however, black rats would have thrived and proliferated, likewise their fleas, and during those times the disease would have been spread by flea bites. Indeed, records in England show that death rates during the colder 1560–1660 period peaked in brief periods of unusual warmth. During England's Great Plague of

1665–1666, temperatures were unusually high—the summers of 1665 and 1666 were 1°C–2°C higher than the average during the 1660s.[40] In London, the epicenter, the plague killed an estimated 100,000 people, one-sixth of the city's population, and caused near-paralysis of civic administration, taxation, protection of property, and the rule of law.[41]

Next, can the timing of the outbreak be explained? Around the turn of the fourteenth century, some triggering disturbance of the natural ecology of the bacterium must have occurred, leading to a spillover into humans. That ecology involves interplay between five species: burrow-dwelling wild rodents and their fleas, black rats, humans, and the *Yersinia pestis* bacterium.

The burrowing rodents, the natural reservoir-host of the bacterium, include marmots, great gerbils, and others. They inhabit vast underground burrow-complexes scattered across the Central Asian steppes (including today's Kazakhstan, adjoining northwest China) and in southern China adjoining the eastern Himalayan foothills. Over the past 2,000 years, other natural reservoirs of plague have been established via trade, warfare, and pandemic spread, and now also exist in northwest India, central and eastern Africa, the western United States, and South America.[42] The rodents' blood-feeding fleas become infected and spread the infection throughout the colony. The relationship between bacteria and wild rodent hosts has a long coevolutionary history that selected for a mutually beneficial less-virulent infection, and so otherwise healthy rodents are rarely seriously sickened by the infection. Not so the black rats or the less frequent human hosts.

Given the complex multistaged ecology of this disease, a new outbreak in a human population requires a *sequence* of influences on several pre-outbreak stages: the environmentally stimulated proliferation of wild rodents (the natural reservoir of the *Yersinia pestis* bacterium) and a consequent high infection rate among the many nonimmune young, then dispersal of infected wild rodents, their subsequent contact with black rats and humans, and finally the disease's geographic spread. Could regional climatic changes have influenced the initial outbreak or subsequent course of this great pandemic?

Likely climatic influences

The circumstantial evidence linking regional outbreaks of plague with changes in climatic condition is abundant. For example, occurrences of human plague in New Mexico have typically followed above-average rainfall in winter and spring.[43,44] Similarly, much of the variation in plague outbreaks in the western United States has mirrored regional fluctuations in climate, particularly increases in rainfall.[45] In China, there is a clear difference in the climatic changes that trigger plague outbreaks between the two distinct outbreak belts in the north and south.[46] In each case, when rainfall levels depart markedly from that region's average rainfall, outbreaks become much more likely.[47] During the Little Ice Age, after the NAO shifted into a negative phase around 1275–1300, Atlantic westerlies progressively moved south, creating more humid conditions in Europe. Eventually, these westerlies reached Central Asia, bringing with them increased precipitation.[48] In relation to the Black Death, these climatic changes in Central Asia and then in China early in the fourteenth century might have set the ecological cascade in motion. How might black rats and their fleas have been infected with plague at that particular time?

Relevant evidence has been building. Ibn al-Wardi, a Syrian writer of the time, wrote that the Black Death came out of "the Land of Darkness" (Central Asia).[49] Reconstruction of Kazakhstan's climate around the turn of the thirteenth century, using diverse proxy indicators, reveals a warmer, wetter, and hence more verdant period.[50] Indeed, research based on Kazakhstan's records on meteorological conditions and bubonic plague incidence for nearly half a century, 1949–1995, has shown how variations in seasonal temperature and rainfall have affected wild rodent numbers and their likely rate of plague infection.[51] The proportion of infected rodents increased in years when spring was warmer and summer was wetter; it is estimated that a 1°C rise in average spring temperature caused a 50 percent increase in the proportion of infected rodents. When the same research team applied their findings to the reconstructed regional climate, they concluded that wild rodents would have thrived and proliferated around the turn of the thirteenth century.[52] This would have been the first step in a typical *trophic cascade*.

The subsequent drier and hotter period in Kazakhstan in the 1320s, with less verdant landscapes, presumably mobilized the now-abundant but hungry wild rodent populations.[53] Local above-ground contacts with black rats and nomad-pastoralists followed. The early fourteenth century was also a time of heightened tensions between Mongolian nomad-pastoralists and agrarian communities in northwestern China, each seeking control over productive land, all within a background context of Chinese resentment of the recently established Mongolian imperial control of China, the Yuan Dynasty. Chinese records show that conflicts in these contested borderlands were closely related to regional climatic fluctuations, and peaked during the generally more verdant period in the early thirteenth century.[54] Stowaway black rats in the saddlebags of Mongol raiders may have carried the plague from the Central Asian steppes into China, where the disruptive floods, falling temperatures, and general social turbulence in the 1330s would have enhanced contacts between rats and humans. And so a human epidemic resulted.[55]

China was one of the busiest trading nations in Eurasia, and an outbreak of bubonic plague in western China or adjoining Central Asia would have spread readily to western Asia and on to Europe via infected black rats furtively accompanying trade caravans and Mongol horse-borne armies. By mid-1346, the Mongols were laying siege to the seaport city of Kaffa on the Black Sea. Bubonic plague broke out first among the Mongols and then in Kaffa, causing terrified European traders to scramble aboard their ships and head for Mediterranean seaports. The plague thus arrived in coastal southern Europe later in 1346.

Other earlier climatic experiences may have helped prepare the ground for this pandemic, which reached Europe just three decades after the Great Famine of 1315–1322. Large epidemiological studies in human populations, backed up by studies in animals, have shown that undernutrition in fetal life and infancy imparts a lifelong predisposition to various disease processes.[56] Many people born during the Great Famine would have carried a weakened immune system into adulthood, making them more susceptible to infection. More generally, the bulk of the European peasantry was in poor biological shape following a series of stressful climatic, environmental, and social experiences. The onset of gradual cooling in Europe, the declining fertility of soils from increased population pressures, and the shift from crop and meat production to sheep-and-wool farming resulted in widespread impoverishment, hunger, enfeeblement, and the abandonment of nonviable villages.[57] The outbreak of the Hundred Years War between England and France in the 1320s added further privation, misery, and hunger.

Then, during the 1340s, continental European populations suffered a decade of climatic extremes, crop failures, and hunger. Crop-stripping plagues of locusts occurred in three successive summers in Hungary, Austria, Bohemia, and Germany, followed by the "millennium flood" of 1342 that destroyed vast areas of crops and in 1344 a severe hot drought that caused harvest failure, famine, and tens of thousands of deaths. This confluence of climate-influenced factors must have increased the spread and lethality of the Black Death and its follow-up outbreaks in fourteenth-century Europe.

Climate and infectious disease: Novel aspects

Might the climate be influenced by infectious disease? Climate scientist William Ruddiman has explored this mirror-image question, proposing that a further, unremarked consequence of the great pandemics in world history was a reduction in carbon dioxide emissions due to population die-off and reduced land-clearing, resulting in global cooling.[58,59] Indeed, the atmospheric record shows clear dips in concentrations of carbon dioxide and methane during the several periods of plague-related depopulation and during the massive virus-mediated depopulation of South America following the Spanish conquest in the sixteenth century—and the temperature record shows concurrent cooling.[60]

Another novel aspect of the climate-and-infection relationship relates to indirect climatic influences on microbial genetic evolution. Within a few decades, a shift in climate may, via natural selection, favor a newly mutant form of a pathogen better suited to the changed conditions. Recall, for example, that around 80,000 years ago, early in the last glacial period, the human-specific body louse that transmits typhus in humans emerged as a new species of louse, able to take advantage of the change in *Homo sapiens*' clothing fashion as the long and cold glacial period got under way. During recent millennia, the body louse has been the disseminator of the serious infectious disease typhus, prone to epidemic outbreaks.

The bubonic plague receded after the 1666 flare-up, muted by the continued cold weather of the latter seventeenth century. It flickered erratically for another two centuries in Europe as rat populations fluctuated, and then began fading out—first in England, Scotland, and

Italy; then, early in the eighteenth century, in France and Russia; and finally in the Baltic ports. Outbreaks remained more common in summer than in winter, but were seemingly not much affected by rainfall. The reasons that the bubonic plague finally disappeared remain uncertain.[61] The Mediterranean black rat (*Rattus rattus*) was displaced by the larger brown Scandinavian rat (*Rattus norvegicus*), which is less susceptible to flea infestation and better adapted to the European cold. Likewise, better domestic hygiene, as well as changes in housing materials and design that provided fewer warm nesting sites for rodents under thatched roofs, may have also contributed: the primary habitat of plague-infected fleas reverted from under-roof to underground.

The crisis of 1560–1650: Plumbing the depths of cold

During the decades of the Little Ice Age's nadir, death rates across Europe were at their highest. Food crises were frequent, and so were the accompanying famines. Conditions were generally cold and harsh; winters were long, snows were heavy, and life was hard. This was evident, too, in the attention paid to the cold, harsh winter landscapes in many paintings by leading European artists.

Europe's climate began to fluctuate in the 1550s, after a steadier and milder half century. In England in the late 1550s, successive colder years caused food shortages, with consequent hunger and undernutrition, while the country was convulsed by civil war. These climatic conditions may have contributed to the spread and high death rate of the severe influenza epidemic that followed in 1557–1558, a regional flare-up of a trans-Atlantic influenza pandemic that killed millions in North and South America during the 1550s.[62]

During the later 1500s, glaciers in Greenland, Scandinavia, and the Alps advanced rapidly. The Arctic sea-ice extended so far south that seafaring Inuit from Greenland landed their kayaks in Scotland. While weather patterns varied erratically from year to year, the spring and summer seasons were unusually cold and wet. During sporadic warmer summers, the bubonic plague often struck again as rat populations surged. Throughout Europe, the growing season was shorter and less reliable, and serious food shortages were common. The combination of plague outbreaks and starvation resulted in high death rates, and population growth plateaued. Weather patterns became more unstable, with severe storms, massive flooding, and many deaths. Coastal storm surges and inundation washed away housing and farmland along the Danish, German, and Dutch coastlines. The Spanish Armada, in 1588, was one spectacular victim of unruly weather.

In Germany, Pastor Daniel Schaller wrote in the local *Theologischer Herold*: "In towns and villages, much weeping and moaning is heard among farming folk that the land has grown weary and even exhausted of bearing fruit, causing great price rises and famine."[63] The environmentally perceptive pastor also recorded that "The waters are no longer as rich in fish as they once were, the forests and fields no longer as abounding in game and animals, the air not as full of birds." This observation accorded with something more widely noted: various ocean fish, particularly the cod stocks in the North Sea and around Iceland, were shifting southward to warmer waters. The cod's liver begins to malfunction as the water temperature drops towards 2°C, and the cod then instinctively migrate in the direction of warmer water.

In 1578, England's Elizabethan government issued the Plague Orders as an attempt to limit the spread of plague and to provide some relief to those now bereft of livelihoods, health, and income. The strictures imposed quarantine on afflicted families and communities, travel was curtailed, and local household taxation regimes were established to alleviate poverty and social unrest. Here were the foundations of the late-Elizabethan Poor Law of 1601. The 1590s were a particularly cold, wet, and socially fractious decade in Europe. In mid-decade, unusually, three consecutive harvests failed, followed by hunger, regional outbreaks of bubonic plague, and misery. Inevitably, some sought profit by hoarding grain and forcing up local prices.

In Europe, the Thirty Years War (1618–1648) was the centerpiece of what historians call the General Crisis of the Seventeenth Century. What began as a religious conflict between Protestants and Catholics within the Holy Roman Empire spilled over and became a war between contending dynasties, the Bourbons and Habsburgs, for domination of Europe. This was a hugely destructive war, often fought in bitter cold, and leaving a trail of trampled farmland, ravaged villages, damaged or pillaged crops and livestock, and emptied royal purses. Around one-third of Germany's population perished, and demographic and social recovery in that region took over half a century.[64] The frequency of famine years doubled, people starved and adult heights declined, the rate of epidemic outbreaks tripled, the numbers of people displaced surged, and child death rates rose.[65] In 1651, towards the end of the English Civil War, the philosopher Thomas Hobbes wrote of the dissolution of all things good in European society—work, culture, travel, trade, public buildings, arts and letters—which were replaced, according to Hobbes, by a "continual fear and danger of violent death."[66]

Many contemporary authorities linked the misery, mayhem, and mortality of the first half of the seventeenth century with supernatural intervention or other extraterrestrial phenomena. A Spanish almanac of 1640 stated that "Whenever eclipses, comets and earthquakes and other similar prodigies have occurred, great miseries have usually followed."[67] Those views, around the time in which the Catholic Church forced Galileo to disavow his thesis of a sun-centered solar system, are understandable. A hundred years later, in the European Enlightenment, the French intellectual Voltaire concluded from his systematic review of the rebellions, wars, food crises, and natural catastrophes of the seventeenth century that they were caused in part by climate change.[68] A new objective rationality was on the rise. Yet it remains surprising, writes historian Geoffrey Parker, that "so far, all historians of the General Crisis have included government and religion in their analysis, but few have considered the impact of the climate."[69] There is now, he points out, a vast volume of climatic, human, and social information available for understanding the mid-seventeenth century: "the human and natural climatic archives show exactly how extreme weather anomalies triggered or fatally exacerbated major political upheavals."[70]

Little Ice Age: What it meant in China and the Pacific Islands

China in the early seventeenth century was beset by severe cold and widespread aridity. Unusual and heavy winter snowfalls occurred in the south. From the 1620s to the early 1640s, the summers were particularly cold, and drought took hold in the north and east of the country, eventually extending to the fertile rice-growing Yangtze Valley region in central-southern China.[71] Chinese government records show that the country's total area of productive farmland fell by around 65 percent between 1602 and 1645 during these cold decades. The annual double-cropping of rice failed in the colder conditions. Yields fell; severe famines followed; internal migrations and social unrest increased.

During 1627–1628, against the background of a growing central governmental fiscal crisis, a severe famine in Shaanxi province in central China resulted in many unpaid and desperate soldiers joining a rapidly growing rural uprising. Rebellious bands, swelling in number, gained control of adjoining territory, including parts of the Yangtze Valley. Government forces managed to hold the line against further advance, but the reprieve was temporary; worse weather was to come. During 1638–1641, China experienced the most severe drought for many centuries.[72] In Shandong province on the upper east coast, the Grand Canal ran dry for the only time in recorded history.

Temperatures plunged further in the early 1640s, during several years of increased volcanic activity and atmospheric shrouding. These cold, dry, and difficult decades in China were punctuated by acute weather disasters, including an extreme flooding of the Yellow River in northern China in 1642, exacerbated by the breaching of levees by angry rebels that caused several hundred thousand deaths. The result of this combination of demographic disruptions and weather disasters, on top of food shortages and chronic rural hardship, was a surge of social disruption and violence, plus outbreaks of smallpox epidemics.[73]

The Emperors' Mandate of Heaven was slipping. In 1644, a full-blown peasant uprising overthrew the three-centuries-old Ming Dynasty.[74] The emperor hanged himself when the rebels forced their way into the Forbidden City, his demise opening the way for horse-borne Manchurians to move south, suppress the rebellion, and establish the Qing Dynasty. Under the new regime, with its cultural legacy of nomadic pastoralism, recovery of farm production was slow, and hardship and hunger persisted in some regions. Helpfully, there were many fewer mouths to feed; indeed, the Qing emperor claimed that during the dynastic transition "over half of China's population perished"—a figure that is consistent with modern estimates of demographic trends in eastern China between 1631 and 1645.[75]

Temperatures and epidemics in China

The vast Chinese imperial archives tell the stories of many epidemic outbreaks over the past eight centuries, catalogued at province level.

During the Little Ice Age in Eurasia, there were 881 epidemics recorded in China, 32 of which afflicted three or more provinces. A chronological listing of epidemics, identified by province, has been compiled in Shanghai from the original two volumes of Ch'en Kao-yung.[76] That listing, matched alongside the record of estimated annual temperatures in central and eastern China, indicates whether the years of epidemic outbreaks were related to cool and warm periods.

It appears, from my analysis of this information, that there was an approximately 35 percent greater probability of an epidemic occurring during colder periods (i.e., temperatures below the 1300–1850 mean temperature in China) than in the warmer periods—and a 40 percent greater probability of a major multiprovince epidemic during colder than during warmer periods.[77] A similar relationship between temperature and epidemics in China has previously been deduced by Gong.[78] The interpretation requires caution. Infectious disease outbreaks can be influenced by either unusually cold or hot weather; both cooling and warming outside the "just right" range can affect the biology, ecology, or social dynamics of infection transmission. However, during the Little Ice Age the "warm" periods are really periods of less cold. No information can be gleaned from this period about the impacts of genuinely hot climatic conditions on the occurrence of epidemic outbreaks.

Climate and conflict in China

The relationship between climatic changes in China, internal conflicts, and the fall of dynasties over the past thousand years is a fascinating topic. China's treasure trove of ancient records includes extensive documentation of all major political and demographic events over the past 3,000 years, stored in the palace archives of successive dynasties. Linking these records of conflict, warfare, and population numbers with newly generated information on climate time-trends in both the northern hemisphere and eastern China has enabled researchers at Hong Kong University to analyze the timing of conflicts since 1000 C.E. for China at large, and for regional conflicts in eastern China since 1500.[79,80] The picture that has emerged shows that most periods of widespread unrest, warfare, and population decline in China occurred during the cooler periods, which accords with other decade-by-decade evidence from eastern China during 1730–1910 that food yields were higher in warmer decades.[81]

The pattern observed over the past thousand years is part of a larger story. Four of the five dynastic changes over the past 11 centuries occurred during or around the end of cooler and drier periods in eastern China characterized by serious food shortages, social uprisings, and increased frequency of warfare.[82] The one exception was the collapse of the Yuan (Mongol) Dynasty in 1368, when the temperature in eastern China, as in Europe, had been rising since the 1340s. Of course, climatic conditions were only one part of the complex causal mix underlying the rise and fall of dynasties.[83] Even so, the link between adverse climate, harvest failures, conflict, and dynastic collapse has been a recurring theme in Chinese political history.

Various dynasties in much earlier times, including the long-lasting Shang, Zhou, and Han dynasties, suffered similar climate-and-food related fates around 1100 B.C.E., 450 B.C.E., and 220 C.E., respectively. The particular climatic conditions that influenced each dynastic collapse were varied,[84] but a weakening of the Asian summer monsoon with its bounty of rain, compounded by a strengthening of the winter monsoon, was a common component of these climatic reversals—and was largely driven by episodic increases in the strength and influence of the El Niño system. The culturally and commercially successful Tang Dynasty collapsed in 907 C.E. after several decades of extreme floods and mass drownings, weather disasters, social disorder, river banditry, and a severe drought-induced famine that caused tens of thousands of deaths from starvation.[85]

The dramatic fall of the Ming Dynasty in 1644, culminating in the emperor committing suicide and the Manchurians quickly taking control, was precipitated by peasant hunger, an unpaid army, unrest, and rebellion.[86] The climatic coup de grâce was a severe drought during 1638–1641—the most severe drought in China for over five centuries, and at its most extreme in the northeast near the seat of government. The resultant food crisis almost certainly played a major role in the popular rebellion and dynastic collapse.[87,88] The population had plummeted by around two-fifths (losing around 70 million) because of the turmoil of wars, starvation, and epidemics that occurred during these climatically and economically miserable few decades.[89]

Little wonder that China's leaders recognize, from history, the political danger lurking in the countryside.

New Zealand and the Pacific Islands

Substantial cooling occurred across the Pacific region from late in the thirteenth century, often referred to as the "A.D. 1300 Event." This coincided with the major climatic transition between the Medieval Warm Period and the Little Ice Age in Europe, China, and North America. It caused a societal crisis in the tropical Pacific Islands, due principally to the combined impacts of cooling and a fall in sea level on food yields.[90] Many communities, from the Solomon Islands in the west to French Polynesia in the east, abandoned their coastal settlements to resettle on inland hilltop sites to avoid both hunger and conflict.

The first people to settle in New Zealand (Māori *Aotearoa*) were from the islands of East Polynesia. They arrived on planned voyages of discovery in the late thirteenth century, during a relatively congenial climatic period toward the end of the northern hemisphere's Medieval Warm. This was presumably a good time to establish food gardens, and the Māori brought with them from warmer, more equatorial climes kumara sweet potato, taro, and gourds. As well as exotic starch-rich vegetables, the Māori had access, at least in the early stages of settlement, to ample supplies of fresh meat, due to the large numbers of seals and large flightless birds such as the moa. This inferred diet accords with skeletal evidence of good bone health and growth in Māori populations in the first 200 years or so after the first landing.

However, the A.D. 1300 Event cooling soon set in, as evident in New Zealand's paleotemperature record—including evidence from stalagmites that New Zealand's temperature dropped by around 1.5°C in roughly 70 years.[91] The region continued to cool during the fourteenth and fifteenth centuries, and, after a brief warmer intermission, temperatures were at their lowest between 1600 and 1750, a period of turbulent weather with stronger westerlies and more frequent storms.

After several centuries of fast-food exploitation, the easily clubbable moas had become extinct by the early seventeenth century. Archaeological studies have unearthed evidence of growth delay in childhood among the Māori, reduced diversity of food sources, and an increase in malnutrition. During the colder seventeenth and eighteenth centuries, there were signs also of more frequent and severe intertribal warfare, marked by the appearance for the first time in the archaeological record of fortified settlements. On occasion, parts of the country that had previously supported large populations were abandoned. For example, at Palliser Bay, on the southern end of New Zealand's North Island, a deteriorating climate from around 1600 was associated with large-scale migrations across Cook Strait to more sheltered sites at the top of the South Island.[92]

The Super El Niño event of the 1780s–1790s

By the mid-eighteenth century, food crises in Europe, China, and North Africa occurred less frequently than before due to advances in crop choices, regional transportation of grain, and provision of credit financing that boosted local communities' purchasing power in larger markets.[93] The major countries of western Europe, particularly the Netherlands, England, France, and Spain, were increasingly buffered against climate-related food crises by their improved internal trading networks, social relief systems, colonial holdings, and participation in international trade.[94] In the Netherlands and England, crop yields were increasing in response to more intensive farming, the planting of nitrogen-enriching plants, and production of animal fodder on previously unused fallow land. Better access to food from afar, more robust health of most populations, and decreased risks of epidemics relieved the precarious dependence on local farm yields. The long-standing relationship between crop failure, high food prices, starvation, and death within small-scale closed agricultural economies was receding. Population size and density increased in towns and cities. World population was heading toward one billion.

In these conditions, "crowd diseases" began to thrive, since organisms enjoyed more-or-less continuous circulation in the urban setting. Indeed, most deaths in seventeenth-century Europe were caused by infectious diseases, not famine, and mortality was not due, on the whole, to sporadic recurrent local epidemics of diseases such as bubonic plague but to established endemic infections such as dysentery, influenza, measles, smallpox, scarlet fever, and louse-borne typhus.[95] (Lice infestation was boosted by the advent of thicker clothing and less frequent public bathing during these cold decades.) By the eighteenth century, tuberculosis was responsible for one in four deaths in Europe, sparing neither young nor old, nobility nor pauper.[96]

The final two decades of the eighteenth century were marred by unusual climatic adversity in several regions of the world. Major volcanic eruptions in Iceland and Japan in 1783 caused temporary cooling around much of the world and darkening of regional skies. In Iceland, the Mt. Laki eruption deposited acid and fluorine, which stunted fodder and crop production, and poisoned sheep and cattle. An estimated quarter of the Icelandic population perished from starvation and fluorine poisoning. Around northern Europe, up to 100,000 extra deaths occurred.

In Japan, Mount Asama erupted and plummeting temperatures led to widespread crop failure, a serious food crisis, and political turmoil.[97] Harvests did not recover until the end of the 1780s, by which time the death toll from starvation and consequent infections and illnesses in Japan had reached over one million.[98]

Some historians argue that a larger and more global climatic force may have contributed to adverse weather in the late eighteenth century. Environmental historian Richard Grove has attributed the food crises that occurred around much of the temperate and tropical world in the 1780s and 1790s to an extreme, slow-forming, and prolonged El Niño event.[99] This, he concludes, was the most severe El Niño of the eighteenth century, spanning much of the period 1785–1795 and with extensive teleconnected influences on regional climatic conditions around the world.[100] The date of the Great El Niño event struck a chord with me as an ardent admirer of Mozart's music. There are many theories as to the cause of the composer's premature death. Might the poor weather conditions have contributed? Records suggest an increase in edema-related deaths among young men in Vienna at that time (Box 8.1).

In the latter half of the 1780s, there were long droughts in Mexico, southern Africa, the Caribbean, and South Asia. During 1785–1786, Mexico experienced a severe famine, recalled as the Year of Hunger and described as "the most disastrous single event in the whole history of

BOX 8.1 Was Wolfgang Mozart's death in December 1791 influenced by the far-reaching Great El Niño event?

In the last week of November 1791, Mozart became bedridden after a frenetic creative period that produced, among other music, the operas *The Magic Flute* and his famous *Requiem*. His hands and feet began to swell, and he had vomiting fits and a fever. He died after an illness of two weeks that had caused edematous swelling of limbs and, ultimately, whole-body bloating. His physician, Dr. Thomas Closset, recorded the cause of death as *hitziges Frieselfieber* ("severe miliary fever"). Buried in an unmarked commoner's grave, Mozart's death has attracted a multitude of cause-of-death sleuths. More than 150 causes of death have been proposed,[101],[102] including influenza, mercury poisoning, exhaustion, autoimmune disorders, and kidney ailments. The majority view is that he died of acute rheumatic fever.

We know that Mozart's death occurred during the most severe years of the Great El Niño event of 1788–1793, the most extreme and sustained El Niño of that century.

A team of Dutch epidemiologists, intrigued by Mozart's death, visited Vienna to record archived details on all deaths registered during the winters of 1790–1791, 1791–1792, and 1792–1793. They compiled monthly data for younger men, older men, and women.[103] In December 1791, there were 47 edema-related deaths in younger men, versus an expected 13 such deaths (based on the two adjoining Decembers).

The weeks before Mozart's death would be the most likely time for weather conditions to have influenced the spread of an infectious agent that might have caused this mini-epidemic. By good fortune, I met Dr. Ernest Rudel, head of the Data and Modeling Division of Vienna's Center for Meteorology, at an international biometeorology conference in 2011. He kindly offered to generate tables of November to January daily temperatures for 1790–1794. These tables show the average daily temperature in the weeks immediately preceding Mozart's illness (8.2°C) was 5.5°C higher than the combined average daily temperature of the adjoining years (2.7°C). Might the unusually high temperatures around the time of Mozart's illness have influenced the brief outbreak of a mysterious infectious disease in younger men? Mozart had been in fluctuating poor health for a month or two, and was probably not in good condition to withstand any potentially lethal infection. A change in weather conditions, in the presence of other causal influences, can often tip the scales and lead to unexpected illness and poor outcomes.

colonial maize agriculture."[104] In the Mexican city of Léon at that time there was a twelvefold increase in the price of corn (maize), a sixfold increase in the death rate, and a halving of the birth rate. Over the following 10 years, prolonged aridity and autumn frosts in Mexico led to persistent drought and famine conditions, culminating in wholesale failures of corn and wheat crops in the mid-1790s.[105] The most severe droughts did not strike Mexico until 1793, indicating that the onset of full El Niño conditions there did not occur until more than two years after the same El Niño event had already caused serious monsoon failure in India—a reminder that the teleconnected impacts of these events are spread widely over both space and time.

Yellow fever in Philadelphia

The far-reaching El Niño of the late eighteenth century brought higher temperatures to North America. In August 1793, an epidemic of mosquito-borne yellow fever broke out in Philadelphia, the temporary national capital.[106] Within months, there were 17,000 cases, one-third of them fatal. Livers failed, bodies turned a sickly yellow, internal hemorrhages caused vomiting of blood, kidneys failed, and death hovered. Philadelphia was beyond the usual northern limit of this tropical disease.[107] However, the city had strong trading connections with the Caribbean, a hotbed of yellow fever. In 1793, the unusually warm and humid weather along the mid-eastern seaboard led to a great proliferation of *Aedes egypti* mosquitoes, the disease's natural vector. Further, infected French refugees from a flare-up of the ongoing slave rebellion in the French colony of Saint Domingue (now Haiti), where yellow fever was endemic, may have brought the disease with them to the eastern coastal cities of America.

Whatever the source, Philadelphia was the unlucky first major victim, suffering about 5,000 deaths, one-tenth of the city's population, between mid-August and early November. Early winter frosts then mercifully culled the mosquito population.

Subsequently, seven of the nine yellow fever epidemics in the United States in the nineteenth century have coincided with El Niño events.[108,109]

First Fleet settlement in Australia (1788–1792): Hostile weather in Sydney Cove

Late in the eighteenth century, England faced several urgent challenges on the colonial and home fronts. The nation was undergoing early rapid industrialization, population growth, and urbanization. People were migrating from country to city as industrial job markets opened up.

However, life in the booming cities was blighted by poverty, petty theft, burgeoning slums, and lethal infectious diseases. The late chill of the Little Ice Age lingered on, and social unrest was threatening.

The British government needed, in particular, a replacement for the recent loss of its American colonies—previously its population safety valve and penal colony. Further, the crisis of His Majesty's overcrowded jails and prison-ship hulks needed resolution, as did the unseemly rise in the numbers of public executions of petty and often penniless thieves.[110] Hence, the government made a bold decision to establish a convict-based colony on the east coast of the southern continent the Europeans called New Holland. James Cook had recently sailed the east coast of the continent, naming it New South Wales and claiming it for the British.

A convoy of eleven ships was assembled in Portsmouth, in southern England. Seven ships were for people, four were for supplies. On January 19, 1788, the First Fleet arrived at Botany Bay, New South Wales. Remarkably, on a journey of eight months and 24,000 kilometers, in turbulent southern seas, and with few experienced deckhands, only 48 deaths (3 percent loss) occurred.

The fleet sojourned in Botany Bay briefly but found it deficient in freshwater and soil and unsuitable as a harbor. An advance party was sent to explore adjoining Port Jackson (today's Sydney Harbour), the spectacular harbor named by Cook as he sailed past without entering in 1770.

For the 1,450 new arrivals, comprising crew, officers, and marines with their families, as well as 751 convicts (543 men, 189 women, and 18 children), plus dozens of sheep, cattle, and horses, the timing of this arrival was unlucky. For the first 10 days, they faced violent winds, lightning, and rain. Lieutenant Clark wrote in his diary: "Thursday 31 January, what a terrible night it was ... thunder, lightening [*sic*] and rain. Was obliged to get out of my tent with nothing on but my shirt to slacken the tent poles."[111] Of much greater consequence, the next five years would expose the struggling colony to unusual climatic extremes, including stormy, torrential, and searingly hot weather. The fleet had arrived just two years before the onset of the extreme El Niño event.

During the first half of 1788, there were some deaths from scurvy (primarily a consequence of shipboard diet) and more from dysentery (often recorded as "flux," in reference to the watery diarrhea). Surgeon Bowes wrote: "Five of the Convict men died last week; there are great numbers ill on shore now, chiefly of dysenteries ... The Dysentery prevails very much on shore now, and many have Dyed of it."[112] There were 27 deaths among convicts that first half year, representing a proportion of 3–4 percent. The weather was persistently stormy and erratic for the next two summers. In the words of one of the colony's diarists: "The rain came down in torrents, filling up every trench and cavity which had been dug about the settlement, and causing much damage to the miserable mud tenements which were occupied by the convicts."[113] These years were apparently the tail end of the ENSO cycle, when rains and storminess typically increase in eastern Australia.

The climate changes—as the Second and Third Fleets arrive

As the Second Fleet arrived in mid-1790 with more convicts, the weather patterns changed. In his book *A Complete Account of the Settlement at Port Jackson* (published and widely read in London in 1793), Watkin Tench, a captain of the marines, wrote that the weather "is changeable beyond any other I ever heard of ... clouds, storms and sunshine pass in rapid succession."[114] Tench could not have known that he was describing the local impact of a particularly strong El Niño event.[115,116] Drawing on his diary records, Tench wrote that the dry conditions were threatening the colony's food supplies:

> Vegetables are scarce . . . owing to want of rain. I do not think that all the showers of the last four months put together, would make twenty-four hours rain. Our farms, what with this and a poor soil, are in wretched condition. My winter crop of potatoes, which I planted in days of despair (March and April last), turned out very badly when I dug them about two months back. Wheat returned so poorly last harvest.

Of course, the harvest deficits should not be blamed solely on the climate. After all, these were settlers with no prior knowledge of Australia's soils, with its organic content and profile of trace element deficiencies that differed greatly from Europe's soils. Further, some of the vegetables and plants brought from England inevitably struggled in a new, harsher and hotter environment.

There was a two-month spike in illnesses and deaths during August–September of 1790. Many deaths were from scurvy and dysentery. The suffering, however, was not equally shared within the colony. Most of the 104 recorded deaths occurred among debilitated survivors of the shamefully mismanaged Second Fleet (which had suffered a 25 percent onboard mortality among convicts). Problems may have been compounded by an acute outbreak of imported infectious disease (most probably typhoid or typhus).[117]

During these years of early settlement, the new arrivals suffered from cholera, dysentery, smallpox, typhoid fever, and venereal diseases. Many convicts were malnourished, even emaciated, upon arrival, and some would have harbored chronic diseases such as tuberculosis and syphilis. Tuberculosis, in the late eighteenth century, was reaching peak prevalence in England, where it was causing the death of 1–2 percent of the population each year. The limited nutrients and stringent rationing of foods in the climatically difficult early years in Sydney Cove and the resulting undernutrition, often severe, would have increased individual susceptibility to infection and delayed recovery from it.

The summer of 1790–1791, between the arrival of the second and third fleets, was hot and dry. Tench commented that at times it felt "like the blast of a heated oven." He describes a vast flight of bats driven before the wind, then dropping either dead or dying, both on the wing and from their trees, "unable longer to endure the burning state of the atmosphere." Other diarists, too, reported seeing masses of dead and dying birds of many kinds. Temperatures over 40°C, never experienced back home in England, caused alarm and suffering.

In March 1791, Governor Arthur Phillip wrote in a letter back to London: "From June until the present time so little rains has fallen that most of the runs of water in the different parts of the harbour have been dried up for several months ... I do not think it probable that so dry a season often occurs. Our crops of corn have suffered greatly from the dry weather." As food supplies dwindled, Governor Phillip tightened food rations, noting that "little more than twelve months back, hogs and poultry were in great abundance, and were increasing very rapidly ... but at this time there was seldom any to sell." Virtually no rain fell for a full year, between July 1790 and August 1791. In November 1791, Governor Phillip recorded that the freshwater Tank Stream, the main local source of water for the harborside settlement, had run dry. Indeed, it remained dry until 1794.

During 1792, the total number of deaths in the settlement was almost three times greater than in 1791.[118] Of the 469 deaths in 1792, nine out of ten were convicts. Extant records indicate that those deaths were not primarily in enfeebled and chronically ill convicts. Most of the convicts, in fact, were reasonably hardy; they were not a random sample of the large and debilitated population in England's overcrowded jails. This Australian venture was no run-of-the-mill British penal settlement; Australia was more than an out-of-sight-out-of-mind dumping ground. This was a deliberate colony-building exercise, and the authorities gave due consideration to individual physical work capacity and to the mix of skills that would be needed.

The episode of scurvy, dysentery, and perhaps typhoid that had followed the arrival of the Second Fleet in mid-1790 was reprised by a less marked but more prolonged rise in deaths after the Third Fleet arrived in late 1791. This, too, was most probably an epidemic of dysentery, perhaps exacerbated by the voyage-weakened and dispirited condition of this third batch of convicts, now struggling to cope with the unusually oppressive heat.

It is difficult to attribute patterns of disease to variations in climate with much certainty, given the variety of hardships suffered by the early colonists. But overall, a fair conclusion is that the extreme weather that prevailed during those first four years of settlement exacerbated the risks of undernutrition, scurvy, dehydration, dysentery, and many other occult infections.

Despite the harshness, infections, and privations of the outward sea passage, most of the early convicts arrived in reasonable health. As the flow of convict consignments increased in the first half of the nineteenth century, meticulous records of age, sex, weight, height, and overall physical conditions were made of all convicts upon embarkation and arrival. The convicts arriving around the turn of the nineteenth century were of similar average height to that of the general English populace. The records on the convicts also showed that, across three successive birth generations in England, spanning a half century from 1770 to 1815, both male and female convicts displayed a progressive reduction in height of 2–3 centimeters.[119] This gradual intergenerational shrinking most probably reflected the general nutritional decline within the English population during early industrialization and an increase in childhood stunting.

Concluding remarks

The story of the Little Ice Age demonstrates the complex ways in which climate can influence many aspects of society. It is a fundamental determinant of health. During the Great Famine of 1315–1322, changes in climate disrupted the food supply, which had consequences for human health and social systems. Too much rain left crops stunted and leached nutrients from the soil. Livestock had little feed, and weakened herds were devastated by disease. In human populations, too, diseases spread more rapidly among the undernourished. Social relationships were strained among desperate, fearful, and hungry populations, leading to disorder and violence. The Black Death in Europe and the wider second pandemic of bubonic plague illustrate how climate conditions can influence the ecologically complex relationships that underlie many infectious disease outbreaks. Climate change interacted with other social and ecological phenomena, causing a chain of follow-on consequences: European populations were already weakened and more susceptible to infection following the Great Famine, increased rainfall in Central Asia may have contributed to a boom in the rodent population, the conditions may have facilitated the transmission to black rats, and recently established trade routes made geographic dispersal possible. Even shorter cycles of warmer and colder seasons might have influenced the spread of the bubonic plague: colder weather aided person-to-person transmission of pneumonic plague, while warmer weather allowed rat populations to increase. It is important to note from the case studies in this chapter that societies and subpopulations vary in their vulnerability; some cope with climatic adversity better than others. England, with its evolving social relief laws, coped better with climatic adversity and food shortages in the seventeenth and early eighteenth centuries than did France, where communities still relied on local voluntarism.

In Philadelphia in the late eighteenth century, changing climatic conditions associated with El Niño extended the range of the warm-climate disease yellow fever north into dense urban populations. Only in recent decades have farmers in the eastern and southeastern regions of Australia come to understand, and cope better with, the fluctuating climatic conditions resulting from the El Niño Southern Oscillation. The prolonged and severe drought in southeastern Australia during the first decade of this twenty-first century, the most severe since European settlement, sorely tested the limits of this coping capacity. Many communities and families suffered; mental health problems increased, and suicide rates in men crept upwards.[120] Meanwhile, as global warming proceeds, and as the drier latitudinal belt between tropical and temperate zones (the southern subtropical ridge) intensifies and extends southward,[121] many of Australia's most productive farming regions in the Murray-Darling Basin—which accounts for two-fifths of the nation's farm-yield—will experience a type of sustained, amplified climatic stress that was only sampled briefly by the First Fleet colonists.

9

Weather Extremes in Modern Times

IN 1816, AGAINST A foreboding climatic background, Mary Shelley wrote *Frankenstein*. She might well have begun: "It was a dark and stormy decade . . ." During the previous year, much of the world had been shrouded by the great ashen veil cast across the skies by the massive Tambora volcanic eruption in April 1815. Europe's 1815–1816 was a cold, gloomy, and tumultuous time.[1] Crops failed and temperatures fell. Bonaparte was consigned to the rocky island of St. Helena, Beethoven entered his more radical and introspective late period, and minor autocratic monarchies around the continent came under increasing political siege as democratic impulses stirred. This chapter examines some of the shorter-term climate shifts and extreme weather events that have occurred over the last two centuries. The disrupted weather following the Tambora eruption, for example, shows how small changes in temperature and rainfall can have major consequences, including failed harvests and epidemic outbreaks. In mid-nineteenth-century Ireland, the failure of the potato crop in wet and relatively warm conditions contributed to food insecurity that devastated the local population. Unusual weather extremes in late-nineteenth-century China, including a period of cooling, facilitated the Third Pandemic of bubonic plague, which spread rapidly through populations already under stress due to harvest failures, conflict, and political turmoil. Such events may intensify in the coming decades as the Earth's average temperature rises and climatic cycles are disrupted and become more variable.

Additionally, the consequences for human population health are amplified by social and political mismanagement and turmoil. We can expect climate change to act as a "force multiplier," exacerbating many of the world's health problems.

From the mid-nineteenth century, the northern hemisphere's Little Ice Age receded as solar activity regained its twelfth-century peak level. The depths of the cold had been reached around 1700 C.E., and the cooling influence of the Siberian High was now receding. The almost year-round ice and snow in northern Europe during those super-chilled earlier times were long gone, and the snowbound, though increasingly grimy, White Christmases of early-1800s Dickensian London were waning.

In the southern hemisphere in far-off colonial Australia and South Africa, gradual warming and increases in regional rains in the mid- to late decades of the nineteenth century eased the opening up of new farmland in some areas (as well as the brutal marginalization of the indigenous occupants). Historical reconstruction of Australia's climate from tree rings, corals, and caves shows that the earlier period 1810–1860, just two decades after initial European settlement, was the coldest time in the past several centuries.[2] After that, as both the temperature and numbers of (mostly British) settlers gradually rose, pastoral and agricultural holdings multiplied and spread. They had yet to learn that this driest and flattest of continents was a land of natural climatic extremes, "of droughts and flooding rains."[3]

Ever optimistic, many would-be farmers pushed on beyond safe climatic-environmental limits. In South Australia in the 1860s, the Surveyor-General had defined the safe limit (Goyder's Line), to the north of Adelaide, beyond which drying and droughts would occur. Cattle grazing might be viable, but growing crops would not. The optimists smiled for a while as crops grew, then smiled less, then suffered and abandoned their now-parched farmland.

Very probably, as our university research team has shown for recent rain-parched times in the eastern state of New South Wales, suicide rates among despairing and nearly bankrupt farmers would have risen in the wake of these unsparing drought years.[4]

In the twentieth century, global temperatures rose. Between 1900 and 2000, Earth's average surface temperature increased by 0.9°C, including a brief cooling period during 1945–1970. The cooling followed a surge of postwar growth in industrial activity and energy generation, causing extremes of local air pollution in the urban and industrial regions of many western countries until Clean Air legislation was introduced during the 1960s and 1970s.

First, though we must return to the early years of the nineteenth century, starting with details of the Tambora story.

The Tambora eruption: The years "without a summer"

I had a dream, which was not all a dream.
The bright sun was extinguish'd, and the stars
Did wander darkling in the eternal space,
Rayless, and pathless, and the icy earth
Swung blind and blackening in the moonless air.

—LORD BYRON, 1816

The eruption of Mt. Tambora on the Indonesian island of Sumbawa in 1815 was a "super-colossal" eruption, one of the most extreme for over a thousand years.[5] The volcano hurled 15 cubic kilometers of ash, sulfate particles, and other acidic aerosols up to 40 kilometers into the skies, casting an atmospheric pall around the world that affected climatic conditions in far-off Europe, northeast America, and maritime Canada.[6] The dismal and darkened year of 1816 became Europe's "year without a summer."

The eruption caused three years of global cooling and erratic weather patterns. This was an unusually volcanic, haze-ridden, and cold decade, since four other major eruptions had occurred during 1812–1814 and there had been a very large eruption in 1809 in the tropics that added to the background cooling trend.[7,8] The Tambora eruption caused a further (but temporary) fall in average temperatures in western Europe by 1.5°C–3.0°C.[9,10]

Overall, that entire second decade was one of bitter cold and widespread misery in Europe, extending further west across Eurasia. Perhaps it was the misfortune of composers Frederic Chopin and Felix Mendelssohn to have been born in eastern and central Europe in 1810 and 1809, respectively, and to have passed through their susceptible early childhood years in unusually cold conditions. Both composers died of tuberculosis in their late 30s. In Vienna in 1822, Franz Schubert wrote his great song cycle *Die Winterreise* (Winter's Journey), imbued with pathos, frozen landscapes, heartache, and tragedy.

In the early 1800s, the landscape paintings by England's J. M. W. Turner, the "painter of light," were distinguished by mists and shrouded skies. London, in the months following the eruption, was treated to sunsets and twilights that appeared orange or red near the horizon and purple to pink above—colors that infused many of Turner's paintings of the time.[11]

Food crises, epidemics, and social unrest in the Tambora decade

Agricultural production in both North America and Europe was an early casualty of the Tambora eruption.[12] The severity and geographic extent of the resultant famines in Europe were of the same order as those in the Great European Famine of 1315–1321. The price of food more than doubled in Germany during 1816–1817,[13] and food riots were widespread, especially in England, France, and Belgium.

The change in climate and the social upheavals that followed triggered outbreaks of epidemic typhus and, in some cases, relapsing fever, another lice-borne disease. These conditions frequently accompany starvation, privation, and cold and damp conditions. As impoverished families huddled together at night in the cold and gloom and the poor crowded together in soup-kitchen lines, the lice that carried the typhus bacteria were able to spread readily. One-quarter of Glasgow's population of 130,000 was struck down and one in ten of the afflicted died, while in London serious epidemics of typhus occurred among the silk workers in Spitalfields and elsewhere.

North America was affected as well by harvest failures and hunger in those cold post-Tambora years, especially the grain-starved northeast of the United States; there was a flow of migrants to other parts of the country. In the spring and summer of 1816, a persistent reddish "dry fog" dimmed the sun, enabling sunspots to be seen by the unprotected naked eye.[14] Snow began falling in early June of 1816 in New York State and Maine. Severe frosts occurred repeatedly during the summer, curtailing the growing season and causing widespread crop damage, harvest failure, and hunger. The strikingly short growing season in 1816 in southern Maine is clearly shown in Figure 9.1.

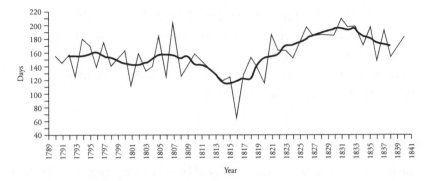

FIGURE 9.1 Length of yearly growing season in southern Maine. In 1816, "the year without a summer," the growing season was half the normal duration. Oppenheimer, "Climatic, Environmental and Human Consequences of the Largest Known Historic Eruption: Tambora Volcano (Indonesia) 1815."[15]
Permission received from Sage.

Chester Dewey, professor of Mathematics and Natural Philosophy at Williams College, Massachusetts, recorded that:

> Frosts are extremely rare here in either of the summer months; but this year there was frost in each of them . . . June 8th, some ice was seen in the morning . . . earth very little frozen . . . wind still strong and piercing from the N.W. Cucumbers and other vegetables nearly destroyed . . . June 10th, severe frost in the morning . . . Ten days after the frost, the trees on the sides of the hills presented for miles the appearance of having been scorched . . . August 22, cucumbers killed by the frost. August 29, severe frost. Some fields of Indian corn were killed on the low grounds, while that on the higher was unhurt. Very little Indian corn became ripe in the region.[16]

In sub-Saharan Africa, the cooling helped trigger an unusually severe drought and food shortages, especially in subtropical southern East Africa. One benefit of this drier period in sub-Saharan Africa was the decline of the deadly infectious disease sleeping sickness. This disease, which along with malaria and other tropical diseases earned equatorial West Africa the label of the "White Man's Grave," is spread by tsetse flies from infected cattle and remains a human health problem in moist equatorial Africa. (It rebounded in the much wetter 1860s.) In northeast China, where cooling and famine were severe, fertility halved—a reminder that severe undernutrition can contribute to reduced libido via effects on sex hormone levels.[17]

Cholera

The first pandemic of cholera occurred between 1817 and 1825. The correlation with unusual worldwide cooling and widespread food shortages, while tantalizing, may have been coincidental. Cholera, caused by fecal-oral transmission of the *Vibrio cholerae* bacterium (via either person-to-person transmission or infected drinking water), thrives in conditions of poverty, crowding, flooding, and warmth. The infection originated long ago in the eastern Gangetic plains of India; there is evidence of an ancient endemic focus of cholera in the Ganges delta, where people sought to placate the goddess of cholera (called by European observers variously Oladevi or Oola BeeBee). In 1816, during increased trading activity and British troop movements around Calcutta, cholera began extending its local geographic range. Spreading north, it reached Canton in coastal southern China in 1820 and then spread rapidly and lethally in China's Yangtze Valley and into Korea and Japan. To the south, it spread via Sri Lanka into Southeast Asia. Aided by British military expeditions seeking to extend the range of colonial control, cholera also spread to the north and west via land routes into Afghanistan and then on to the Middle East.

Six other pandemics have followed, and the seventh—the largest and longest yet, originating in Indonesia in 1961—is still with us. The earlier pandemic outbreaks of cholera all originated in India in the nineteenth century, in 1817, 1826, 1845, 1862, 1881, and 1899. The timings of those six pandemics, while not obviously related to particular trading activities, troop movements, crowding, and displacement, show a suggestive influence of periodic variations in monsoon rain patterns, particularly in the first half of the nineteenth century[18]—a relationship that accords with the observed association of cholera outbreaks with both severe floods and droughts in southern India during the first half of the twentieth century.[19]

It was, incidentally, the search for the causes of cholera outbreaks by British colonial surgeons, meteorologists, and Indian scholars in the mid-nineteenth century that produced early clues about the existence of the Pacific-based El Niño Southern Oscillation (ENSO) cycle. Recent ecological studies have shown that the cholera bacterium, while quiescent, naturally stays within the outer layer of tiny algal species in coastal and estuarine waters, and that when those waters either warm or become nutrient-laden, as happens in the coastal waters of Peru, Bangladesh, and South Africa during El Niño events, the cholera bacteria multiply, enter the local marine food-web, and are eventually ingested by humans dining on fish.[20]

Midcentury: Blighted potatoes, bubonic plague

Potatoes in the morning, potatoes at noon,
And if I rose in the night it would still be potatoes.
—GAELIC RHYME (TRANSLATED) FROM THE EARLY 1800s

Potatoes were introduced to Ireland in 1590 after being brought to Europe by Spanish conquistadores from their place of origin in the South American Andes. By the early nineteenth century, the potato had become the staple crop in Europe's poorest regions. In 1840s Ireland, potatoes accounted for nearly half of the national food crop. Indeed, they were almost the only food source for over three million wretchedly poor rural Irish families. One could in fact stay healthy on a (sufficient) diet of potatoes. If savings allowed, the peasantry sometimes added salt, cabbage, and fish and drank a little buttermilk. Potatoes are rich in protein, carbohydrates, minerals, and several key vitamins—and many poor, landless Irish peasant families actually ate more healthily than their counterparts in England and much of Europe, who depended on bread as the main staple.

Vincent van Gogh's early somber-toned oil painting *The Potato Eaters* depicting an 1880s family in a humble setting in village Netherlands captures this dietary dependence of the peasant families on potatoes. The domestic setting of this Dutch scene—familiar to van Gogh from his own rural childhood—was several cuts above the thatched and unlit hovels in which many poor rural Irish peasant families lived in the 1840s, the decade in which the devastating Irish Potato Famine occurred.

Tragically, another factor had heightened the vulnerability of Irish peasant families to starvation. As crowding and hunger increasingly pressed on the land-starved peasantry, they switched from "apple" potatoes to "lumper" potatoes. The knobbly lumper potato had higher yields, but it lacked taste and stored badly, and in Ireland's moist conditions the lumper was easy prey for fungal infection.

The notorious Potato Famine in 1846–1849 had several modest connections with climatic factors. Conditions in northern Europe during much of the 1830s and early 1840s had been cold and wet, and crop yields were low. The cold was reinforced in 1835 when the massive volcanic eruption of Cosigüina in Nicaragua dumped ash over much of Central America and caused temperatures around much of the world to drop by 0.75°C for a half decade. Much of Ireland's large and impoverished rural population was chronically undernourished. Then in mid-1845 a fungus, *Phytophthora infestans*, was apparently imported into Europe in a shipload of diseased tubers from either North or South America (an unresolved issue). As bad luck would have it, 1846 was a relatively warm and humid year in Europe, with moist southerly winds and copious rain on the Atlantic fringe, including Ireland. Under these conditions, the fungus, once introduced, could spread quickly, causing crops to wilt and potatoes to rot. The "potato blight" was widespread in Europe. Ireland, warmer and moister than other countries courtesy of the heat-releasing Gulf Stream, was particularly vulnerable.

Three decades before Robert Koch's exposition of the germ theory, the infectious nature of this fungal blight was not understood. In England, there was dispute between a learned professor of botany and a pastor-naturalist over whether the disease was due to waterlogging (professor) or to the damage caused by the fungus-like growth (botanist).[21] The latter proved much closer to the truth and, for the learned professor, a professional blight.

As the potato blight took hold and spread, Ireland's annual potato harvests halved in the worst years from 1846 to 1849, although temperatures were now a little cooler. This was a terrible blow to those already starving. Since dirt-poor rural families could not afford alternative foods, even if they were available, a tidal wave of severe hunger and starvation swept the country.

The famine was exacerbated by the free-market ideology that prevailed (though selectively applied) in Britain. The government, keen to have its cake and eat it, refused to constrain the free market in its ideologically correct *export* of grains from Ireland by English landlords of Irish lands, but was reluctant to repeal the protectionist Corn Laws that forbade corn *imports* from outside Britain. By the time the Corn Laws were repealed, at least a million poor rural Irish had died from starvation and from the typhus epidemic that swept through the soggy, crowded countryside and the now crowded urban slums. Typhus thrives in damp and dismal conditions, when people crowd together and families may huddle together. The lice that carry the typhus bacteria thrive in thick layers of unwashed clothing, and they infect humans primarily via feeding on their blood.

Cholera, too, broke out in many workhouses and urban districts. Overall, approximately 10 times as many poor Irish died from lice-borne disease as died from frank starvation.

In the following decades, up to half of the surviving Irish population emigrated, many going to America and Australia. The famine left in its wake an enduring trail of psychological, social, and political upheaval: mental depression, a lowered birth rate, family fragmentation, hostility to the English and to their Queen Victoria ("the famine Queen"), a labor force depleted in the middle age range, and a loss of national pride and purpose.

The latter nineteenth century

Since the early 1860s, thermometers have been used systematically on land and at sea to measure Earth's surface temperature, and they tell us the average global surface temperature has risen by around 1.2°C in the last 150 years. At first, warming represented the final dissipation of the centuries-long Little Ice Age. But fossil fuel–powered industrial activity gathered momentum, especially in Europe and North America, and discernible human-induced warming entered the climate change equation.

Serious food shortages and famines became less frequent in the industrializing countries as trade, transport, and energy-powered technologies changed the agricultural and commercial landscapes. Even so, in the 1860s a series of cold waves and heavy rainfall caused serious harvest failures and hunger uprisings in much of Europe. Higher prices constrained the poor more than the rich; according to Mulhall's *Dictionary of Statistics*, Fellows of the Royal Society on average were 3.9 inches taller and 21 pounds heavier than "burglars and other convicts."[22] During the 1870s, and then again in the 1890s, extreme droughts and hotter temperatures occurred in China, South Asia, and the southern hemisphere. The droughts were associated with failures of the great Asian summer monsoon system and with unusually strong El Niño events and their desiccating swathe extending westward from the western-equatorial Pacific region. There was a 1.5°C increase in sea-surface temperature as warm surface waters and moisture-laden air flowed from the western to eastern Pacific to pile up and drench Peru, while disseminating drying conditions to the immediate and distant west.

These late-nineteenth-century droughts caused huge numbers of deaths, particularly in the populous countries of India, Brazil, and China.[23] The famines of 1876–1878 were due to a combination of ENSO-driven falls in regional rainfall and the impact of colonial policies enforcing the integration of local food markets into the increasingly global market. The greatest human toll from the famines occurred in situations where the state authorities lacked the will or capacity to provide food aid to famine victims.[24] In China, the Qing Dynasty had been seriously enfeebled, both politically and economically, by the nineteenth-century military defeats and exploitative trade demands imposed by European maritime powers, while in India the British Raj was preoccupied with overseeing a profitable extractive colonial enterprise.

In India, the colonial rulers initially withheld famine relief on the premise that Malthusian "natural checks" should apply to such stricken populations. Famine deaths would bring population numbers back into line with food supplies. In 1877, the British Viceroy of India, Lord Lytton, pronounced that the Indian population had a tendency to increase faster than food is produced—a clear Malthusian view. In line with free-trade orthodoxy, his colonial regime continued to export India's grain to Europe, even during the crisis years of 1877–1878. Here were shades of the Irish potato famine debacle. As a marginal concession, make-work public projects were instigated in which skin-and-bone laborers were fed a meager daily diet. Meanwhile, across the country, millions of deaths occurred from starvation and infectious diseases.

China and bubonic plague: The Third Pandemic

A few days following the death of the rats,
Men pass away like falling walls
    —POEM BY SHI DAONAN, "Death of Rats" (China, 1792)

Extreme weather, including disrupted rainfall patterns, in eighteenth-century China contributed to the outbreak of bubonic plague. The region was already enduring economic and social instability. Wars, opium addiction, and famines, along with widespread internal rebellion, had weakened both the populace and their Manchurian Qing imperial rulers. The size of the national population had declined by around one-fifth in the preceding 20 to 30 years.

The onset of colder temperatures in China beginning in the late 1840s resulted in widespread harvest deficits of 10 to 25 percent, regional crop failures, and food crises.[25] Hunger intensified the peasants' agitations for land tenure reform, spawning broadly based protest movements, the Muslim Nian rebellion in the west and the Taiping rebellion in the east. The Taiping rebellion, led by an aggrieved candidate who had failed the civil service examination and who claimed to be the younger brother of Jesus of Nazareth, quickly gained momentum. In the 1860s, a large army of Taiping protesters seized control of towns and cities throughout the Yangtze Valley and then the Central Plain, including the prize city of Nanking. The rebellion was only quelled when European military powers backed up the teetering Qing dynasty.

European reinforcements were self-interested. The British, German, and French governments were seeking concessionary footholds, trading rights, and profits in China, and demanded that the Chinese imperial court receive international diplomatic missions. Relations went from bad to worse. During the 1840s and 1850s, the Chinese and British had fought two Opium Wars, the result of an attempt by the British to balance their international trade books by imposing on China a profitable import from their India dominion, opium. As opium addiction had spread, affronting Chinese cultural values and sapping local economies, Chinese authorities had grown angry and the Opium Wars followed.

Against this turbulent background, unusual weather extremes in the 1860s perturbed the usual self-contained subterranean circulation of the plague organism among wild rodents. A local epidemic appeared first in the 1860s in Yunnan, China's most southwestern province, when cooler temperatures and lower rainfall mobilized hungry plague-infected wild rodents.[26] Meanwhile, much further north in China the conditions were wetter and warmer—local conditions that were conducive to plague outbreaks specifically in that northern environment.[27]

The plague spread widely across the southern region and then into northern China. The southern and northern belts of infected areas (Figure 9.2) match closely the known habitat areas of infected rodent populations. The critical climatic conditions that influenced plague activity within its wild host populations differed between these two regions, reflecting differences in evolutionary adaptation between northern and southern rodent-and-climate ecology.[28] Outbreaks in the normally dry north increased with *increasing* rainfall, while outbreaks in the normally wet south increased with *decreasing* rainfall.

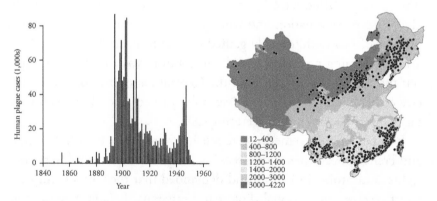

FIGURE 9.2 Bubonic plague, the Third Pandemic. Annual incidence in China, ca. 1850–1950 (graph), and map of outbreak sites in annual rainfall zones. Note the fivefold difference in rainfall between the northern and southern plague belts. Xu et al., "Nonlinear Effect of Climate on Plague during the Third Pandemic in China."[29]

Permission received from PNAS.

In 1894, after bubonic plague had ravaged China for three decades, including 70,000 deaths in Guangzhou Province in that one year, the disease then spread globally via shipping from Hong Kong. Serious outbreaks followed in North America, Australia, and elsewhere. India, however, bore the brunt of this pandemic spread. From the time the disease arrived by sea in 1896, repeated outbreaks occurred in India until the 1920s. Typically, those outbreaks were strongly influenced by seasonally varying temperature and rainfall, declining in intensity as rainfall peaked during the monsoon and then increasing as conditions became drier and warmer—until temperatures of around 27°C were reached, when case numbers began to dwindle before ceasing at around 30°C.[30] Indonesia's first case of plague occurred in 1910 in a tiny railstop in East Java, where a shipment of grain from India's neighbor Burma was unloaded and a hapless local householder fell ill from the disease.[31]

Bubonic plague is still very much with us this century. Greater global connectedness has hastened integration of the disease into wild rodent ecology in many parts of the world. Changes in land use, in the location of settlements and recreational behaviors, in human mobility, and in the climate all contribute to the spread of plague.[32] Most modern outbreaks occur in locations with an average annual temperature within the range of 24–27°C.[33] As the Earth's average temperature rises in the coming decades, this temperature band will slowly extend to higher latitudes, extending the range of bubonic plague into new areas.

Twentieth century: Approaching the present

Early in the twentieth century, weather disasters and climatic extremes continued as ever to influence rates of human disease, disability, and death. Later in the century, temperatures began to rise more sharply, especially from the mid-1970s, as the influence of human actions in altering our planet's terrestrial, atmospheric, and ecological systems assumed increasing dominance. Some scientsits argue the Earth has entered a new epoch, the Anthropocene, or Age of Humans. But the usual mix of nonclimatic and climatic influences on human health makes it difficult to tease out the effects of single causes. Since rising temperatures and extreme weather events from the mid-twentieth century onwards are well documented and often discussed (for example, in IPCC reports), the concluding sections of this chapter highlight some key themes: food crises, infectious disease, floods, and other weather disasters.

Climate and food crises in the modernizing world

Food shortages, sometimes famines, continued into the twentieth century. Some were largely due to the seismic social upheavals of revolution, as in Cuba, and national policies in the wake of European decolonization. Some, especially in South Asia and later in sub-Saharan Africa, were strongly climate-related. Other famines, such as in Stalinist Russia and Mao Zedong's China, were due to ideological and political mismanagement. The famine of 1932–1933 in the Ukraine, known as Holodomor (extermination by hunger), which killed seven million people, was thinly disguised genocide by Stalin.

In the Indian subcontinent, the extremity of the 1943 Bengal Famine was largely due to social and political circumstances, argues economics Nobel laureate Amartya Sen.[34] The famine caused two million deaths in a regional population of 60 million, many of them in Calcutta. Cooler weather and a devastating tsunami on India's east coast in October 1942 had reduced cereal grain yields, and much of the saleable harvest in 1943 was simply not available to the poor. Their food "entitlement" under social contract principles was precluded, says Sen, by poverty, caste barriers, and deficient hunger alleviation policies. Further, the authorities were more concerned with protecting the free-market economy and with the military threat foreshadowed by the Japanese bombing of Calcutta in December 1942. Grain merchants hoarded much of the grain against both the possibility of war with Japan and the rumored possibility of a fungal blight affecting the next rice harvest.[35,36]

In Midwest America in the 1930s, already struggling with the Great Depression, the regional climate became drier. The subtropical jet stream with its associated subtropical ridge, typically located around 30°N, had temporarily changed its sinuous course. Traveling east, its usual northward excursion over the American mainland temporarily reached further inland, as did the aridifying subtropical ridge, meaning that most of the southern United States was deprived of rain.[37] During 1931–1936, a prolonged drought resulted, coming on top of the chronically eroded, overplowed soils. This generated huge dust storms in the American and Canadian prairie lands. The farmland and human catastrophe was most extreme in Oklahoma and Texas, where the exhausted bare land turned into the famous Dust Bowl.

Hunger, physical stresses, and mental health disorders were rife, and many farming families fell seriously ill from malnutrition and dust-inhalation pneumonia. Thousands of families (the "Okies") abandoned desiccated farms and headed west to California, seeking work.

China's last great famine, the worst of the twentieth century, was the "Great Leap Forward" famine of 1959–1961.[38] This caused around 30 million deaths (almost 1 percent of the world's population) and had a visible impact on the graph of global population growth rate (Figure 9.3). Here, if further evidence were needed, is testimony to the killing power of famines—in this case, a famine attributed by the Chinese government to adverse climatic conditions, but by many others to inefficient and corrupt regional agricultural management and food distribution.[39] China's large southern neighbor India, for many centuries a focal point for food crises and famines, has managed to avoid a major famine since the State of Maharashtra's disaster in 1972–1973, when 130,000 deaths were attributed to starvation.

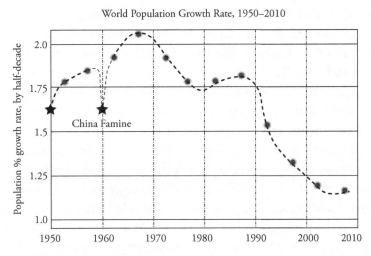

FIGURE 9.3 Time trend in the annual percentage growth rate of world population: actual 1950–2010 (Stars added to UN data). UN, Population Division of Dept of Economic and Social Affairs, *World Population Prospects. The 2012 Revision.*[40]

During the second half of the twentieth century, the center of gravity of food crises, famines, deaths, and refugee flows shifted progressively to sub-Saharan Africa, where temperatures are already uncomfortably close to the thermal limits of important crops. For instance, a meta-analysis of more than 20,000 field trials in Africa found that each degree day corn (maize) spends over 30 degrees centigrade, the yield is reduced by 1 percent, and if water supplies are limited the impact of heat is almost doubled.[41]

Starvation in Ethiopia in 1985 was described by the BBC as a "biblical famine . . . the closest thing to Hell on Earth."[42] The famine began in the early 1980s, the product of a combination of climatic adversity, corrupt strong-arm government, weak administration, and superimposed warfare. In 1982–1983, a moderately strong El Niño event caused a marked decline in rainfall in Ethiopia. By that time, the quasi-Marxist Communist military junta called the Derg was ruling Ethiopia following their toppling of Emperor Haile Selassie after that government's failure to ameliorate the great drought-induced hunger of 1972–1973.[43] The new rulers collectivized farms and sought to suppress secessionists in the northern provinces. During the early 1980s, farm yields stagnated, the agricultural system deteriorated, and then drought descended. Failed harvests led to a rapidly worsening famine in which nomadic pastoralists suffered particularly. As starvation spread, an estimated 300,000 deaths resulted and famished refugees poured into the Sudan and Somalia.[44]

In the Republic of North Korea, the most centrally controlled state of all in modern times, the great famine of 1995–1996 was triggered by an extreme weather event, when a torrential deluge occurred in little over a week. Two-fifths of the nation's rice fields were submerged, and local production of this basic food item halved. Half a million people were displaced, and as hunger took hold, death rates surged. Further flooding in 1996 compounded the country's difficulties, as did a prolonged drought in 1997 in areas close to the Chinese border. This run of climate misfortune severely tested the country. It is difficult to be certain, but a careful weighing of information indicates there were very likely 600,000 to 1,000,000 famine-related deaths in North Korea between 1995 and 2000.[45] Predisposing factors included overuse and degradation of much of the farmland, economic mismanagement, corruption, and despotic government.[46] The collapse of the Soviet Union meant that North Korea lost its major trading partner and sponsor. The end of cheap oil and grain from the Soviet bloc was a particularly severe blow. Other causes of vulnerability included fragile transport networks, a primitive health-care system, and arrangements for distributing food that even in times of plenty favored the military and government officials rather than rural populations.

Famines cast a long shadow. There are serious consequences, for instance, for health and life expectancy among famine survivors, and even among their offspring. Although not all the evidence from co-horts of those born during a famine is consistent,[47] much of the human and animal experimental research has shown that major nutritional deficits in infants, fetuses, or indeed in their mother's early-life history impart lifelong increases in the likelihood of various serious disease processes: type 2 diabetes, high blood pressure, coronary artery disease, and others. But this story appears to extend further. Children from China's Anhui province, for example, conceived during the great famine of 1959–1961 were found to be at twice the risk of experiencing schizophrenia compared with children conceived before or after the famine.[48] Elsewhere in China, those who were one year old at that famine's onset had on average 2 percent lower adult height, 6 percent lower weight, 3 percent less completion of schooling, and 7 percent lower participation in the workforce than the general population. Intriguingly, in the next generation born in and around the 1980s, children of a parent born during the great Chinese famine were smaller and lighter than average.[49]

Infectious diseases in the twentieth century

As the twentieth century progressed, human populations grew more rapidly, crowded into cities, cleared more land for agricultural expansion (and to produce feed grains for enlarged herds of livestock, as meat eating became more widespread), traveled more widely, over-used antibiotics, and relied increasingly on surgical procedures that involve direct biological contact between individuals—blood trans-fusion, organ transplants, and intravenous fluid infusion. In all of this were many new ecological disruptions, and hence new niches for infectious agents to exploit. Later in the century, changes in regional climates began altering further the landscape of possibilities for the emergence and spread of infectious diseases and for prolonged seasonal transmission.

Many of the more interesting contemporary examples of climate change and infectious disease relate to vector-borne infections and to other infections with complex environmental and ecological determinants. However, globally, much of the impact of climate change will be on more prosaic but higher-burden infectious diseases, especially diarrheal diseases in children. Cholera persisted through the last century, and the ongoing seventh pandemic, beginning in the 1960s with a new strain (El Tor) of the bacterium, has become the longest ever. This disease will almost certainly remain a prominent source of climate-related outbreaks in many poor populations. Severe floods and other weather disasters will continue to cause local cholera epidemics and surges in other diarrheal diseases. Systematically recorded colonial data from southern India in the period 1900–1940 shows that outbreaks of cholera in the populous Madras Presidency occurred repeatedly in response to the two extremes of surface water conditions—droughts and heavy flooding.[50] During droughts, the more concentrated surface-water supplies contain heavier doses of the bacteria, while during flooding, water supplies and basic sanitation systems are often disrupted and displaced people crowd together. There has been a marked increase in abundance of *Vibrio* organisms (including the subtype that causes cholera) in the North Sea since 1980, matching the rise over the same period in sea surface temperature.[51]

Malaria remains a focus of climate-related concern. Close analysis of cases of malaria in the highlands of Ethiopia and Colombia found a shift to higher altitudes in warmer years.[52] This is a sign that rising temperatures will pave the way for the disease to enter densely populated high-elevation regions in Africa and South America unless steps are taken to control the vector and treat promptly any cases that do occur.

During the latter half of the twentieth century, as many countries became wealthier and better informed, their rates of malaria declined. However, the reappearance of locally transmitted malaria in Greece in association with high levels of immigration, economic hardship, and cuts in health services demonstrates that the disease remains a threat wherever the climate is suitable.[53]

Floods and weather disasters

Floods are always with us. Some, such as the annual Nile floods and, historically, the annual floods in southern Mesopotamia, all bearing nutrient-rich silt, are life-sustaining. Many are life-terminating. Recall the period of coastal storm surges and inundation in the late sixteenth century, during the coldest phase of the Little Ice Age, which washed away housing and farmland along much of the Danish, German, and Dutch coastlines. In mid-1931, when China experienced heavy and prolonged summer rain, the immediate death toll from floods in the Yellow River and Yangtze River basins was approximately half a million. The eventual death toll from the ensuing famine and spread of infectious diseases, particularly during 1932–1933, was well over six million.[54]

The world is now experiencing an upturn in the frequency and intensity of floods, fires, heat waves, and other weather disasters.[55] In contrast, there has been virtually no increase in geological disasters (earthquakes, tsunamis, volcanic eruptions).[56,57] Since 1923, there has been an increasing frequency of very large hurricane-associated ocean surges on the United States Atlantic coast, with Hurricane Katrina–magnitude events occurring twice as often in warm years as in cold years.[58] Katrina, which was a category 3 hurricane when it crossed the Gulf Coast of the United States in August 2005, caused almost a thousand deaths. Black Americans were two to four times more likely to be killed by the storm as whites, and mortality rates were highest among the elderly (75 years and over).[59]

Emergency services were better prepared seven years later when Sandy, the largest storm ever seen in the North Atlantic, struck the east coast of the United States. But nevertheless the hurricane caused about 230 deaths, $50 billion damage to property, and long-term health problems, including those caused by toxic molds in flooded houses and mental health disorders.[60]

Heat waves have emerged as a significant hazard, as shown by the extreme events that affected France in 2003 and Russia in 2010. These periods of heat wave were described as "three-sigma events," meaning temperatures so high were extremely unlikely to occur in the historic record, and the impacts were also unprecedented, amounting to tens of thousands of excess deaths.[61] Though less well documented, extreme heat causes even greater mortality in low-income tropical countries such as India, where the effects of record high temperatures in the last five years have been aggravated by power failures and water shortages.[62]

Concluding remarks

And so we reach the present. The second decade of this century is experiencing an uptrend in the rate and severity of weather disasters, as mentioned earlier. It is now nearly certain that this increase exceeds historical decade-to-decade variation and that this is, in large part, due to climate change and the greater amount of energy now embedded in the climate system.

These extreme events consistently draw the attention of the general public and the media. Increasingly often, human-driven climate change is invoked as the explanation for the severity of such events. And yet the longer-term toll of adverse health impacts of climate change will most probably be dominated by the consequences of ongoing shifts in the prevailing (or regionally average) climatic conditions. Weather-related seasonal mortality, for example, is not just influenced by short-term fluctuations such as severe cold and heat waves, but also by gradual changes in climate. During the four decades since the 1960s that Australia experienced rising average temperatures, the ratio of summer to winter deaths increased.[63] The influence of climate change is evident in the recent Australian mortality record. These longer-term changes in climatic conditions will affect crop growth and yields, livestock health, water availability, the mobility and activity of infectious agents, stability of coastal settlements, population displacements, and tensions over dwindling environmental resources.

So the journey from two million years ago through to this century is now completed. Climate change has shaped human evolution and both challenged and nurtured human social and cultural systems, especially during the expansion of agriculture and urbanization in the Holocene. In the next chapter, that long and variegated history of the relationship of climate change to human well-being, health, and survival will be reexamined to seek out key themes, patterns, and messages from the past.

10

Humans throughout the Holocene

WHAT DOES THE STORY about human experiences of past natural climatic changes tell us in broad terms? At the least it points to the types of risks to the health, survival, and social stability of diverse populations likely to result from this century's human-driven climate change. This chapter and the next look to that future, and to how humankind might avert the looming environmental and social crises. In preparation, paleoclimatologist Raymond Bradley argues that we should seek "insight into the nature and magnitude of past regional [climatic] anomalies and their human impacts by examining the Holocene paleoclimate record."[1] This chapter asks whether there are discernible patterns in the long story of past climate change and population health adversity. Have particular conditions and factors determined vulnerability to those adverse impacts? On this basis, how vulnerable are today's populations, rich and poor, tropical and polar, to the climatic stressors and associated risks anticipated in this twenty-first century?

An overview of the Holocene's 11,000-year climatic experience, a mere sliver of the total *Homo sapiens* experience of climate variation, provides a good entry point. This latest of the nine interglacial periods during the past million years has been the one in which anatomically and behaviorally modern humans began exerting increasing control over the environment and its carrying capacity by shifting toward growing crops, herding animals, managing water flows, and building settlements. That agriculturally based era is shown near the top right in

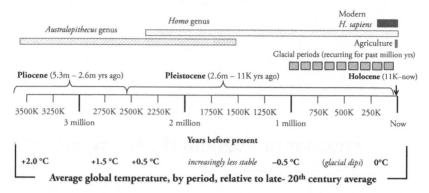

FIGURE 10.1 Temporal background to climatic influences on evolution of human biology and culture. Holocene is at far right.
Source: Author.

Figure 10.1, a moment in geological time. Throughout the Holocene, the climate was, as ever, naturally variable, and the gods of rain, sun, wind, and warmth held great sway over harvests, hunger, and viability.

The Holocene climate experience

We will summarize here the main points from earlier chapters as the basis for a discussion of patterns in the past which may point to future vulnerability and implications for human well-being.

The Holocene has been a period of *relative* climatic constancy, with an average global surface temperature of 15°C. Changes in temperature and rainfall spanning several centuries have occurred, as have many shorter-term fluctuations. Rises and falls in regional decadal-average temperatures have rarely exceeded 2°C, and most have been within a 1°C–2°C range. Temperatures in Europe differed by about 1.5°C between the peak Medieval Warm and the Little Ice Age nadir five centuries later. In China, a similar, though lesser, temperature difference occurred between those periods. In parts of the southern hemisphere, too, cooling occurred particularly during the thirteenth and fourteenth centuries.

Acute climatic changes, due to major volcanic eruptions, super El Niños, or other abrupt changes in regional atmospheric circulation systems, have often caused extremes of weather, social upheaval, destruction of infrastructure, and crisis. Multidecade events, such as prolonged droughts or the northern hemisphere's extreme cold of the early 1600s, caused hardship, misery, increased illness and death, and social unrest. Much longer regional climatic shifts characterized the Holocene Climatic Optimum, the European region's Roman Warm, the prolonged drying in the American southwest from the twelfth to fifteenth centuries, and the Little Ice Age in Eurasia. These all exerted enduring influences on social, cultural, and technological changes, affecting agricultural systems, undernutrition and starvation, epidemic outbreaks, displacement of populations, and destabilization of political regimes.

Humans, as products of evolution's central "survive the present" criterion, lack a strong hardwired instinct to act on behalf of the distant future and on behalf of species and ecosystems that are far away and unfamiliar. Earlier societies typically made *reactive* responses to climatic adversity, and often did so too late. They had limited information and awareness that critical environmental limits were being breached and limited formal knowledge of the workings of ecosystems and the climate system, and accordingly they had little forecasting capacity that could support *proactive* planning. So they did little more than build city walls, irrigation systems, and granaries to buffer them against any future floods and droughts. Today we know much more about the workings of the planet, the biosphere, and the climate system, as well as the mounting pressures on those systems. This knowledge, teamed up with our cerebral capacity for abstract thought and imagining the future,[2] should enable us to respond in more proactive fashion.

The past as analog for the future

Taking the extra step to get from historical examples to insights about the contemporary world is always tricky. Many a statesman and not a few scholars have slipped up when attempting this step . . .

—JOHN R. MCNEILL[3]

The first point to note is that we face a change in global climatic conditions far greater and faster than anything in recorded human history. At most, following the cooler seventh and eighth centuries in the North Atlantic region, an enduring 1°C–2°C warming during the ninth and tenth centuries C.E. enabled the colonization of Iceland and Greenland by the Norse Vikings. Yet the world may be heading towards an average global warming of up to 4°C by later this century, with even larger rises in polar and subpolar regions. The projected rate of global heating outstrips anything in the geological record; the last time the planet's temperature rose by 4°C was 56 million years ago, but that change occurred over thousands of years, not over a single century.[4]

As the temperature rises, most risks will veer upwards, often steeply; in terms of impact a rise of 4°C does not always equal 2°C plus 2°C. The World Bank grasped this nettle in 2013, issuing a report on the potentially dire consequences of a world that could be 4°C hotter by later this century.[5]

There is, nevertheless, much we *can* learn from the Holocene experience. Overall, cooling and drying episodes outnumbered heating episodes and caused most of the human adversity, but excursions beyond the "just right" comfort-zone temperatures, *either* by warming or cooling, impair biological function and well-being. This applies widely: to human health and survival in the face of extremes of heat or cold, to mosquito survival and rates of development of the malaria pathogen inside the mosquito gut, to plant photosynthesis and crop yield, and, within the built environment, to the structure and safety of transport modalities (such as road traffic, railway lines, power lines), and also to rainfall and humidity. For many health-related outcomes, such as the ease of cholera transmission, food yields, and the reproductive capacity of plague-transmitting fleas, both too much and too little rain increase the risks to health. Whereas warming in Central Asia around 1300 C.E. helped trigger the bubonic plague pandemic, it was rapid cooling during the late 530s that prompted the spread of the disease and then, in Constantinople, ignited the explosive Plague of Justinian.

In summary, the historical record indicates that most of the natural world's species, ecosystems, and regional physical processes that our societies and biological health depend on are sensitive to rather small changes in climatic conditions. Those plants, animals, and ecosystems must take the climate as it comes; if stress, disease, deaths, displacement, and extinction follow, that is part of the uncompromising currency of life and survival in the natural world. We humans, however, have partially insulated ourselves with cultural assets and technological skills that can mitigate or adapt to—but not eliminate—our vulnerability to the vicissitudes of climate and environmental changes.

Past patterns of climate and health

Historians, writes Niall Ferguson, "cannot (and should not even try to) establish universal laws of social or political 'physics' with reliable predictive powers."[6] Ranged against that view is the folk wisdom that if we do not learn from history we are doomed to repeat it. Indeed we can derive general conclusions from sets of similar experiences that are likely to apply in similar situations in today's world and its immediate future.

Over most of the Holocene, average life expectancies were less than half of today's figure, and most deaths resulted from starvation, infectious diseases (especially since the onset of large-scale human crowding in cities around 2500 B.C.E.), and the immediately ensuing impacts of violence and warfare. Hence for much of the period covered by this book the scope of health impacts available to study has been limited to the realm of the "Four Horsemen of the Apocalypse," famine, pestilence, warfare, and conquest, plus the impacts of weather disasters. Many of the less dramatic or more acute health impacts are more difficult to glean from the historical record. These include deaths from heat waves, exacerbations of diarrheal diseases, fatal and nonfatal injuries from extreme weather events, and mental illness following weather disasters, farm failures, loss of livelihoods, and bereavement. All these afflictions must have been significant causes of suffering and early death.

The *main general conclusion* to be made about climatic impacts on health and survival during the Holocene is this: whether in the Arctic, temperate regions, or the tropics, the climatic comfort zone that sustains food and water supplies, stability of ecosystems, and other basic needs is confined within a narrow range of temperatures and a particular pattern of seasonal rainfall. Outside that comfort zone, stresses mount, biological function is compromised, and human health is impaired.

Human societies, trading on accrued experience, understanding, and cultural accessories, can tolerate brief and extreme variations in climate. Similarly, each of us as an individual organism can cope with quite a wide range of weather on a day-to-day and week-to-week basis. However, while today's better-resourced societies may side-step many of the early risks from climate change, the ecological systems and species that our societies depend on are the Achilles heel. Those ecosystems and their species, finely tuned by evolution to their prevailing climatic conditions, lack an additional protective veneer and are therefore more vulnerable than humans to changes in the environment.

Climate and food yields

The concept of the "just right" comfort zone has particular relevance to food production, as we discussed in Chapter 1. Both cooling and warming relative to the familiar long-term seasonal average, as well as either excess or insufficient levels of rainfall, can cause harvest losses and nutritional deficits. The climate-food-nutrition linkage has long been the foremost determinant of human well-being; indeed, during the first half of the Holocene most of the evidence linking changes in climate with human health refers to that relationship. Given its political ramifications, the climate-nutrition axis has frequently attracted the attention of scribes, priests, and potentates. There were signs of chronic under-nutrition in the bones of early Neolithic farmers and in the skeletal remains and middens in the drying Sahara of 5,500 years ago, while early written records from Mesopotamia and the Nile Valley provided for the first time contemporary observations of floods, droughts, crop yields, death rates, and conflicts. Many of those early clues have come from Eurasia, Mediterranean North Africa, and, to a lesser extent, southern North America, where corn-growing Amerindian communities around 10,000 years ago replaced the earlier Clovis preagrarian culture. But so far, there is negligible information from elsewhere; we know little about pre-Mughal India, most of Africa before 1800 C.E., and the Americas before 1500 C.E.[7]

Periods of moderate warming above the Holocene average temperature, accompanied by good seasonal rainfall, typically enabled expansions of agriculture and population coupled with fewer recorded health crises. This was so in Mesopotamia, the early imperial "Western" Roman Empire, and mid-stage Medieval Europe, and was the basis for territorial expansion by groups such as the Norse Vikings (ninth and tenth centuries C.E.) and the Mongolian peoples (twelfth and thirteenth centuries).

But those societies often pressed their environmental support base too far, mostly in attempts to maximize food production. This left them vulnerable to the recurrence of cooler temperatures when crop yields declined and stressed livestock did not produce as expected. The tidal pattern of good times and bad mirrors the familiar boom-and-bust cycles intrinsic to species everywhere; *Homo sapiens* is no exception.

Drought, says economic historian Cormac Ó Gráda, is the most frequent and fundamental cause of the famines that led to the great die-offs in history and the decline and collapse of various civilizations. What causes famines is never straightforward, since climatic, environmental, social, and political factors are all involved, but the ground truth is that crops and livestock depend on favorable climatic conditions and weather patterns. Ó Gráda writes:

> Most of the worst famines on record have been linked to either too much or too little rain ... In the later Roman period, drought was the main factor in three cases out of four.... In prerevolutionary China, drought was twice as likely to cause famine as floods. This was particularly so in wheat-growing regions ... Drought was also responsible for the massive Bengal famine of 1770, which may have resulted in millions of deaths ... The catastrophe in northern China in the late 1870s came in the wake of exceptional droughts in 1876 and 1877.
>
> In temperate zones, cold or rain, or a combination of both, were more likely to be the problem. The Great European Famine of 1315–17 was the product of torrential downpours and low temperatures during summer 1315 ... The Great Frost of 1740 led to *bliain an áir* (the year of carnage) in Ireland in 1740–41 ... In Kashmir, the great flood of 1640–42 wiped out 438 villages "and even their names did not survive."[8]

Of the Four Horsemen, famine has been the most potent cause of catastrophic mortality. On rare occasions, warfare has been almost as disastrous, as with the murderous conquests by Genghis Khan late in the Medieval Warm as he swept westward through Central Asia. Conquest, via enslavement, persecution, impoverishment, and cultural destruction (as with the American Indian and Australian Aboriginal populations in recent centuries), erodes the morale, strength, and health of the conquered people. Pestilence can, in extreme cases, kill 30 to 50 percent of a population (as occurred with the Plague of Athens, the Antonine Plague in Rome and its hinterland, the Plague of Justinian, and the Black Death in Europe). Famine, though, with its impact over most of the Holocene on human health and survival and on social coherence and political stability, has been the dominant destroyer of population morale, viability, and of civilization at large.

Food crises have also influenced the major historical movements of populations. Much of the ebb and flow of people across the Eurasian landscape over the past four millennia was driven by the varying rhythms of climatic changes, agricultural opportunities, and food shortages. This included migrations by the northern Mesopotamian pastoralists (ca. 2200 B.C.E.), "Sea Peoples" (ca. 1200 B.C.E.), Baltic farmers (ca. 300 B.C.E.), the Germanic "barbarian" invaders of the Roman Empire (third to fifth centuries C.E.), the Mongol hordes (end of Medieval Warm), and the one million migrating famine-afflicted Irish (mid-nineteenth century). This flow of peoples is human ecology writ large. It underscores what is most basic and sought-after for well-being, health, survival, and social stability—farmland, water supplies, and food.

Climate and conflict

We may not usually think of war and conflict as "public health" issues, and yet both the logic and the evidence for doing so are strong. Civil and military organizations are now generally aware that climate change is a "threat multiplier" likely to exacerbate existing resource vulnerability, especially in weak states, and that it therefore looms as a threat to both human security and national security.[9,10,11,12] In 2003, the United States' Pentagon issued a report stating: "As abrupt climate change lowers the world's carrying capacity aggressive wars are likely to be fought over food, water, and energy."[13]

The human history of the Holocene shows an association between climatic adversity and the outbreak of violent conflict.[14] The combined analysis of 60 different original studies, spanning information gleaned from the past 10,000 years across all major world regions, led researchers to conclude: "Deviations from normal precipitation and mild temperatures systematically increase the risk of conflict."[15] Civil wars in sub-Saharan African countries were more likely in periods when the temperature was above average or well below average.[16] In studies of 54 countries affected by El Niño events, conflict became much more likely as the strength of the event increased, especially during times of drying, cooling, and floods. The authors concluded that the primary causes of climate-related conflict were detrimental effects on food yields and freshwater supplies.

Over the past six to seven millennia, adverse climatic and other environmental influences on food yields have been a recurring source of harvest failures and hunger and the main cause of deaths and conflicts.[17] Temperature affects the length of growing seasons, soil moisture affects plant growth, and temperature and humidity influence the occurrence of infections and pest infestations. For today's predominantly agrarian-based societies, as in the past, temperature and rainfall are crucial influences on farm yields and hence on social and biological well-being. The world's major staple crops have been developed by repeated selective planting of higher-yielding variants of plant species over many centuries. These descendant species are finely attuned to their historically prevailing ranges of temperature and soil moisture and are likely to falter or fail outside that just-right range. Given the usual "Goldilocks" limits on the range of climate conditions compatible with good food yields and general ecological stability, quite small changes in climate beyond that range have been a recurring cause of violence, starvation, epidemics, and population displacement.[18,19]

The paths by which climate change influences the occurrence of conflict and warfare are diverse and dependent on their interplay with combinations of other factors.[20,21] A common argument, often made by economists, is that when climatic adversity causes economic downturn, people may see more economic value in joining the conflict than in struggling with their normal livelihoods. Climate-related economic downturn can also weaken and destabilize a state, rendering it vulnerable. History shows that when food, water, and access to raw materials are in short supply, the ensuing social disorder, infectious disease, and traumas of conflict bear most heavily on population groups that are most disadvantaged by ethnic and economic circumstances. During the decline of the Indus Valley's Harappan civilization around 1900 B.C.E., against a background of increasing rainfall uncertainty, falling food yields, and rising hunger, death rates from local violence and infectious disease were highest in the socially and economically marginalized communities.[22]

Syria's protracted civil war during this century's second decade has shown how coexistent economic stress, religious rivalry, environmental degradation, and regional climate change can easily coalesce in a bitter and destructive conflict. Frequent conflicts over water have occurred historically in this Middle Eastern region because of its natural water scarcity, the introduction of irrigated agriculture, and the tensions of religious and ethnic diversity. The severe and prolonged drought in rural northeast Syria that immediately preceded the uprising in 2011 had turned much of this ancient heartland of the Fertile Crescent into an arid and barren desert.[23] Irrigation systems became degraded, aquifers ran dry, sandstorms proliferated, and the landscape became littered with abandoned villages. Grazing animals largely disappeared as the region's herders lost 85 percent of their livestock. And when the drought's demoralizing human impact was exacerbated by an ill-judged shift in the Assad government's agricultural policy, levels of poverty, anger, and hostility rose, and tens of thousands of aggrieved northerners headed for the cities.[24,25]

During internal armed conflict, most of the health damage results not from combat itself but from social disorder, population displacements, and the breakdown of healthcare systems and essential services, such as programs for vaccination and for maternal and child health. Indeed, in modern conflicts, such as that in Afghanistan, health professionals, hospitals, and other infrastructure have been deliberately targeted and disease prevention programs obstructed.[26] Many deaths then result from starvation, malaria, and childhood infectious diseases.[27]

Deteriorating climatic and environmental conditions, along with the desire to escape from the tethers of poverty and illiteracy, will almost certainly increase displacements and migrations over the coming half century and beyond. In my own region of the world, early relocations have begun to occur from some of the low-lying Pacific islands, with movements between islands, between some island states and, by negotiation, to New Zealand. South Asian and Middle Eastern refugees who reach Australia via Indonesia, a chronic source of great public consternation, policy confusion, angry debate, and national shame, may well be dwarfed by future migrations from many parts of Asia as population pressures, extremes of weather, sea-level rise, shortage of fresh water from rivers and aquifers, and climate-impaired crop yields become an intolerable joint burden. Will vote-seeking political slogans exploiting xenophobic fears look naive and selfish within a few decades, or will divisive rhetoric and behavior only increase?

As the tempo of climate change and weather extremes has begun to rise, the prospect of future conflict is commanding more attention from scholars, governments, armed services, and nongovernmental organizations.[28] Among the concerned commentators, some foresee an environmentally disrupted, resource-depleted, and conflict-prone world ahead—a world of heightened inequalities, perhaps a "fortress world" of privileged minority enclaves.[29,30] Such a world, they conclude, could arise if humankind stays on its current trajectory.[31,32,33] Beyond the violent and more acute consequences of conflict for human health and survival, the full spectrum of human health consequences of climate-influenced migration is not yet well understood. There is already in hand plenty of analog knowledge about the health consequences, both adverse and beneficial, from displacements and migrations due to other circumstances: warfare, political or religious persecution, and so on. We need better understanding of the health experiences of actual "climate/environment migrant" groups and populations before, during, and after the move, along with knowledge of the health consequences for the occupied, dispossessed, or hosting populations.[34] This topic will increase in importance as a focus of research and policy input as the worldwide number of people on the move, either as migrants or as climate refugees, increases over the coming warmer decades.[35]

Climate and infectious disease

Contagion was not understood until the late nineteenth century, and during most of the Holocene, epidemics were presumed to be caused by a displeased deity, the transient conjunction of planets, or malodorous emanations known as miasmas. Hence, other than the triggering of distinctive epidemics, such as bubonic plague outbreaks, typhus epidemics during famines, and malaria's fluctuations in geographic range in northern Europe during the Medieval Warm and ensuing Little Ice Age, there is little known about changes in climate and outbreaks of specific infectious diseases. There is virtually no information on patterns of infectious disease, even of the broad-brush kind, from pre-colonial sub-Saharan Africa and the pre-Columbian Americas. This continental gap has special significance, because the prevailing types of infection differ between temperate and tropical zones.

Most temperate-zone infections—measles, mumps, whooping cough (pertussis), smallpox, chicken pox, and others—are acute "crowd diseases." They originate from animal sources and become fully human-adapted, without need of insect vectors or animal hosts. Because they induce lasting immunity in infected individuals, they need a self-renewing crowd of hundreds of thousands of people. These diseases became more common mid-Holocene as animal-human contacts in and around towns and cities became more common, and successful venturer-microbes managed to establish a foothold in the larger urban populations in the Middle East and Mediterranean regions.

In contrast, most tropical-zone infections are less acute, do not cause lasting immunity, rely on an insect or other vector organism for transmission, and usually have a natural wild host animal. Humans are merely incidental hosts. Such diseases do not depend on continuous person-to-person circulation within the human species for their survival. Examples (with vectors and/or natural hosts) include yellow fever (mosquitoes: monkeys), sleeping sickness (tsetse fly: cattle and ungulates), Chagas Disease or "South American sleeping sickness" (triatomine bug: guinea pigs), Ebola virus (uncertain host: bats, monkeys), and malaria (mosquito).

Changes in climate beyond the limits of the just-right zones have affected the occurrence of many infectious diseases, particularly those spread by vector organisms, such as mosquitoes. For example, the immature malarial plasmodium cannot mature and proliferate in the mosquito if temperatures are too low, and the mosquito cannot survive long if temperatures are too high. The flea that spreads the plague bacterium cannot reproduce if conditions are too hot or too cold, too humid or too dry. Nor is it simply a matter of the biological consequences of temperature, rainfall, and humidity. Historically, both cooling and warming have influenced the outbreak and spread of infectious diseases by affecting human living conditions and social processes, as have excessively wet or dry conditions. The pandemics of bubonic plague in the sixth, fourteenth, and nineteenth centuries C.E. in Eurasia have shown how regional climatic changes can trigger spill-over of the rodent-hosted disease into human populations by influencing biological, ecological, demographic, and social factors. More generally, human susceptibility to infection has often been heightened by climate-related nutritional stress, displacement, crowding, and hygienic decline.

Finally, remember that the major prevailing epidemic diseases have changed over time. For example, plague, typhus, and smallpox are no longer numerically important concerns in most of the world, although annual numbers of bubonic plague cases have risen a little in recent decades. Today, concerns focus on the likely influences of climate change on diseases such as malaria and other mosquito-borne viral infections (dengue, West Nile virus, chikungunya, Japanese encephalitis, Zika, and others), for which there are no certain identification in the historical record.

Temperature extremes

Extremes of heat and cold, in their own right, afflict and kill people. The Little Ice Age in Europe and China brought many more cooling episodes than warming episodes. Major volcanic eruptions have superimposed many acute cooling events over the past two millennia, either global or regional in their impacts. We, though, are moving into a warming world. Does history tell us much about the consequences of living in an unusually hot environment or experiencing extreme heat events?

There are no volcano-equivalent events that have caused a year or two of warming, although briefer extremes of heat have often occurred as part of natural climate variability. Information about historical impacts of acute hot episodes is therefore rare. Examples of specific episodes of hotter temperatures and adverse health outcomes include:

- Europe, summer 1344: hot drought conditions reduced the grain harvest, causing local famines that killed many tens of thousands. This prelude to the Black Death, which broke out three years later, would have further weakened the biological and social resilience of the population to infection.

- Hotter temperatures in Central Asia for several decades around 1300 C.E.: warming and drought acted as trigger for the outbreak of bubonic plague, subsequently spreading into China and then Europe. The analysis of twentieth-century data from Kazakhstan showed that a 1°C rise in springtime temperature typically causes a sharp increase in the rate of occurrence of bubonic plague in humans, mediated by the increased numbers and activity of infected wild rodents.

- Philadelphia, 1793: A critical increase in local temperature and humidity helped boost mosquito numbers, and the notorious epidemic of Yellow Fever resulted, with 5,000 or more people dying over three months from among the city's 50,000 population.

- The Great (bubonic) Plague of London broke out after an extremely hot year, 1665—during the coldest century of the Little Ice Age. Were black rat populations in London stressed and disturbed, or did they simply thrive and multiply in the warmer and moister conditions? We do not know.

- Europe in 2003 and Russia in 2010: Extremes of summer heat and unusually lethal heat waves in each of those countries also caused 30 percent losses in grain harvests, with corresponding rises in food prices. There were, as a consequence, food shortages and nutritional deficits in poorer communities in grain-importing countries.

How globally representative is Eurasia's history?

The climate-and-health experiences of Europe and China, while the best-documented, do not necessarily directly apply to other regions. But note that before the nineteenth century, both Europe and China were essentially agrarian societies in which most of the population were poor rural peasants and tenant farmers. So those historical experiences within Eurasia's mid-latitude region are of relevance to much of today's low-income world.

During the past millennium in Eurasia, the fourteenth and seventeenth centuries were times of climatic extremes, human misery, famines, epidemics, and increased mortality. In *A Distant Mirror: The Calamitous 14th Century,* Barbara Tuchman argued that historians had long struggled to fit that century into their historical analyses, given the prevailing assumption of postmedieval human progress. The Great Famine of 1315–1322, the outbreak of the Hundred Years War, and the midcentury Black Death catastrophe which killed over one-third of Greater Europe's population were the deepest dips on the historical skyline.

The climatic fluctuations of the first half of the fourteenth century amplified Europe's agricultural and economic woes and exacerbated the violence and suffering that accompanied profound social changes. For example, official Christian beliefs and proscriptions no longer tallied with actual conduct, even within the Church itself, and these increasingly blatant discrepancies, together with the manifest inability of the Church to stay the hand of the Black Death, led to a loss of moral authority. Tuchman writes: "When the gap between the ideal and real becomes too wide, the system breaks down. Legend and story have always reflected this; in the Arthurian romances the Round Table is shattered from within."[36]

Some respite occurred from the 1360s through to the 1430s, and then climatic and environmental conditions deteriorated again until warmer and steadier conditions reappeared at the end of the fifteenth century. Between the 1570s and 1590s, the temperature began its longest descent of the Little Ice Age, crops failed, hunger became widespread, and witch-burning escalated. The historians' General Crisis of the Seventeenth Century was now on Europe's doorstep.[37] The nadir of the cooling occurred in the early 1600s, exacerbating hunger, physical stunting, epidemic outbreaks, social unrest, displacement, and the chaotic and destructive Thirty Years War. The bubonic plague flared up again in the mid-1600s, including the Great Plague in London. Far to the east of Europe, China's Ming Dynasty fell in 1644 after two dire decades of cold, rain, and crop failures, capped off by an extreme four-year drought.

That Eurasian experience of climatic vicissitudes provides the basis for the next section, a discussion of the factors that influence population vulnerability to climate change.

Historical population vulnerability to climate change

*In looking for evidence of climatic impact in the course of history, it is
sensible to look most at the marginal areas near the poleward and arid
limits of human settlement and activity, for it is there that vulnerability
is likely to be greatest.*

—HUBERT LAMB, 1995[38]

Terms like *vulnerable, susceptible, resilient,* and *coping capacity* all
float around in the same verbal soup. Strictly, "vulnerability"—as
used by the climate change science network—refers to something
more than "susceptibility." A hemophiliac may be very *susceptible* to
serious injury in a sword fight, but would actually have a low *vulnerability* if the opponent's sword was blunt and if the hemophiliac
was wearing protective armor and had good sword-avoidance agil-
ity. The *vulnerability* of a human group to climate change therefore
reflects the threefold combination of the magnitude of the external
(climate-related) exposure, the population's inherent susceptibility
(biomedical profile, robustness of infrastructure and social-cultural
resilience), and the population's capacity to make adaptive, risk-less-
ening changes—physical, technical, behavioral, and organizational
changes.

The preceding chapters have shown many different combinations of these three aspects of vulnerability. The Maya civilization
had learned how to build large, elevated, well-watered food gardens
in otherwise water-insecure environments. Among the many Maya
centers, those that had the most commodious underground water-
storage caverns experienced less actual water shortage during the late-
stage droughts than did other centers. While major civilizations in the
Americas were suffering from severe droughts and drying during the
ninth century C.E., the Norse Vikings were colonizing Iceland and
Greenland as the North Atlantic region warmed. In time, though,
their susceptibility to climate change would become apparent, largely
because their resilience as a settler society was compromised by reluc-
tance to adapt their habits of living, eating, and clothing to suit the
Greenland environment, the northern climate, and the available marine
food species. When long-term cooling arrived in the thirteenth and
fourteenth centuries, along with a sea-ice blockade and dwindling

connections with the Norwegian homeland, the Greenland settlements disappeared.

Urban and village communities were relatively powerless to take adaptive action to halt the ravages of the first (Justinian) and second (Black Death) ravages of bubonic plague. The rural village of Eyam in Derbyshire, England, on the urging of the village rector, altruistically shut itself off from the rest of the regional population, hoping to confine the plague that had already arrived there in 1665.[39] Over three-quarters of the Eyam population perished. The high death rate was probably a result of the villagers remaining in contact with *Yersinia pestis* bubonic plague bacteria transmitted by flea vectors from a local rodent reservoir. The story is told poignantly by Geraldine Brooks in her historical novel *Year of Wonders*.[40] It was not until the last decades of the nineteenth century that the cause and transmission of infection became generally understood. Previously, when epidemics were commonly viewed as the result of divine retribution, cosmological conjunctions, or malodorous miasmas, there was little a community could apparently do. Great bonfires were lit in cities and towns around Europe to purge the air of the (Black Death) Plague. More practically, the Venetians in the late fourteenth century realized that newly arriving merchant ships should be confined at anchor for a quarantine period, given that plague epidemics often occurred in Europe's ports soon after ships had docked and unloaded. This crucial adaptive response lessened Venice's vulnerability to future outbreaks of plague (and would have done so more effectively if the disease had not also established itself on land).

While the nature and processes of infection were not understood, a great deal was known everywhere about the growth, yields, handling, and distribution of food. During the eighteenth century, food crises declined in frequency in the world's more technologically advanced countries, though more rapidly in some than in others. Better roads, canals, markets, and storage facilities eased inequalities and geographic disadvantage. At the same time, social resilience was increasing on a broad policy front. For example, England (with social relief laws) coped better with climatic adversity and food shortages than did France (which still relied on local voluntarism).

One factor that long contributed to past vulnerability to climate-related health risks was a limited understanding of how nature's complex systems work, and why, therefore, they are prone to being perturbed, weakened, and inadvertently changed. Knowledge was based mostly on the day-to-day experiences of specific circumstances. Hence people did not necessarily expect a repeat of past adverse outcomes if and when the same environmental conditions recurred. The 1930s Dust Bowl episode in the American plains states is illustrative; it highlighted a frontier farming mentality that led people to ignore the known cyclical drought experiences of the previous two centuries and, instead, to blithely misuse and erode the land.

In some situations, cultural and social dependence on sophisticated technology and infrastructure restrict the options for adaptation. Low-lying flood-endangered coastal urban populations cannot pack up their city and relocate inland; nomadic pastoralists can, if the hinterland is not occupied by others. Among the Pacific Islands, relocation in response to rising seas is less feasible today than it was around 1300 C.E., when migrations occurred during a period of cooling and falling sea levels that reduced coastal food yields. Today's larger populations, fixed settlements, and lack of vacant land make migration more difficult. Many populations are increasingly vulnerable.

Climatologist Hubert Lamb, as noted above, recommended studying human historical experiences in *marginal areas* of settlement where vulnerability is high. We can learn a great deal from the experience of the Norse Vikings, constrained by cold, ice, and custom; from the Maya, who creatively found ways to feed themselves in difficult and water-insecure environments; and from the dire experiences of the Scots, Finns, Swedes, and others near the Arctic Circle when temperatures fell and crops failed during horror-filled decades in the Little Ice Age. We can also learn about risks to food security and human health from the experience of the nomad-pastoralists who lingered too long in the Sahara as it turned from green to brown six millennia ago; from the Harappan civilization in today's northern Pakistan when, 4,000 years ago, that region and adjoining parts of Central Asia began drying as the Southwest Asian summer monsoon began to drift south and inland river flows receded; and from the Sahelian fringe-dwellers (such as those in Mali, Chad, and Niger) in more recent times as monsoon systems shifted and rains failed.

For infectious disease outbreaks and the cascade of injuries, illnesses, mental disorders, and deaths that accompany warfare, looking to marginal areas is not particularly helpful for understanding vulnerability. Epidemic outbreaks over the past two millennia have occurred across a wide climatic and geographic landscape. More important is how sensitive the ecology, biology, and sociology of infection transmission are to a change in climate in the context of changes in the size and density of settlements, patterns of unhygienic urban living, undernutrition and weakened immunity, and the social disorder and poverty that have often followed famines.

In short, we would do better to include all types of populations in our synthesis, whether close to or far from the geographic margins. And so to the question about the vulnerability of today's populations to the increasing climatic stressors and resultant health risks during this coming century.

Implications for this century

A familiar but naive argument is that modern populations are different from their predecessors; they have knowledge, experience, superior technology, interconnectedness, and more flexible governance. Well, it is not that simple—partly because one-third of the world's population live in conditions more like the past than the modern, high-income, urban present. In many ways, higher-income societies *are* better placed to cope than were our predecessors, but in many other ways they are at a disadvantage.

Contemporary vulnerability can be assessed on the three criteria listed earlier: the external exposure, the intrinsic susceptibility of the population, and its adaptive capacity.

The extent of climate-related exposure

The two most important dimensions of climate exposure are magnitude and duration. The hotter, wetter, or drier a region becomes, the more intense the exposure, and the longer those climatic conditions last, the greater the pressure on affected populations. The severe heat wave in Europe in August 2003 showed how the physiological coping capacity of many, especially older, people was exceeded by heat overload due to both the extent of heating and its duration. Few in Europe had experienced such high temperatures, and authorities were unprepared. Most of the elderly who died had little support or means to respond to the increasing temperatures—younger members of families were away for summer holidays, and many medical professionals and carers were also on leave. Those living on higher levels of apartment buildings died because hot air rose and became trapped overnight. Heat disrupts sleep, amplifying the effects of heat exposure during the day.[41] High temperatures cause stress on the human body as it battles to regulate its core temperature at 37°C. The heart is forced to work harder as it pumps blood away from internal organs and toward the skin, and dehydration increases the risk of strokes or heart attack. More than 70,000 people died in Europe as a result of the heat wave.[42] The probability of more severe and frequent heat waves is increasing. The severity of the heat wave in Russia in 2010 that killed an estimated 55,000 people was the worst recorded anywhere in the world for at least the last 30 years.[43] Researchers suggest that climate change has made Europe's summer heat waves two to four times more likely.[44] Deaths from heat waves in modern and wealthy cities are a wake-up call.

The best-documented historical example of the impact of severe, long-term climate change comes from the seventeenth-century crisis in Europe.[45] Declining temperatures, morale, social cohesion, and food security had rendered much of Europe's population hungry, hopeless, and helpless. Violence, banditry, wanton destruction, and warfare escalated, dominated by the chaotic Thirty Years War of 1618–1648. The frequency of famine years doubled, people starved and adult heights declined, the rate of epidemic outbreaks tripled, the numbers of people displaced surged, and child death rates rose.[46] Not only was the level of climatic adversity unusually great, but this was a society that had lost much of its resilience, internal trust, moral guidelines, and authority structures; it was unusually susceptible.

Sustained climate changes have been seriously damaging, even in previously well-adapted cultures. The multicentury changes in climate that impinged on the Sumerians, the Harappans, the Classic Maya, and (less overtly) the Western Roman Empire significantly influenced the fate of those societies. Invariably, on this extended scale, the change in climate was part of a multifaceted process of environmental, social, and political decline. This is the *longue durée* influence of climate on which Fernand Braudel focused his assessment of the waxing and waning of historical influences on human societies and their affairs.

Changes in climate of several decades duration have often caused suffering and increased death rates; just as important, for the purposes of this book, are the subsequent periods of recovery. Examples include the occasions when sustained falls in the annual Nile flow caused Egypt extremes of hunger and hardship, followed by top-to-bottom social and political reorganization, and the Great Famine in Europe in the early fourteenth century.

The inherent susceptibility of the exposed population

A population's intrinsic susceptibility has biological, infrastructural, and sociocultural aspects. Many of today's populations have metabolic profiles that are, in various ways, less robust than in the past. This is largely due to changes in ways of living associated with the advent of wealth, urbanism, consumerism, an "industrial" food system, cheap fossil energy, reduced physical activity, and an aging population.[47],[48] Average blood pressures are higher, arteries silt up, diabetes is more common, and levels of blood cortisol (the main stress hormone) are chronically raised. An example of inherent susceptibility applies during major heat waves when the elderly, those with higher blood pressure and underlying heart disease, and the seriously overweight are much more likely to die or to require hospital treatment.

Earlier populations had simpler, nonurbanized living patterns and greater reliance on natural (though often micronutrient-deficient) foods, and their lives were shorter anyway. Modernizing populations are therefore, on average, more biologically susceptible to many climate-related stresses. The accelerating rise in many underlying health disorders in lower-income countries, as Westernized economic development and urbanization occur,[49] foreshadows further increases in heat-related death from heart attacks, strokes, and the complications of diabetes in a warmer world.

The *infrastructural* profile also influences susceptibility. When all is going well, infrastructure is supportive and enabling, but in times of overload and damage it often becomes a source of inflexibility and crisis. The OECD has warned that large modern electronically hyperintegrated cities have an increased vulnerability to many external stressors, as was evident when Superstorm Sandy lashed New York and crucial support systems failed.[50] Multisystem connectivity makes for a greater vulnerability of the community.[51] Modernized urban living now depends on a complex infrastructure based on extensive automation and electronic connectivity—time-saving in the good times (if you can remember log-in passwords), but very vulnerable to disruption. The national electricity grid has been described as "the glass jaw of American industry." If a catastrophic weather disaster or a combination of several disasters disabled the grid, much of the American continent could be crippled for weeks—lacking supplies of fresh food, water, and automotive fuels; without internet communications; and with limited basic services such as hospital and nursing-home care. Further, the hyperconnectivity of the modern world means that while information flow and the possibility of mutual assistance is enhanced, that interdependence among populations and countries precludes being insulated from distant shocks. The global financial crisis in the early years of this century underscored the modern world's vulnerability to a "one down, all down" process.

Heightened extremes of heat will disrupt many aspects of infrastructure. In industrialized and urbanized societies, the risk foci are many; they include overloaded electricity grids and hence power failures, the buckling of roads and rails (perhaps obstructing emergency travel), and outbreaks of wildfires in the urban fringes. During the extreme heat wave of summer 2009 in Adelaide, South Australia, several hundred train services were canceled because of buckled rails, while emergency refrigeration morgue facilities were hired to cope with the acute excess of heat-related deaths. For the first 98 percent of the Holocene, there were no risks to sealed roads, rails, and refrigeration.

In the Arctic region, much infrastructure is now at risk from the background warming that is proceeding there twice as fast as in the world at large. Northern Norway has warmed by 2°C in the past 40 years. Permafrost and tundra (literally the *infra*structure) are thawing, building foundations have been disturbed, and electricity transmission pylons and poles have tilted or fallen. Meanwhile, herds of reindeer, precious sources of protein, hides, and livelihoods, are more likely to drown while crossing over lakes with thin ice cover, while traditional hunters (and recreational hikers and campers) may be immobilized when the normally solid permafrost terrain underfoot turns to mud.

International food trade is now part of the global economy and, in theory, a bulwark against food insecurity. The importation of food is now widely taken for granted, especially by those who produce far less than they consume, including Singapore, the Netherlands, and the Middle East oil states. Yet this is an obvious focus of vulnerability, illustrated by the cascade of effects following extraordinary prolonged summer heat in western Russia in August 2010 that led to soaring world food prices and abrupt reduction in wheat supplies to major client countries.

Often overlooked in this discussion is the fact that climate change is not the only contemporary threat to future food trade and security. Population numbers are still increasing, food-producing systems on land and sea are under stress from soil degradation and loss, biodiversity losses are occurring on a global scale, there are water shortages on many continents, and our oceans are over-fished. Efficiency-driven culling of the ancient diversity of strains and types of many plant foods (such as potatoes, wheat, and bananas) has lowered the resilience of agricultural systems and reduced options for the future. Finally, a population's susceptibility reflects its level of *social and cultural resilience*. Societies with coherent and stable structure, mutual trust, shared and well-applied knowledge, efficient information dissemination, good government, and community-level capacities are better placed to cope with climatic threats or crises. Societies such as that of North Korea—authoritarian and inflexible, with an uninformed (and bizarre) leadership and a wretchedly poor and cowed population—are highly vulnerable. Research by our university team during 2010–2012, with local collaborators in Cambodia, Vietnam, and Fiji, assessed the types of rural community processes and governance arrangements best able to respond to the threats from climate change to water security. Community-based governance, open decision-making, and well-informed risk-management strategies were evidently important to maintaining community cohesion, resilience, and effective adaptive action.[52]

Susceptibility is also affected by the increasing physical connectivity between populations, and in this sense the world is becoming more contagion-enabled. Several millennia ago, novel infectious diseases were mostly confined within regions. Then came a long period of exchange and equilibration as physical connections via trade and military campaigns intensified between city-states and, later, between civilizations across Eurasia and then across the oceans.[53] If human-driven climate change alters wind patterns and surface-water locations, the migration paths of various wild birds will change—and so too the patterns of spread of avian influenza, West Nile virus, and tick-borne encephalitis may then alter. In a world with continuing increases in livestock numbers, density of poultry production, and nearby urban populations, climate-related influenza epidemics could occur more readily.

A closing comment on vulnerability is that much of the world's living space is now effectively full, with little vacant hinterland available for relocation. Increases in population displacement are likely to exacerbate geopolitical instability and intercommunity tension, in addition to the relocation itself being a cause of poor nutritional, physical, and mental health in many of the displaced groups.[54]

The capacity to adapt

Finally, the third leg of the vulnerability stool: the extent to which a community or population can take *adaptive action* to lessen climate change–related risks to health. In the wake of a quarter century of procrastination and befuddlement by national governments, a surge of research and assessment into the options, funding, and implementation of adaptive strategies is under way. Having deferred effective mitigation (abatement) of human-induced climate change, we now have no alternative but to rely on adaptation strategies to a greater extent than would otherwise have been necessary. There are, however, limits to what can be done to stave off the escalating and multiplying risks as climatic conditions pass thresholds and patterns of damage and system disruption become more widespread, weakening social institutions and effective governance. Heroic technological responses may be imaginable, but before long they could become inadequate or simply not feasible.

Countries and populations differ greatly in their capacity to adapt to climate change. Part of the international moral dilemma is that the poorer populations of the world, who have contributed relatively little to the current excess load of greenhouse gases in the atmosphere, will generally be at greatest risk by dint of where they are and what resources they have. Although poorer and geographically vulnerable communities may have good adaptation ideas, they commonly lack the necessary resources for major projects. Even so, most of these societies and cultures have traditional, though often dormant, adaptive capacities—such as mixed cropping, hillside terracing, and nutrient recycling—that have been sidelined by modern economic development orthodoxies and international funding programs. These, if reactivated, may well help the rich world learn a thing or two about sustainable strategies, if not about specific methods, in relation to increasing the resilience of agricultural practice and some types of water management.

The Pacific region provides a good example of how economic and technological change can affect adaptive capacity. During the past twelve centuries, Pacific Island communities have been vulnerable to the vagaries of climate change, particularly droughts and, during the prolonged cooling trend around 1300 C.E., the fall in sea level. Seven centuries ago, the combination of droughts and sea-level fall seriously disrupted some Pacific societies. However, the greater basic physical vulnerability in the past was often matched by high resilience, flexibility, and coping capacity that enabled those societies to recover and reorganize quite quickly.[55] The arrival of Europeans in recent centuries has had mixed effects on coping capacity across the Pacific. Long-distance migration, remittances from abroad, and information and expertise flowing through international connections have all been positive forces in many respects. On the other hand, the imposition of alien systems of resource management and negative aspects of economic and cultural globalization have reduced social cohesion, cultural identity, and overall resilience.[56] Pacific Island populations are part of the increasingly homogeneous global family, characterized by fixed settlements, high-density populations, a declining natural resource base, and import dependence (including reliance on health-damaging refined and processed foods).

To recap: the vulnerability of a human population reflects a complex mix of circumstances and characteristics. Vulnerabilities to adverse health impacts and to social disruption in this century will change as exposures to increased levels of unfamiliar and adverse climatic conditions increase. Demographic, economic, political, infrastructural, and other changes will modulate the impact of future climate regimes, sometimes lessening it, sometimes amplifying it. We are heading into unknown territory.

Concluding remarks

The rear view mirror of history offers a view of the world's climate as two-faced: sometimes friendly, sometimes fearsome. Both sustained and abrupt changes in climate have left their mark on food production, epidemic outbreaks, and conflict, leading in each case to deaths, injuries, and a trail of disease. Those events, often dramatic, readily find their way into the historical record. In contrast, before the mid-twentieth century we know little about the health impacts of heat waves, the mental health impacts of climatic adversity, and lower-profile infections such as dysentery. Nor, of course, are there explicit event-based records of the *benefits* to health, fertility, and longevity during periods of benign climate conditions. Those must be inferred from skeletal remains and records of births and deaths, where available.

The *extent* of a change in climate is very often as important as the actual *direction* of the climatic change. Human cultures, their built environments, and, most important, the natural resource base on which they depend have been shaped over several thousand years to the Holocene climate that prevails in that region. Both significant cooling and warming, acting via biological, ecological, and social impacts, may endanger food yields, contribute to infectious disease outbreaks and spread, affect water quality and availability, and exacerbate social disruption, impoverishment, and displacement.

The greatest recurring health risk has been from food shortages. Societies everywhere have faced periodic hunger, undernutrition, deaths, and social unrest. The historical evidence of climatic influences on infectious disease epidemics is less strong than for hunger and undernutrition, and it spans a shorter period of calendar time, from around 4,000 years ago. Even so, in a future warmer world the range, rates, and seasonal duration of many infectious diseases are likely to increase as bacteria and vector organisms (mosquitoes, fleas, etc.) multiply faster—up to a temperature limit that threatens their survival.

Some will argue that humans have coped with past climate changes and can therefore do likewise in the future. Two points should be understood. First, historically, many populations and whole societies did *not* cope well; much misery, impoverishment, illness, and early death resulted, and some societies went into hastened decline. Second, the future climate and environment will not be a simple variation on the Holocene experience. Given our current trajectories, within a century or so this planet is likely to have a distinctly different climate, landscape, and biotic profile from the world that we and our *Homo sapiens* ancestors have known.

So how do our prospects stand? What do we need to think about and do differently? The final chapter examines where trends are now pointing, the awkward context within which these challenges must be addressed, and the prospect of a sustainability transition—assuming we can renegotiate the now-menacing dénouement of the Faustian bargain.

II

Facing the Future

Trend is not destiny.

—RENÉ DUBOS[1]

MANY CIVILIZATIONS HAVE COME and gone over the past 6,000 years; some declined rapidly, some lingered, and a few renewed and rebuilt. These rise-and-fall cycles have been variously attributed to the typical increase in complexity of a society over time as procedural solutions to successive layers of problems accumulate and eventually stifle purpose and productivity,[2,3] or to heightened social stratification, inequality, and consequent uprisings.[4] But beyond the city walls are other explanations. Many societies have overexploited and degraded their natural environmental base; in other cases, natural changes in regional climates and environments have impaired harvests, caused water shortages, mobilized epidemics, or fomented political disorder.

In the twenty-first century, populations around the world face unprecedented but broadly foreseeable changes in climate on a global scale, with impacts compounded by other environmental and demographic pressures. We cannot predict the consequences for human populations, but they may be dire—especially if *runaway* climate change occurs. The modest warming that has occurred since the mid-1970s, associated with increased severity of weather disasters, is already affecting human health and safety, via heat waves and other extreme weather events, physical injury, child undernutrition, changes in infectious disease ranges and seasonality, mental trauma and depression, and population displacement and lost livelihoods.

Can we find another, safer way forward? Our elaborate primate brain with its unique higher-cognition planning capacity enables us, when pushed, to imagine alternative futures and to behave flexibly and seek transformative changes.[5] But other human foibles and frailties intervene. These include the widespread assumption of unlimited economic growth, an instinct to retain current social and cultural structures, and the limitations of rapid-turnover democratic government. Structural impediments also persist: the continuing poverty and illiteracy of several billion people; the heterogeneity of cultures, beliefs, and political systems; and modes of scientific research not yet well suited to studying complex environmental and social systems. These make the task ahead more complex, but not impossible.

To make headway will depend on people and communities understanding climate change in terms closer to home. Talk of emissions, trajectories, scenarios, ocean acidification, targets, and timetables does not connect with daily lives. People actually experience debilitating episodes of heat, they worry about their children's physical safety, they fret about parched crops or gardens, and they wonder if infectious microbes might multiply in the local water supply. But climate change, in the abstract, does not automatically command attention and trigger alarm, it does not activate our inbuilt "fight or flight" response, and you cannot see it through the bedroom window. In short, climate change does not register on the personal "risk thermostat."[6] We naturally respond to external threats that are immediate and visible, have an obvious cause, and will affect us directly. But climate change is invisible, long-term, and unfamiliar, and its causation is both complex and unsettling—*we* are the cause.

So how to proceed? Via education and informed public discussion, human-induced climate change must be understood as an unintended consequence of Western (and increasingly global) civilization's deep-set belief that continued material growth is both normal and inherently good. And yet it is humankind's energy- and environmentally-intensive economic activity that now jeopardizes the foundations of future human well-being, health, and survival. The word "survival" sounds dire. Yet some foresee a decline in human life expectancy and social order within the next century or two, or even a widespread collapse of civilization.[7,8,9,10] Martin Rees, former president of Britain's prestigious Royal Society, has written that he thinks the odds "are no better than fifty-fifty that our present civilization on Earth will survive to the end of the present century."[11] These are not idle speculations.

Awareness of the fundamental risks to human health from climate change will exert leverage on decisions about taking effective action. For most people, ensuring good health and survival has higher priority than protecting property, maintaining local economic growth, or sustaining tourism. In 2013, the United States Congress debated a Climate Change Health Protection Act, seeking a coordinated national action plan. Introducing the bill, Congresswoman Lois Capps stated: "One of the most troubling and immediate impacts of climate change is its harmful effects on public health." A greater shared recognition of the human face of climate change will infuse new energy into the case for radical and urgent international action.

Exponential impossibilities

Human life should grow, not quantitatively through the conquest of nature, but qualitatively in co-operation with nature.

—RENÉ DUBOS[12]

The idea that progress equates to material growth is deeply embedded in post-Renaissance Western culture and in the now-dominant economic system, market capitalism.[13] The increasing connectivity and interdependence among countries has disseminated this assumption globally. But we also see more clearly the extent of the negative externalities, the environmental, ecological, and social damages, that market-driven extraction, production, consumption, and waste disposal cause.

Yet growth remains alluring, and all but the world's poorest enjoy its immediate benefits, comforts, and convenience. The economic orthodoxy is that a dynamic and growing economy based on market capitalism generates profits and new wealth; otherwise, investors would be reluctant to inject more capital. In basic terms, increased production needs more workers, and increased consumption needs more consumers. Henry Ford astutely paid his production-line producers enough for them to also become consumers.

The assumption of continuing economic growth has been described by historian John McNeill as the ideological sheet anchor of the twentieth century for both capitalism and communism.[14] "Economic thought did not adjust to the changed conditions it [growth] helped to create," he argues, and so mainstream economics continued to legitimate and thus, in effect, cause great and mostly damaging ecological change.[15] Growth, however, need not be defined by crude physical measures of total production; it can be measured by an increase in the actual *value* (including quality, utility, and durability) of goods and services produced, or material well-being achieved, for each unit input of labor or capital. This can be achieved by making technological improvements that change how we do things and so increase the value created per unit input of labor and capital.

The more radical idea of a no-growth Steady State economy has been mooted by some economists.[16] They argue not for a socially static self-denying economy, but instead a cultural shift from excessive, energy-intensive material consumption to a value-adding economy based on recycling, a low-carbon footprint, and a broader view of human needs and opportunities.

Herman Daly compares the challenge we face today versus that of the past: "The [erstwhile] *empty* world assumption guarantees that the newly expanded production will always be worth more than the natural wealth it displaces. But what may well have been true in yesterday's empty world is no longer true in today's *full* world. This is an upsetting prospect for growth economists—growth is required for full employment, but growth now makes us collectively poorer."[17]

This no-growth idea has had some illustrious prophets and proponents. The pioneering economist Adam Smith in the 1790s foresaw the eventual coming of a "stationary state" economy.[18] Half a century later, John Stuart Mill wrote:

> *The increase of wealth is not boundless... At the end of what they term the progressive state lies the stationary state ... a very considerable improvement on our present condition... A stationary condition of capital and population implies no stationary state of human improvement.*[19]

Note that last sentence. Curbing physical and material growth and environmental depletion does not mean a future of joyless hair-shirt living. Rather, the change should create space for more fulfilling and equally shared activities, and for material comforts produced within a low-throughput recycling economy.[20] Continued exponential material growth is simply not logically possible; it must eventually result in a collision between the current politically irresistible force of economic growth and the immovable reality of a finite Planet Earth.

When, in 2007–2008, the interconnected global economy imploded, the global financial crisis became the overwhelming preoccupation of national governments[21]—to the neglect of the greater looming environmental crises. We witnessed attempts to reactivate the twentieth-century economic growth model, but "in a 21st century world subject to fundamentally different biophysical constraints."[22]

The demon lurking beneath conventional market accounting is the "externalizing" of off-market costs, which ignores the huge, often longer-term, economic costs from environmentally damaging impacts, such as greenhouse gas accumulation, ocean acidification, and species extinctions.[23] Such specious accounting may suit an imaginary bubble society that dispatches its wastes into space, but not real-world societies that depend on the continuing viability and productivity of the wider environment of which they are part. Yet that model largely persists.

Reframing "progress" as if humans mattered

We are facing both an opportunity and a duty to rethink what progress really means and to build stronger and more inclusive visions for the future.

—ÁNGEL GURRIA, Secretary-General, OECD, 2009[24]

The global economic crisis has focused minds on restoring growth. But does growth necessarily mean progress? What about factors which growth depends on, such as the environment or happiness?

—JOSEPH STIGLITZ, *OECD Observer*, 2009[25]

For most national governments and their constituencies, the concept of gross domestic product (GDP) remains on the "progress" pedestal; this measure of average per-person income and wealth is assumed to tell all. But the pedestal is becoming precarious. This one-dimensional measure, developed during the 1930s Great Depression when it was important to stimulate economic activity, misses something important. Research on the sources of happiness and life satisfaction tells us that the personal reward from ever-escalating, often compulsive, consumption is a mirage.[26] Our most basic psychological needs, as social beings, include a sense of personal autonomy, good interpersonal relationships, self-acceptance, and opportunities for personal fulfilment. Perversely, a culture that extolls the endless pursuit of money and possessions impedes fulfilment of these personal and social needs.

Even as a measure of national economic performance the GDP is defective, since it is the sum total of *all* market transactions, whether they enhance or reduce welfare. The more frequent the need to remove industrially contaminated topsoil, the greater the GDP. Worse, many monetary transactions, such as marketing logged forest timber, take no account of the extended real costs to the environment, local communities, and future generations. Conventional market economics assigns such off-market costs to some imaginary noneconomic realm, present or future. And so externalities, including the costs of climate change such as reduced life expectancy and greater healthcare costs, are conveniently left off the balance sheet.

A more meaningful assessment is a broader people-based measure of inclusive wealth and social progress.[27] This idea has been expounded by economist Partha Dasgupta, who argues the need for a measure of the true social worth of a society's capital assets—not only manufactured and infrastructural assets, but also the stocks of human, knowledge, and natural and environmental capital.[28] This idea now resonates widely.[29,30] "If more of us valued food and cheer and song above hoarded gold, it would be a merrier world," as J. R. R. Tolkien's hero Bilbo Baggins lamented.

There is a long-running debate about what constitutes "the good society." Aristotle and Buddha claimed progress was above all a matter of *balance*, including ethical and moral qualities. Today, measures of "genuine progress" include the Inclusive Wealth Index, the UN's Human Development Index, and the Genuine Progress Indicator.[31] The Inclusive Wealth Index measures the full value of a country's capital assets beyond financial and manufactured capital, and includes natural and environmental, social, and human capital assets, as proposed by Dasgupta.[32] The Human Development Index combines measures of life expectancy, literacy, educational attainment, and per-person income—but not environmental conditions. The Genuine Progress Indicator measures the actual *economic welfare* generated by economic activity, not just total economic activity as in the GDP.[33] This difference is highlighted by the fact that, while the globally averaged GDP has tripled since 1950, the actual economic *welfare* estimated by the Genuine Progress Indicator peaked in the mid-1970s.[34]

The worldwide diffusion of consumerism, driven by rising wealth in many lower-income populations and by commercial contrivance, is a major contributor to humankind's global environmental footprint by dint of the energy embedded in the relentless production of "goods" such as new-model mobile phones and coffee-making machines. But if awareness of escalating environmental damage reorients consumer preferences, industry, perhaps grudgingly, will follow suit in the pursuit of profitability.[35]

Skepticism, doubt, and denial: Disputing the messenger

There are none so blind as those who will not see.

—OLD QUOTE, based on Jeremiah 5:21

The evidence assembled by the United States government to justify the Iraq War in 2003 was speculative and fabricated. But President George W. Bush and his advisors were determined to believe it. Conversely, the scientific evidence on human-driven climate change is orders of magnitude more thorough, coherent, and corroborated than the spurious "weapons of mass destruction" claim. But with climate change, many people and politicians want *not* to believe it. The prospect is unsettling, a challenge to the economic status quo and to its ideological base, and a political high-risk zone. Denial also avoids personal conflict. As John F. Kennedy noted, belief in myths allows "the comfort of opinion without the discomfort of thought."[36]

But there are deeper layers of difficulty in seeing and accepting. Biological evolution, favoring immediate survival-enhancing action, has made little provision for responding to nonurgent and gradual changes. Harvard psychologist Daniel Gilbert cautions us that although climate change may threaten our futures, it does not threaten this afternoon's plans:

> *The human brain is exquisitely sensitive to changes in light, sound, temperature, pressure . . . and just about everything else. But if the rate of change is slow enough, the change will go undetected . . . Because we barely notice changes that happen gradually, we accept gradual changes that we would reject if they happened abruptly.*[37]

Much public discussion of climate change has been confused and adversarial, fraught by misunderstandings, genuine difficulty in accepting unusual and unsettling science, and deliberate doubt-fostering denialism.[38] Doubt entails uncertainty; denial entails rejection.[39]

There are, of course, huge financial, ideological, and political vested interests in the economic status quo—a global economy in which later-stage market capitalism, operating under loosened regulatory controls, is rapidly increasing the world's wealth divide.[40] Those who are deeply immersed in the economic machine cannot comfortably come to terms with the implications of climate change.[41] The fostering of doubt is, regrettably, made easier by the professional caution of most scientists, who avoid the public spotlight lest they politicize their science. Within the wider community, there is a widespread form of "soft denial." Who wants to be the bearer of bad news, the person who sounds dogmatic despite the complexities of the science, the person who dampens coffee break chatter? Even so, many people can see that climate change is a big and urgent deal for government, community, and planet. In the words of Australian singer-songwriter Emma Tonkin:[42]

> *Deep down we know . . .*
> *The waiting storm . . .*
> *We are beginning to fray . . .*
> *The silence is gone . . .*
> *Deep down we know*

As understanding of climate change and its adverse impacts grows, priorities and values will change.[43] Max Planck, the eminent atomic physicist, wrote: "A new scientific truth does not triumph by convincing its opponents and making them see the light, but rather because its opponents eventually die, and a new generation grows up that is familiar with it."[44] Climate change will not pause, however, to wait for generational change. In Barbara Kingsolver's *Flight Behavior*, set in a conservative Appalachian community, the central figure, Dellarobia Turnbow, says to a visiting ecologist with expert knowledge about human-driven climate change: "I'm not saying I don't believe you. I'm saying I can't."[45] Her words capture in microcosm this inability to see the world differently; the idea of "climate change" simply does not fit the worldview of her community.

Climate change as heresy

Climate change may constitute the fourth great scientific heresy of the past millennium. The Copernican, Darwinian, and Freudian "heresies," in the sixteenth, nineteenth, and early twentieth centuries respectively, each confronted prevailing assumptions about the centrality of humans in the grand scheme of things. Copernicus and Darwin hesitated to publish their theories lest the punitive might of religious orthodoxy be provoked. Their discoveries relegated humankind, long considered the pinnacle of God's creation, to living on a minor planet circling around a small sun, with humans merely a recent bud on one branch of an ever-evolving tree of life. Freud argued that humans were not agents of free will and rational judgment, but functioned at the behest of a tripartite brain dominated by subconscious inhibitions, guilt, and parental influence.

Climate change confronts us differently. It does not gainsay the physical pre-eminence of humans on this planet but highlights the dire consequences of their overwhelming domination of the natural world. The three earlier heresies revealed the inconvenient truth that we humans are *less* central to the cosmos, less permanent and less autonomous than had long been assumed, but the climate change story tells us that, actually, we have become hugely *more* powerful than our predecessors could have imagined. For many, the idea that humans are now so powerful, that we are disrupting the Earth system itself, is a threat to fundamentalist religious views, quasi-scientific theories, and material vested interests.[46,47]

During a debate on climate change policy in the United States Congress in 2010, several Republican senators invoked God's covenant with Noah after the Flood that such a catastrophe would not be repeated: "I will not again curse the ground any more for man's sake ... While the earth remaineth, seedtime and harvest, and cold and heat, and summer and winter, and day and night, shall not cease... This [the rainbow] is the token of the covenant which I make between me and you and every living creature that is with you, for perpetual generations."[48] That divine covenant, the senators argued, precludes climate change forever. Amusing at one level, this senatorial absurdity alarms at another. It is yet another dismal example of the dismissal of scientific rationality.

Other impediments to action: Poverty, governance, and modes of research

By 2013, the world's population had already spent half the "safe" global budget of human-generated carbon emissions for 2000–2050—around 500 billion metric tons of carbon dioxide out of the upper limit of 1000 billion metric tons necessary to prevent warming above 2°C. The likelihood of our staying within that budget is shrinking each year. If a serious reduction in international emissions does not begin until 2020, an estimated actual reduction (not just slowing) of around 10 percent *every year* will then be needed through to midcentury.[49]

Any such fast-track approach to a climatically stable future would face impediments. Humankind remains beset by widespread persistent poverty and disadvantage, outdated governance structures, shortsighted social goals, and a mode of science not well attuned to solving "wicked" problems: that is, complex, world-scale, environmental, population, and social problems.

Poverty and the development agenda

Worldwide, the numbers of people living in extreme poverty is declining, but even so the rich-poor income gap has widened in most countries. In Pope Francis's words, "While the earnings of a minority are growing exponentially, so too is the gap separating the majority from the prosperity enjoyed by the happy few." This imbalance, he said, "is the result of ideologies that defend the absolute autonomy of the marketplace and financial speculation."[50] In 2013, a mere 1 percent of the world's families owned almost half of the world's total wealth.[51] The statistics of disparity are multiple and inevitable under market capitalism, according to Thomas Piketty.[52] Meanwhile, in poor, high-fertility countries such as Nigeria, Timor Leste, northern India, and Pakistan, the number of people living in urban slums and shantytowns is increasing. Poor and vulnerable communities are at the greatest overall risk to health and physical safety from a change in climate and extremes of weather.[53]

Further, poverty itself is both a consequence and a cause of climate change. Poverty is heightened, for example, by extreme weather events that damage local crops, while changes in temperature and rainfall, especially in poorly irrigated regions, reduce yields, undermine livelihoods, often create tensions, and erode community cohesion and facilities such as microcredit. Poverty also contributes to climate change. Tree-felling to generate farmland or firewood increases local environmental drying, soil carbon loss, and CO_2 emissions from burning vegetation. Household reliance on low-grade biomass fuels in several hundred million poor unventilated dwellings not only poisons the indoor air but releases methane, climate-active black carbon particulates, and carbon dioxide into the air.[54] Squeezing an extra hunger-relieving crop into the annual family-farming cycle gradually exhausts the soil and its capacity to absorb carbon.

A misplaced view persists in some quarters that attending to the climate change agenda will divert funds from economic development in low-income countries. This "it's either one or the other" argument was invoked by ExxonMobil's CEO at the 2013 annual shareholder meeting when, according to media reports, he commented dismissively: "What good is it to save the planet if humanity suffers [from reduced oil supply]?"[55] The counter view is that an environmentally low-impact way of living need not compromise the needs of economic and social development.[56] And better accounting of all the off-market costs from collateral environmental damage makes it plain that poverty reduction, socioeconomic development, and environmental sustainability do in fact lie on the same path.[57,58] As environmental scientist John Williams says: "We cannot continue to foster one story that assumes an infinite planet and is framed around the paramount need for economic growth while maintaining the other story around the paramount need to protect an increasingly fragile natural world. [We need a] new story ... to empower a transition to a society that lives within the means of a finite planet and improves global wellbeing at the same time."[59,60]

Renovated governance

Our response so far is just like that before the Second World War, an attempt to appease. The Kyoto agreement was uncannily like that of Munich, with politicians out to show that they do respond but in reality are playing for time.

—JAMES LOVELOCK[61]

For several thousand years, government has been mostly about local food security, urban infrastructure, defense (or attack), taxing, and the peace and security of the populace. Climate-related food shortages, hunger, and starvation have played critical roles in eroding political authority, as in Mesopotamia, the Classic Maya culture, dynastic China, and others. Over the past century, industrializing societies have sought to constrain *local* environmental problems, particularly the toxicological and microbial hazards from industrial effluent, unhygienic conditions, and crowding. But this has not equipped modern societies to deal with disruption of the Earth system itself—a disruption that must be tackled on a broad front, within the context of other momentous global changes. These include increases in population size and human mobility, the West-to-East shift in the economic and consumer center of gravity, and the recent growth of private multinational corporate power. These are issues beyond the control of individual nation-states.

In this second decade of the twenty-first century, there is a manifest lack of clear global leadership at a time when an unprecedented internationally coordinated stewardship of the Earth system is needed.[62] The current patchwork quilt of independent national governments, specialized international agencies, and commercial and civil organizations—all differing in character, constituency, geographic scope, and topic focus—is ill equipped to act coherently to redress these great and damaging global environmental changes.[63,64]

Intergovernmental cooperation is hampered by the embedded legacy of Europe's Treaty of Westphalia in 1648, agreed to after the chaotic Thirty Years War. Westphalia guaranteed independent and equal status for sovereign countries, an undertaking that, for all its benefits, has nurtured an enduring ethos of narrow national self-interest.[65] In this worldview, climate change is primarily framed in terms of *state security*; it is therefore constructed as a threat to national economies, sovereignty, and borders. It would be much better if it were understood primarily in *human security* terms—as a fundamental threat to human well-being, health, safety, poverty alleviation, and conflict avoidance.[66]

The UN system, forged during the first half of the twentieth century, despite its collectivist charter, is poorly equipped to tackle this century's great task of restabilizing the Earth system and assisting the worldwide reframing of social and cultural objectives.[67] Beyond the systemic problem of sectoral fragmentation and territoriality, UN decision-making is constrained by its requirement for consensus. It treads a difficult line between recognizing the self-interest of member states and making policy that is in the collective good. Climate change, biodiversity losses, water sharing, oceans management, and fisheries protection are all, in principle, supranational "global commons" issues. In practice, they often remain at the mercy of short-sighted national self-interest.

At the time of editing this book, world leaders met for the 2015 United Nations Climate Change Conference (COP 21), held in Paris over two weeks from November to December 2015. Many participants, media commentators, and scientists hailed the "historic" meeting for setting a new path for the future. Of the 195 delegate nations, 186 agreed to reduce or limit increases in greenhouse gas emissions. They set a general goal to limit warming to 2 °C above preindustrial levels, and agreed they should aim for 1.5 °C . It was a strong signal that future economies should be low-carbon.

The spirit of the agreement was positive, but the pathways by which reductions in emissions would be made were not outlined. Christiana Figueres of the UN Framework Convention on Climate Change warned that even if the voluntary pledges were fully implemented, the world would still be on track for close to 3 °C warming by 2100.[68] This is an important distinction when a degree or so could be the difference between retaining much of the Greenland ice sheet or losing it.[69] Some climate models have shown that it is possible to keep to warming of 2 °C if there is rapid and substantial investment in renewable energy technologies such as solar, wind, and geothermal. Other models require massive increases in bioenergy and carbon capture and storage, both of which are fledgling technologies which have not been successful on a large scale.[70]

While the Paris meeting made it clear that governments and investors should plan for the shift to a decarbonized world economy, it was also apparent that governments are preparing for a future marked by conflict, forced migration, and economic damage in vulnerable areas. The president of the European Commission, Jean-Claude Juncker, warned the conference that climate change could "destabilize entire regions and start massive forced migrations and conflicts over natural resources."[71] Staff working for the UN High Commissioner for Refugees said, "Climate-related displacement is not a future phenomenon . . . it is a reality; it is already a global concern."[72] The Paris agreement acknowledged developing nations would suffer "loss and damage" but made provisions to bar any possibility that wealthy nations or corporations could be held liable. Little was said about the threats climate change poses to human population health.

While international entities and national governments delay, civil society and popular movements push for action and accountability on climate change. The People's Climate March, staged in New York in September 2014, drew over 300,000 participants, making it one of the largest single climate protests in history. In the lead-up to the 2015 United Nations Climate Change Conference in Paris (COP 21), more than 600,000 people marched in rallies across the globe, urging world leaders to make meaningful commitments to addressing climate change. Local social movements opposed to intensive fossil fuel development have emerged in many countries. Opposition to the Keystone XL pipeline, which would have delivered tar-sands oil from Canada to the Gulf of Mexico, became the "Blockadia" movement. Like Australia's rural-based "Lock the Gate" protests against new development of coal seam gas and coal mining on prime agricultural land, it broke all the stereotypes of environmental demonstration: it began in conservative inland states, with farmers and ranchers joining urban environmentalists and North American indigenous groups to fight the construction of the pipeline. Similar movements have formed in Southeast Asia, Bangladesh, India, and elsewhere.[73]

Other actors are looking for ways to mitigate and adapt to climate change. Enlightened private-sector organizations eager to stave off foreseeable adverse climatic impacts on their industry (e.g., insurers) bring useful professional skills, management experience, and worldly wisdom.[74] The cultural sector—design, arts, architecture, museums, and humanistic scholarship—is making important contributions to ideas for living in a changed world of the future. If we are leaving the Holocene behind us—the epoch in which our cultural systems and civilizations evolved—how do we make sense of this change, and what does it mean to live in the Anthropocene? These questions move debate and understanding of future possibilities beyond just a set of economic or technological choices to matters of justice and equity.[75] The Haus der Kulturen der Welt (HKW), an art and ethnographic museum in Berlin, hosted the scientific meeting of the International Commission on Stratigraphy in October 2014. The museum also runs an "Anthropocene Curriculum" bringing together scientists, practicing artists, designers, museum specialists, and humanities scholars. The HKW recognizes that the Anthropocene demands debate that is both cultural and scientific. China is home to the world's first Museum of Climate Change, an outreach initiative launched in 2012 by the Chinese University of Hong Kong, with support from philanthropic donors. The Deutsches Museum in Munich, the world's largest museum of technology, launched a multimillion-dollar exhibition in 2014, *Willkommen im Anthropozän* (Welcome to the Anthropocene: The Earth in Our Hands), which will travel around the world, and the National Museum of Australia in Canberra has plans to collaborate with the Deutsches Museum to develop an exhibition on understanding Australia's place in the Anthropocene.

At the local level, cities and towns offer opportunities for a smaller scale of activity incorporating the processes of "deliberative democracy"—an important aspect of the Transition Town movement.[76] For example, building and housing standards, public transport, road design, green space, recreational facilities, water conservation, and urban food production are related to abating climate change, and are well suited to city-level discussion and decisions. More creatively, so are urban agriculture and urban food foraging.[77] The digital revolution enables cities to share knowledge and management information on resource flows, waste outputs, greenhouse emissions, and their social and environmental impacts.[78]

In Melbourne, Australia, the city council has recently planted many thousands of trees, mainly to lessen the urban heat island effect. Large tracts of public parks are being excavated for high-volume underground storage of urban storm water to help offset future droughts and water shortages. Local governments everywhere have a role to play in monitoring vulnerability to future extremes of temperatures, rainfall, storms, and rising seas, and seeking climate-proof land-use planning decisions.[79] However, retrofitting cities to adapt to a twenty-first-century climate and its many impacts will not be easy. Flood-prone New Orleans is hostage today to ill-informed decisions made a hundred years ago to build huge riverbank levees, now seen as high-maintenance structures that obstruct better solutions.

Modes of research

Our current modes of scientific research have strong historical roots in the "classical" experimental method that disaggregates the real world into its parts, enabling each to be studied in isolation. But weighed against the complexity of today's global-scale problems, such as human-driven climate change and its likely impacts, mainstream science is not well prepared to engage in systems-based thinking and analysis. That deficit is heightened by the compartmentalized structure of the research realm. Different disciplines, while *potentially* capable of mutual enrichment of thought, understanding, and method, feel most secure within their own small fenced-off areas.

Another downside of the classical model of science, simple in its experimental logic, is that it sustains an unrealistic public assumption that "good" science gives specific, certain, and unambiguous results. Yet the real-world *whole* is much more complex and unpredictable than the sum of its disaggregated *parts*, the latter being the main focus of classical experimental science. Systems-based analysis elucidates the internal workings of a complex system such as the climate system, and can identify the optimal intervention points for stabilization or restoration. But there is no single answer, no bottom-line $E = mc^2$. The complexities and uncertainties of real-world systems cannot be swept under the carpet of randomization of compared entities and orthodox statistical tests.

A radical idea that some scientists support as a solution to anthropogenic climate change is geoengineering, or more precisely, climate engineering. This would entail large-scale technological intervention in Earth systems to prevent or ameliorate adverse climate change. Some of the proposed planetary interventions include spraying a solar-reflecting veil of fine particulates into the upper troposphere, creating an array of mirrors to orbit the planet and reflect solar radiation away, seeding the oceans with fine iron dust to stimulate CO_2-absorbing algal growth, adding lime to the oceans to reduce their acidity and maintain their CO_2 uptake capacity, and removing CO_2 from the atmosphere with vast vacuum cleaner–like devices. Public discussion has lagged far behind such adventurous engineering ideas, even though long experience highlights the potential for unexpected, disastrous, and perhaps irreversible consequences—such as a reduction in global rainfall that would result from dimming the incoming sunlight.[80] Many of today's questions are too big and complex, with too many ramifications for human health, to be left just to scientists and engineers: "science is not the only voice in the room, and workable solutions need to account for a plurality of voices."[81]

Environmental stewardship: Paths to sustainability

The future, it has been wisely said, is not somewhere we are going; it is something we are making. A transition to an environmentally sustainable and socially equitable future would be the fourth great transition in the way that humans have lived in relation to the environment.[82,83] The earlier transitions entailed radical shifts from hunter-gathering to early farming, then to city-dwelling, and then to industrialization. Each unfolded over time as experience and ideas evolved. However, transforming our environmentally damaging way of living to accord with the limits of Earth's biocapacity will entail a different type of transition: global, deliberate, and rapid.

That will require a clearer and more widely shared understanding of how the human species fits within the global ecosystem as participant, not plunderer—an understanding captured well by the word "biosensitivity."[84] The future challenge for societies is to act primarily as environmental stewards, not owners or masters.[85] Second, there is urgent need for social investment in energy-efficient and environmentally benign technologies. Many examples are already on the horizon, such as making biofuels from urban waste, or using solar-powered nano-semiconductors to split water into hydrogen fuel and oxygen.[86]

The alternative, hard to imagine, would be an overheated and biotically impoverished world with many uninhabitable areas, food and water shortages, great loss of species, disrupted ecosystems, and damaging weather disasters. All those changes would exacerbate existing health inequalities and amplify social tensions, conflicts, and the displacement of people. If a divided "enclave world" of the privileged and the abandoned were to emerge, then uprisings and conflicts might follow.[87,88] Recognition of the great and growing threats from population pressures, climate change, and other global environmental changes has prompted the UN to set future-oriented Sustainability Development Goals for 2016 and beyond that seek both to reduce existing well-being and health deficits and to align human practices with Earth's resource limitations to secure the future for oncoming generations.[89]

Stewardship: More than just treading lightly

It is our species that has disrupted the planet's complex operating system, and we now know that our priorities and practices must change. The responsibility that we should now exercise is momentous and epochal.

It is worth spending a moment recalling that story. Almost four billion years ago, self-replicating molecules emerged within Earth's primordial chemical cauldron, perhaps triggered by flows of energy and chemical gradients on the ocean floor.[90] Natural selection then kicked in, and biological evolution responded to continuing change and opportunity. Today, after the most recent two million years of climate-influenced evolutionary branching within the *Homo* genus, only one *Homo* species survives.

Homo sapiens now occupies most of the planet's land surface. What is our moral responsibility as Earth's now-dominant species? Can we avoid unraveling further the self-renewing solar-powered ordering of the living world? That maintenance of order occurs despite the universal tendency toward *entropy*—the tendency for ordered material and its embedded energy to degrade to randomly distributed molecules and heat?[91] Our escalating use of energy is causing entropy more rapidly than the resultant degraded energy, low-grade heat, can be dissipated via radiation outward into space. And so the planet necessarily warms towards a new thermal equilibrium in which order and entropy (disorder) can be kept in balance. Meanwhile, the problem escalates as humankind ramps up the use of energy, particularly by burning fossil fuels.

Effective stewardship requires appropriate responses on many fronts. There is an old sailors' saying that "We do not control the wind, but we *can* set our sails." Currently we do not know the size of Earth's stocks of many of the assets that we depend upon, including fossil fuels, groundwater, species diversity, and the oceans' capacity to buffer increased acidity. Nor do we know what unexpected changes in Earth's biography lie ahead. Effective stewardship will at least buy us some time. Curbing global warming as soon as possible is an obvious priority, and early gains could come from reducing emissions of the shorter-lived but potent heat-trapping pollutants, principally methane, tropospheric ozone, hydrofluorocarbons, and black carbon.[92] Besides, these short-lived pollutants carry less political baggage and vested interest than does fossil-fuel-generated carbon dioxide.

Advancing beyond fossil-fuel dependence will not be easy. Previous energy revolutions, from harnessing water power to burning fossil fuels, have initially taken a heavy toll on human well-being, including the uprooting of communities. Early industrializing populations suffered severely from the "four Ds": disruption, deprivation, disease, and death.[93] Progressive, politically nonpartisan, social and political management will be crucial for a more benign transition.

Food security, the sine qua non

Food security and sufficiency will be central to achieving a sustainable future world, but not everyone will be well fed en route. The prospects for food production matching population growth and rising consumer demands for animal protein look precarious—especially given competition from biofuels, and the ongoing increase in energy-inefficient production of animal feed instead of staple foods for local populations.[94]

Indeed, food shortages are likely to be this century's defining crisis resulting from climate change, acting in concert with other systemic environmental changes.[95,96] While today's wealthy populations now require smaller land areas to feed each person than were required 3,000 years ago,[97] this century's vastly greater and still-expanding population and intensified methods of food production will, if persisted with, lead to a crisis of food insufficiency, along with cumulative soil damage caused by nonsustainable practices.

Radical changes in food systems will be a prerequisite to a shift to sustainable living. That, in turn, assumes a change in public understanding and preference—a tall order, given the central role of food in cultures everywhere. Does the solution lie primarily with transformations to large-scale industrial farming or to smallholder farming, or both? Current industrial agriculture has high immediate productivity, but very low levels of energy efficiency and resilience; it uses 10–15 units of input energy to produce one energy-unit of food and contributes one-quarter of total global greenhouse gas emissions.[98] Meanwhile smallhold farming in many regions has been achieving gains in both yields and resilience via more diversified and mixed farming at small to medium scale.[99,100] The World Bank's Agriculture for Development program has focused on enhancing the productivity of smallhold farms and their integration into global markets, but the latter clashes with a counterstrategy for smallhold farmers to concentrate on enhancing locally suited methods and foods oriented primarily to local markets.[101,102]

Is agriculture to be primarily a commodified handmaiden of global economic growth, or a basic resource for local food provision and ecological sustainability?[103] There is plenty of evidence that a shift to sustainable farming practices can both increase food yields and reduce greenhouse emissions.[104] Increasing the efficiency of livestock production is often more effective in reducing greenhouse gas emissions than sustainable efficiency gains in food-crop yields; together, emissions reductions of around one-quarter can be attained. In the northern United States, the reduction of chemical fertilizer use in row-crop agriculture reduces greenhouse emissions and environmentally polluting nitrogen runoff while improving soils and crop yields.[105] Restoring soil carbon (biosequestration) as part of climate change abatement improves soil microbial activity, organic structure, and hence crop yields. Using seed drills instead of ploughs, planting cover crops, and leaving postharvest crop residues can all transform soil from a carbon source to a carbon sink. In sub-Saharan Africa, local "evergreen agriculture" is boosting food yields by integrating the "Four Fs": nitrogen-fixing fertilizer, fruit trees, fodder trees, and food-cropping patterns.[106]

Genetic modification (GM) of plants, still environmentally controversial, faces continuing problems.[107] Drought-resistant, heat-resistant, and pest-resistant GM seeds will increasingly be needed, but farmers cannot afford the seeds produced by big transnational corporations. State-funded research subsidies, on contractual terms, could cut development costs and retail prices of GM seeds. It would be better, though, if state-funded research institutions were to include GM research in their portfolio of development strategies, perhaps via public-private partnership. Whatever else, social ownership of the enhancement of world food production, especially GM research, is a prerequisite for a world of equitable and sustainable access to healthy food.

Outside the genetics laboratory, nature often responds equally well to quasi-natural interventions. The discovery that a specific strain of nitrogen-fixing bacterium found in sugarcane can colonize the cells of major crop plants means that those plants may be able to capture growth-enhancing atmospheric nitrogen, a feature currently restricted to just a few plants.[108] Improving the soil's organic content and hence its microbial profile reduces its reliance on chemical fertilizers, pesticides, and herbicides.[109] And locked away out of sight are the world's high-security seed banks, an insurance against future need. This resource also *could* be actively used to help restore the genetic diversity of food crops and their resilience to changes in climate, particularly since four-fifths of the world's diet now comes, somewhat precariously, from a narrow band of a dozen selectively bred flowering plant species (including wheat, barley, oats, rye, rice, corn, potatoes, beans, and tomatoes).

Final comments

This book began with the question: why write about *past* climate change and its impacts? Climate scientists aim to learn, from recent and historical information, about the climate system's *sensitivity*: what temperature rise occurs with a doubling in carbon dioxide concentration? Equivalent historical information from human experiences of past natural changes in climate can tell us about the *sensitivity* of human populations to different types and durations of climate change. That should alert us to the *fact* of climate threats to population health and social stability, and will provide some *guidance* as we act to arrest human-driven climate change this century.

Two metaphors appear throughout the book. One is the Faustian bargain, first applied in earlier chapters to the move from hunter-gatherer existence to agriculture. This shift meant that the long-term fate of humankind depended on a new way of living, one which led, step by step, to increased food supplies, and the creative efforts of stratified and specialized societies. What followed also were population growth, amplified social inequalities, a new regime of infectious diseases, and escalating energy requirements.

The second framing metaphor is the Goldilocks "just right" zone. This applies not just to the temperature of porridge. Humans thrive in the familiar, middle zone. Deviations (such as unusual heat, cold, drought, or heavy rains) stress social, ecological, and physiological systems and sometimes push them past their limits.

But the Faustian bargain has another dimension. Urbanism and the upscaling of human activity after the agricultural revolution led to the utilization of more and more energy for a specific purpose. This was to expand the Goldilocks Zone so that humans could survive and prosper in formerly marginal environments (including climates that were previously "too hot" or "too cold").

At first, the only energy sources were fire and human effort. Domestication of animals increased power outputs, but the quantum leap in human adaptability was achieved by combustion of coal, oil, gas, and other biomass. In other words, continued expansion of the Goldilocks Zone was hitched to the burning of fossil fuels.

This is the modern form of the Faustian bargain: by choosing to utilize fossil fuels to expand the Goldilocks Zone, humans unwittingly agreed to alter Earth's climate. We are now trapped in a bizarre paradox of our own making. We burn fossil energy to expand our reach (and the Goldilocks Zone), meanwhile spewing out pollution that slowly shrinks the zone and makes it less predictable.

What makes our current predicament so different from past adaptations to changing climate? In the past, human ways of life were not the cause of climate shifts. And until the agricultural revolution, we were not wedded to ways of life that made expansion of the Goldilocks Zone imperative. It is easy to say the circuit breakers will come from energy sources that are climate-neutral, and from new ways of living that do not press so hard on environmental limits. But putting these aspirations into practice will be difficult.

The world community has wasted much of the past 25 years arguing about climate science, defending national interests, fretting about threats to our growth-dependent economies, and being distracted by the global financial crisis—the "tyranny of the contemporary."[110] Meanwhile, at least five million extra deaths have occurred worldwide, particularly in children, that are reasonably attributable to recent climate change. And yet we find it difficult to look beyond the climatic threats to polar bears, Amazonian forests, and coral reefs. Climatic threats also apply to *Homo sapiens*—another animal species, albeit one with a unique cultural and technological veneer that might confer some early protection against climatic stresses.

Historically, climatic influences have often been beneficial, maximizing food yields and social stability and physical progress. But more often, changes in climate have been detrimental, contributing to food shortages, famines, infectious disease outbreaks, weather disasters, and conflicts over resources that foment social disorder and topple regimes. The historical story also gradually reveals the perils of the unconstrained pursuit of material progress, a central goal of the original Faustian bargain. That mindset has caused misjudgments and grief over the centuries. The Maya, for example, became obsessed with building ever-grander edifices, just like the leaders of today's Dubai. Little attention was paid to adapting their practices to increasingly evident adverse changes in climate and environment. Now humankind is on a treadmill, attempting to produce more food (including more and more red meat), electricity, safe drinking water, houses, and consumer goods for an ever-growing population, and doing so within a largely deregulated market-based economic system that routinely discounts long-term environmental damage.[111]

Here are the three overarching conclusions from this book:

First, natural climate change has been an *ever-present*, though often unremarked upon, influence on patterns of health, disease, survival, population displacement, conflict, and social and political stability. It has acted overtly and abruptly or as a subtle background shift in conditions.

Second, all species and ecosystems have evolved to survive and thrive within a particular climatic niche. If conditions change substantially, whether hotter or colder, wetter or drier, then stability, productivity, and even viability decline rapidly. Risks to human health then follow.

Third, our modern ideas and scientific methods are encumbered with outmoded legacies that impede thinking, understanding, and taking rational action. We are perturbing large and complex ecological, environmental, and social *systems* and therefore must understand and respond to these changes within that real-world systems frame.

The advent of human-induced climate change also throws into higher relief the moral dilemma relating to intergenerational responsibility.[112,113] Earlier generations in Western societies were comfortable in their assumption that they would, as a matter of course, bequeath a positive legacy to the next generation. But our generation cannot take comfort from the changes that are now unfolding around us. We may bequeath a world that, two or three generations hence, is very much hotter and more climatically erratic than the world that we know.

Can we combine pessimism of anticipation with optimism of the will?

Reversing those high-risk trends looks less plausible with the passage of each year as the required annual rate of emissions reductions rises; it is a daunting task for a species that has never before needed to take global-scale and radical action on behalf of the future. Perhaps we can take heart from past radical international about-turns, such as the abolition of transoceanic slavery in the nineteenth century or, more recently, the cessation of major ozone-destroying industrial gases. The wild card in today's situation is that radical changes in energy, in its sources, genera-tion, and use, will affect *everyone*, not just those engaged in a particular trade or industry (Box 11.1). Neither of those examples of earlier about-turns entailed risks to the health, safety, and survival of people every-where. And that should help swing the argument about whether, how, and when to respond to climate change.

Beyond the untapped full potential of our high-performance pri-mate brain is another potential trump card. For the first time, human minds can connect electronically and instantly around the planet, sharing ideas, information, and action plans. In contrast, it was im-possible for the Sumerians to advise the Harappans, 5,000 kilometers to the east, on how to counteract drying and soil salination. Today's planetwide connectivity offers new leverage for collective understand-ing and progressive action (Box 11.1). Despite our modern scientific knowledge and understanding, we too are just a transient part of the grand continuing geological and historical procession. Our ways of living, producing, organizing, and tackling daunting problems will become a substrate for future historians to compare the experiences of the early twenty-first century with those of earlier societies facing natural climate changes. Some societies coped, some were weakened, and some collapsed. If, in the future, the records show that effective

---

BOX 11.1 Message on Climate Change to World Leaders

"Human-induced climate change is an issue beyond politics. It transcends parties, nations, and even generations. For the first time in human history, the very health of the planet, and therefore the bases for future economic development, the end of poverty, and human wellbeing, are in the balance. If we were facing an immi-nent threat from beyond Earth, there is no doubt that human-ity would immediately unite in common cause. The fact that the threat comes from within—indeed from ourselves—and that it de-velops over an extended period of time does not alter the urgency of cooperation and decisive action."

Signed by over 4,000 scientists worldwide, July–August 2014

international agreement on halting human-driven climate change could not be reached, short-term material growth remained the primary goal, and a global climatic-environmental crisis ensued, the conclusion would be clear to future scientists: the label *sapiens* was little more than hubris.

Yet our species has not been tested collectively, globally, like this before. Human ingenuity and imagination may flourish as never before. Alfred Russel Wallace, Darwin's younger contemporary naturalist-evolutionist, wrote in light of his often-harrowing experiences in the Amazon "It is the struggle for existence ... which exercises the moral faculties and calls forth the latent sparks of genius."[114] If we learn from the past, understand the present, and imagine a better and more sustainable future, we may yet call forth such sparks to ignite corrective action and light the path to a sustainable way of living on a finite planet.

NOTES

Preface

1. John Darwin, *After Tamerlane: The Rise and Fall of Global Empires, 1400–2000* (London: Penguin, 2008), p. 6.
2. Joerg Friederichs, *The Future Is Not What It Used to Be: Climate Change and Energy Scarcity* (Boston: MIT Press, 2013).
3. "Epidemiology" refers to the study of the distribution and determinants of health disorders in human populations and studies of optimal preventive interventions at population level. Despite the word's appearance, it has no special alignment with infectious disease "epidemics."
4. Donella H. Meadows, Gary Meadows, Jorgen Randers, and William W. Behrens III, *The Limits to Growth* (New York: Universe Books, 1972); Barry Commoner, *The Closing Circle: Nature, Man, and Technology* (London: Random House, 1971); Paul R. Ehrlich and Anne H. Ehrlich, *Population, Resources and Environment: Issues in Human Ecology* (San Francisco: W. H. Freeman, 1970).

Chapter 1

1. IPCC, "Summary for Policymakers," in *Climate Change 2013: The Physical Science Basis: Contribution of Working Group 1 to the Fifth Assessment Report of the Intergovernmental Panel on Climate Change*, edited by Thomas F. Stocker et al. (Cambridge and New York: Cambridge University Press, 2013), available at http://www.climatechange2013.org/images/uploads/WGI_AR5_SPM_brochure.pdf.
2. Anthony J. McMichael, "Health Impacts in Australia in a Four-Degree World," in *Four Degrees of Global Warming: Australia in a Hot World*, edited by Peter Christoff (London: Routledge, 2014), pp. 155–171.

3. IPCC, *Climate Change 2013.*

4. Richard A. Muller, Judith Curry, Donald Groom, Robert Jacobsen, Saul Perlmutter, Robert Rohde, Arthur Rosenfeld, Charlotte Wickham, and Jonathan Wurtele, "Decadal Variations in the Global Atmospheric Land Temperature," *Journal of Geophysical Research: Atmospheres* 118, no. 1–7 (2013): 5280–5286, doi: 10.1002/jgrd.50458.

5. James Kantor, "Scientist: Warming Could Cut Population to 1 Billion," *New York Times*, March 13, 2009, available at http://dotearth.blogs.nytimes.com/2009/03/13/scientist-warming-could-cut-population-to-1-billion/?_r=0.

6. Rachel Warren, "The Role of Interactions in a World Implementing Adaptation and Mitigation Solutions to Climate Change," *Philosophical Transactions of the Royal Society A*, no. 369 (2011), 217–241, doi: 10.1098/rsta.2010.0271, p. 234.

7. Shaun A. Marcott, Jeremy D. Shakun, Peter U. Clark, and Alan C. Mix, "A Reconstruction of Regional and Global Temperature for the Past 11,300 Years," *Science* 339, no. 6124 (2013): 1198–1201, doi: 10.1126/science.1228026.

8. Neville Brown, *History and Climate Change: A Eurocentric Perspective* (London: Routledge, 2001).

9. An early book on this topic, from 1985, is T. M. L. Wigley, M. J. Ingram, and G. Farmer, eds., *Climate and History: Studies in Past Climates and Their Impact on Man* (Cambridge: Cambridge University Press, 1985). However, that book contained very limited information about impacts on human health and survival.

10. John L. Brooke, *Climate Change and the Course of Global History: A Rough Journey* (New York: Cambridge University Press, 2014).

11. Jessica L. Blois, Phoebe L. Zarnetske, Matthew C. Fitzpatrick, and Seth Finnegan, "Climate Change and the Past, Present, and Future of Biotic Interactions," *Science* 341, no. 6145 (2013): 499–504, doi: 10.1126/science.1237184.

12. "IPCC—Intergovernmental Panel on Climate Change," World Meteorological Organization, http://www.ipcc.ch/.

13. Anthony J. McMichael and E. Lindgren, "Climate Change: Present and Future Risks to Health and Necessary Responses," *Journal of Internal Medicine* 270, no. 5 (2011): 401–413, doi: 10.1111/j.1365-2796.2011.02415.x.

14. Andrew T. Guzman, *Overheated: The Human Cost of Climate Change* (Oxford: Oxford University Press, 2013).

15. Richard Levins and Richard C. Lewontin, *The Dialectical Biologist* (Cambridge and London: Harvard University Press, 1985).

16. Geoffrey Parker, "Crisis and Catastrophe: The Global Crisis of the Seventeenth Century Reconsidered," *American Historical Review* 113, no. 4 (2008): 1053–1079, doi: 10.1086/ahr.113.4.1053.

17. In today's more integrated multicultural world, the religion-neutral acronyms B.C.E. and C.E. are used to refer to "before the common era" and the "common era." They replace the Christian B.C. and A.D.

18. Jared Diamond, *Guns, Germs and Steel: A Short History of Everybody for the Last 13,000 Years* (London: Random House, 1998).

19. OECD, *Future Global Shocks: Improving Risk Governance* (Paris: Organisation for Economic Cooperation and Development, 2012).

20. Paul Coombes and Keith Barber, "Environmental Determinism in Holocene Research: Causality or Coincidence?" *Area* 37, no. 3 (2005): 303–311, doi: 10.1111/j.1475-4762.2005.00634.x.

21. Gustaf Utterström, "Climatic Fluctuations and Population Problems in Early Modern History," *Scandinavian Economic History Review* 3, no. 1 (1955): 3–47, doi: 10.080/03585522.1955.1041146.

22. Fernand Braudel, *Civilization and Capitalism, 15th–18th Century: The Structures of Everyday Life; The Limits of the Possible*, trans. Siân Reynolds (New York: Harper and Row, 1981).

23. Emmanuel Le Roy Ladurie, *Times of Feast, Times of Famine: A History of Climate since the Year 1000* (New York: Doubleday, 1971).

24. Julia Adeney Thomas, "Comment: Not Yet Far Enough," *American Historical Review* 117, no. 3 (2012): 794–803, doi: 10.1086/ahr.117.3.794. She writes (p. 802): "Understanding the history of our altered physical world will require us [historians] to readapt our magpie tendencies and steal some eggs from the nests of scientists . . . In the 1960s and '70s, social history was energized by the collateral disciplines of economics and sociology; then, with the linguistic and cultural turns in the 1980s and '90s, anthropology, cultural studies, and literary theory reoriented our outlook. Now it is to our more distant intellectual cousins in biology, chemistry, physics, and related fields that we must turn."

25. Parker, "Crisis and Catastrophe."

26. Brooke, *Climate Change*.

27. Ciara Raudsepp-Hearne, Garry D. Peterson, Maria Tengö, Elena M. Bennett, Tim Holland, Karina Benessaiah, Graham K. MacDonald, and Laura Pfeifer, "Untangling the Environmentalist's Paradox: Why Is Human Well-Being Increasing as Ecosystem Services Degrade?" *Bioscience* 60, no. 8 (2010): 576–589, doi: 10.1525/bio.2010.60.8.4.

28. Jonathan A. Foley, Ruth DeFries, Gregory P. Asner, Carol Barford, Gordon Bonan, Stephen R. Carpenter, and F. Stuart Chapin, "Global Consequences of Land Use," *Science* 309, no. 5734 (2005): 570–574, doi: 10.1126/science.1111772.

29. Tedros A. Ghebreyesus, Mitiku Haile, Karen H. Witten, Asefaw Getachew, Ambachew M. Yohannes, Mekonnen Yohannes, Hailay D. Teklehaimanot, Steven W. Lindsay, and Peter Byass, "Incidence of Malaria among Children Living near Dams in Northern Ethiopia: Community-Based Incidence Survey," *British Medical Journal* 319, no. 7211: 663–666, doi: 10.1136/bmj.319.7211.663.

30. E. A. Malek, "Effect of Aswan High Dam on Prevalence of Schistosomiasis in Egypt," *Tropical and Geographical Medicine* 27, no. 4 (1975): 359–364.

31. Carlos Corvalan, Simon Hales, and Anthony J. McMichael, *Ecosystems and Human Well-Being: Health Synthesis; A Report of the Millennium Ecosystem Assessment* (Geneva: World Health Organization, 2005), available at http://www.millenniumassessment.org/documents/document.357.aspx.pdf.

32. Anthony J. McMichael, "Globalization, Climate Change, and Human Health," in *New England Journal of Medicine* 368 (2013):1335–1343, doi: 10.1056/NEJMra1109341. Anthony John McMichael, "Climate Change and Global Health," in *Climate Change and Global Health*, edited by Colin D. Butler (Wallingford and Boston: CABI, 2014).

33. Chris Funk, Shraddhanand Shukla, Andy Hoell and Ben Livneh "Assessing the Contributions of East African and West Pacific Warming to the 2014 Boreal Spring East African Drought," *Bulletin of the American Meteorological Society* 96, no. 12 (2015): S77–S82.

34. A. Park Williams and Christopher Funk, "A Westward Extension of the Warm Pool Leads to a Westward Extension of the Walker Circulation, Drying Eastern Africa," *Climate Dynamics* 37, no. 11 (2011): 2417–2435, doi: 10.1007/s00382-010-0984-y.

35. Aiguo Dai, "Drought under Global Warming: A Review," *Wiley Interdisciplinary Reviews: Climate Change* 2, no. 1 (2011): 45–65, doi: 10.1002/wcc.81.

36. Anthony John McMichael, "Globalization, Climate Change, and Human Health," *New England Journal of Medicine* 368 (2013): 1335–1343.

37. Robin A. Weiss, "Apes, Lice and Prehistory," *Journal of Biology* 8, no. 2 (2009): 20, doi: 10.1186/jbiol114.

38. World Bank, *Turn Down the Heat: Why a 4°C World Must Be Avoided* (Washington, DC: World Bank, 2012), available at http://www-wds.worldbank.org/external/default/WDSContentServer/WDSP/IB/2012/12/20/000356161_20121220072749/Rendered/PDF/NonAsciiFileNameo.pdf.

39. See Figure 1 in Anthony J. McMichael, "Insights from Past Millennia into Climatic Impacts on Human Health and Survival," *Proceedings of the National Academy of Sciences of the United States of America* 109, no. 13 (2012): 4730–4737, doi: 10.1073/pnas.1120177109.

40. Darrell Kaufman, Nicholas McKay, Thorsten Kiefer, and Lucien von Gunten (The PAGES 2k Consortium), "A Regional View of Global Climate Change," *Global Change* 81 (2013): 18–23, available at http://www.igbp.net/news/features/features/aregionalviewofglobalclimatechange.5.64c294101429ba9184d44a.html.

41. Hubert H. Lamb, *Climate, History and the Modern World* (London: Routledge, 1995).

42. Oscar Branson, Simon A. T. Redfern, Tolek Tyliszczak, Aleksey Sadekov, Gerald Langer, Katsunori Kimoto, and Henry Elderfield, "The Coordination of Mg in Foraminiferal Calcite," *Earth and Planetary Science Letters* 383 (2013): 134–141, doi: 10.1016/j.epsl.2013.09.037.

43. P. D. Jones, T. J. Osborn, and K. R. Briffa, "The Evolution of Climate over the Last Millennium," *Science* 292, no. 5517 (2001): 662–667, doi: 10.1126/science.1059126.

44. Rudolf Brázdil, Christian Pfister, Heinz Wanner, Hans von Storch, and Juerg Luterbacher, "Historical Climatology in Europe: The State of the Art," *Climatic Change* 70, no. 3 (2005): 363–430, doi: 10.1007/s10584-005-5924-1. The sequence of changes in the quality of written information over the past millennium for Western and Central Europe is as follows. Before 1300: Reports of (natural) weather disasters and of climatic variations that had social and economic consequences. 1300–1500: Regular reports on types of summers and winters (less regular reports on spring and autumn). 1500–1800: Systematic descriptions of monthly weather, and to some extent daily weather also. 1680–1860: Records of instrumental (thermometer) measurements made by isolated individuals. First short-lived international network observations. From 1860: Records of instrumental observations made within the framework of national and international meteorological networks.

45. Paul J. Crutzen and Eugene F. Stoermer, "The 'Anthropocene,'" *Global Change Newsletter* 41 (2000): 17–18. See also Will Steffen, Paul J. Crutzen, and John R. McNeill, "The Anthropocene: Are Humans Now Overwhelming the Great Forces of Nature?," *Ambio* 36, no. 8 (2007): 614–621, doi: 10.1579/0044-7447(2007)36[614:TAAHNO]2.0.CO;2, and Will Steffen et al., "The Anthropocene: Conceptual and Historical Perspectives," *Philosophical Transactions of the Royal Society A* 369, no. 1938 (2011): 842–867, doi: 10.1098/rsta.2010.0327.

46. For an alternative understanding of the Anthropocene, see William F. Ruddiman, "The Anthropocene," *Annual Review of Earth and Planetary Sciences* 41 (2013): 45–68, doi: 10.1146/annurev-earth-050212-123944.

47. Edward O. Wilson, *The Social Conquest of Earth* (New York: Liveright, 2013).

48. Vaclav Smil, *Harvesting the Biosphere: What We Have Taken from Nature* (Cambridge: MIT Press, 2013). Today the total biomass of humans is around 125 million metric tons, while their livestock weigh another 300 million tons, compared with just 10 million tons for all wild vertebrate animals.

49. Andrew Y. Glikson, *Evolution of the Atmosphere, Fire, and the Anthropocene Climate Event Horizon* (Dordrecht: SpringerBriefs in Earth Sciences, 2013), doi: 10.1007/978-94-007-7332-5_1.

50. Robert L. Sherlock, *Man as a Geological Agent* (London: H. F. and G. Witherby, 1922), p. 343.

51. Erle C. Ellis, Jed O. Kaplan, Dorian Q. Fuller, Steve Vavrus, Kees Klein Goldewijk, and Peter H. Verbur, "Used Planet: A Global History," *Proceedings of the National Academy of Sciences of the United States of America* 110, no. 20 (2013): 7978–7985, doi: 10.1073/pnas.1217241110.

52. William F. Ruddiman, Michael C. Crucifix, and Frank A. Oldfield, eds., "The Early-Anthropocene Hypothesis," special issue, *The Holocene* 21, no. 5 (2011).

53. Royal Society Science Policy Centre, *People and the Planet* (London: The Royal Society, 2012).

54. Will Steffen, "The trajectory of the Anthropocene: The Great Acceleration," *The Anthropocene Review*, 2/1 (April 1, 2015), 81–98, doi: 10.1177/2053019614564785.

55. United Nations, Department of Economic and Social Affairs, Population Division, *World Population Prospects: The 2015 Revision, Key Findings and Advance Tables*, Working Paper No. ESA/P/WP.241 (New York: United Nations, 2015).

56. Global Footprint Network, *Atlas of Global Ecological Footprint, 2010* (Paris: GFN, 2010).

57. Anthony J. McMichael and Colin D. Butler, "Promoting Global Population Health While Constraining the Environmental Footprint," *Annual Review of Public Health* 32 (2011): 179–197, doi: 10.1146/annurev-publhealth-031210-101203.

58. Based on figures in Stephen V. Boyden, *The Biology of Civilisation: Understanding Human Nature as a Force in Nature* (Sydney: UNSW Press, 2004).

59. Vaclav Smil, *Energy in World History* (Boulder: Westview Press, 1994).

60. Smil, *Energy in World History*.

61. Clive Ponting, *A New Green History of the World* (London: Vintage, 2007).

62. Boyden, *Biology of Civilisation*.

63. More formally, a system is a naturally integrated complex of processes interacting with one another and (within limits) able to self-stabilize via homeostatic feedback processes.

64. Ludwig von Bertalanffy, *General System Theory: Foundations, Development, Applications* (New York: George Braziller, 1968). See also International Society for the System Sciences, "Ludwig von Bertalanffy (1901–1972)," January 1999, www.isss.org/lumLVB.htm.

65. Paul Shepard and Daniel McKinley, *Subversive Science: Essays toward an Ecology of Man* (Boston: Houghton Mifflin, 1969). Why "subversive"? The ideas of ecologists were often focused on how prevailing patterns of economic development and military activity in the 1960s would damage the natural environmental resource base upon which societies depend. This argument was not welcomed by powerful military-industrial interests (particularly in the United States) at the height of the Cold War. Ecologists who advocated decreasing economic pressures on natural systems were seen by many as left-wing challengers to the status quo. Their science was regarded as "subversive."

66. Stephen Boyden, *Western Civilization in Biological Perspective: Patterns in Biohistory* (Oxford: Oxford University Press, 1987).

67. Johan Rockström, Will Steffen, Kevin Noone, Åsa Persson, F. Stuart Chapin III, Eric F. Lambin, Timothy M. Lenton, et al., "A Safe Operating Space for Humanity," *Nature* 461, no. 7263 (2009): 472–475, doi: 10.1038/461472a.

68. Rockstrom et al., "Safe Operating Space."

69. Sources: Yaha Abawi et al., "Relationship between El Nino-Southern Oscillation and the Incidence of Malaria in the Solomon Islands," presentation to AusAID and the Australian Government Bureau of Meteorology (Suva, 2009); Lachlan McIver, Rokho Kim, Alistair Woodward, Simon Hales, Jeffery Spickett, Dianne Katscherian, Masahiro Hashizume, et al., "Climate Change and Health in Fiji: Environmental Epidemiology of Infectious Diseases and Potential for Climate-Based Early Warning Systems," *Fiji Journal of Public Health* (2012): 7–13.

70. Paul Davies, *The Goldilocks Enigma: Why is the Universe Just Right for Life?* (London: Penguin, 2004). In the children's fable "Goldilocks and the Three Bears," a young girl intrudes into the three bears' house and chooses what's "just right" for her from each of several sets of items. She ignores the ones that are too extreme (large or small, hot or cold, etc.) and chooses the one in the middle that is "just right."

71. Johann Wolfgang Goethe, *Faust/Part One*, trans. Philip Wayne (Harmondsworth: Penguin Books, 1975), p. 92.

Chapter 2

1. Zita Martins, Mark C. Price, Nir Goldman, Mark A. Sephton, and Mark J. Burchell, "Shock Synthesis of Amino Acids from Impacting Cometary and Icy Planet Surface Analogues," *Nature Geoscience* 6, no. 12 (2013): 1045–1049, doi: 10.1038/ngeo1930.

2. Li Li, Christopher Francklyn, and Charles W. Carter Jr., "Aminoacylating Urzymes Challenge the RNA World Hypothesis," *Journal of Biological Chemistry* 288, no. 37 (2013): 26856–26863, doi: 10.1074/jbc.M113.496125.

3. Sean A. Crowe, Lasse N. Døssing, Nicolas J. Beukes, Michael Bau, Stephanus J. Kruger, Robert Frei, and Donald E. Canfield, "Atmospheric Oxygenation Three Billion Years Ago," *Nature* 501, no. 7468 (2013): 535–538, doi: 10.1038/nature12426.

4. Andrew Glikson, "Fire and Human Evolution: The Deep Time Blueprints of the Anthropocene," *Anthropocene* 3 (2014): 89–92, doi: 10.1016/j.ancene.2014.02.002.

5. Peter B. deMenocal, "Cultural Responses to Climate Change during the Late Holocene," *Science* 292, no. 5517 (2001): 667–673, doi: 10.1126/science.1059287.

6. Andrew P. Schurer, Simon F. B. Tett, and Gabriele C. Hegerl, "Small Influence of Solar Variability on Climate over the Past Millennium," *Nature*

Geoscience 7, no. 2 (2013): 104–108, doi: 10.1038/ngeo2040. The sun is actually a relatively weak influence on Earth's temperature, despite some popular beliefs and deliberate misrepresentations.

7. John Tyndall, *The Forms of Water in Clouds and Rivers, Ice, and Glaciers* (London: Henry S. King, 1872).

8. See also a simple illustrated explanation at http://www.sarmento.eng.br/ Documentos/14149824-How-the-Sun-Affects-Climate-Milankovitch-Cycles .pdf.

9. Adapted from Wikimedia Commons graphs, with annotations and dashed lines by author, http://en.wikipedia.org/wiki/File:Orbital_variation.svg.

10. Named after the pioneering British meteorologist George Hadley.

11. Wenju Cai, Agus Santoso, Guojian Wang, Evan Weller, Lixin Wu, Karumuri Ashok, Yukio Masumoto, and Toshio Yamagata, "Increased Frequency of Extreme Indian Ocean Dipole Events due to Greenhouse Warming," *Nature* 510, no. 7504 (2014): 254–258, doi: 10.1038/nature13327.

12. Michael H. Glantz, *Currents of Change: El Niño Impact on Climate and Society* (Cambridge: Cambridge University Press, 1996).

13. Tom Griffiths, "The Roaring Forties," in *A Change in the Weather*, edited by Tim Sherratt, Tom Griffiths, and Libby Robin (Canberra: National Museum of Australia Press, 2005), pp. 152–164.

14. Jean-René Vanney, *Histoire des mers australes* (Paris: Fayard, 1986), p.22, cited in Griffiths, "The Roaring."

15. Wenju Cai, Tim Cowan, and Marcus Thatcher, "Rainfall Reductions over Southern Hemisphere Semi-Arid Regions: The Role of Subtropical Dry Zone Expansion," *Scientific Reports* 2 (2012): 702, doi: 10.1038/srep00702.

16. Seth Westra, Lisa V. Alexander, and Francis W. Zwiers, "Global Increasing Trends in Annual Maximum Daily Precipitation," *Journal of Climate* 26, no. 11 (2012): 3904–3918, doi: 10.1175/JCLI-D-12-00502.1.

17. Jesse Kenyon and Gabriele C. Hegerl, "Influence of Modes of Climate Variability on Global Temperature Extremes," *Journal of Climate* 21, no. 15 (2008): 3872–3889, doi: 10.1175/2008JCLI2125.1.

18. Curtis C. Ebbesmeyer and W. James Ingraham Jr., "Pacific Toy Spill Fuels Ocean Current Pathways Research," *Earth in Space* 75, no. 37 (1994): 425–430, doi: 10.1029/94EO01056.

19. Donovan Hohn, *Moby-Duck: The True Story of 28,800 Bath Toys Lost at Sea and of the Beachcombers, Oceanographers, Environmentalists, and Fools, Including the Author, Who Went in Search of Them* (New York: Viking, 2011).

20. William Burroughs, ed., *Climate: Into the 21st Century* (Cambridge: Cambridge University Press, 2003), p. 174.

21. Andrew G. Turner and H. Annamali, "Climate Change and the South Asian Summer Monsoon," *Nature Climate Change* 2, no. 8 (2012): 587–595, doi: 10.1038/nclimate1495. See also Moetasim Ashfaq et al., "Suppression of South Asian Summer Monsoon Precipitation in the 21st Century," *Geophysical Research Letters* 36 (2009): L01704, doi: 10.1029/2008GL036500.

22. Arathy Menon, Anders Levermann, and Jacob Schewe, "Enhanced Future Variability during India's Rainy Season," *Geophysical Research Letters* 40 (2013): 3242–3274, doi: 10.1002/grl.50583.

23. Sharon Nicholson, "A Climatic Chronology for Africa," 75–81, 251–254, cited in James Webb, *Desert Frontier: Ecological and Economic Change along the Western Sahel, 1600–1850* (Madison: University of Wisconsin Press, 1995), pp. 4–5.

24. See James C. McCann, "Climate and Causation in African History," *International Journal of African Historical Studies* 32, no. 2/3 (1999): 261–279, and his discussion of George Brooks, *Landlords and Strangers: Ecology, Society, and Trade in West Africa*, 1000–1630 (Boulder: Westview Press, 1993).

25. See Hubert H. Lamb, *Climate, History, and the Modern World* (London and New York: Methuen, 1982).

26. McCann, "Climate and Causation."

27. "Disaster Statistics," Insurance Council of Australia, 2015, available at http://insurancecouncil.com.au/statistics.

28. Dim Coumou and Stefan Rahmstorf, "A Decade of Weather Extremes," *Nature Climate Change* 2, no. 7 (2012): 491–496, doi: 10.1038/nclimate1452.

29. Aiguo Dai, "Increasing Drought under Global Warming in Observations and Models," *Nature Climate Change* 3, no. 1 (2013): 52–58, doi: 10.1038/nclimate1633.

30. IPCC, *Climate Change 2013: The Physical Science Basis, Contribution of Working Group I to the Fifth Assessment Report of the Intergovernmental Panel on Climate Change* (Cambridge and New York: Cambridge University Press, 2013), doi: 10.1017/CBO9781107415324.

31. Thomas Litt, Christian Ohlwein, Frank H. Neumann, Andreas Hense, and Mordechai Stein, "Holocene Climate Variability in the Levant from the Dead Sea Pollen Record," *Quaternary Science Reviews* 49 (2012): 95–105, doi: 10.1016/j.quascirev.2012.06.012.

32. The physics of the warming process is intriguing. The various greenhouse-gas molecules absorb outgoing long-wave (infrared) radiation energy from Earth's surface, store it transiently via "origami-like" bending of the molecule, and then relax the bending and rerelease the radiant energy—but this time in all directions. And so the atmosphere warms—as it has been doing, erratically but upward, since early in the industrial revolution.

33. Actually, we have had two great strokes of atmospheric luck. The ozone in the middle atmosphere (stratosphere) filters out most of the biologically damaging part of the incoming solar energy—the shortwave ultraviolet (UV) radiation that damages DNA and exposed tissues (such as skin and eyes) and impairs plant growth. That ozone is a product of early life on Earth when, for the first three billion years of life, all organisms were confined to the UV-protective environment of seas and waterways. As aerobes evolved, releasing a novel gas, oxygen, into the atmosphere, slowly the level of oxygen in the lower atmosphere built up, some of it being

converted into ozone (O_3), which migrated to the stratosphere. Thanks to that ozone, within the past half billion years it has become safe for aerobic life forms, plant and animal, to live on land.

34. Higher-resolution ice cores are now untangling this contentious question of leads and lags between CO_2 and temperature during the transition out of the last glacial period and into relatively rapid warming. The cores show that temperature leads CO_2 at the beginning of the rewarming process (triggered by the ending of the 100-year Milankovitch cycle that underlay the preceding long cooling trend). But that is only for a short time; as CO_2 is released by warming oceans, it is soon in the driver's seat and leading temperature for the rest of the transition. This accords well with our current understanding of the feedback processes.

35. Ayako Abe-Ouchi, Fuyuki Saito, Kenji Kawamura, Maureen E. Raymo, Jun'ichi Okuno, Kunio Takahashi, and Heinz Blatter, "Insolation-Driven 100,000-Year Glacial Cycles and Hysteresis of Ice-Sheet Volume," *Nature* 500, no. 7461 (2013): 190–193, doi: 10.1038/nature12374.

36. Spencer Weart, "The Carbon Dioxide Greenhouse Effect," *Discovery of Global Warming*, March 2015, available at http://www.aip.org/history/climate/co2.htm.

37. Carmel Doyle, "Irish Scientist John Tyndall: Climate Change Visionary in 1861," Silicon Republic, September 26, 2011, available at http://www.siliconrepublic.com/clean-tech/item/23757-irish-scientist-john-tyndal.

38. John Tyndall, *Contributions to Molecular Physics in the Domain of Radiant Heat* (London: Longmans, Green, 1872).

39. Iain McCalman, *Darwin's Armada* (London: Pocket Books, 2010), pp. 354–358.

40. PALAEOSENS Project Members, "Making Sense of Palaeoclimatic Sensitivity," *Nature* 491, no. 7426 (2012): 683–691, doi: 10.1038/nature11574.

41. The name "greenhouse" has stuck even though the physics of heat-trapping by greenhouse gases differs from the heat-capturing process of a glass-paneled garden greenhouse.

42. Gilbert N. Plass, "The Carbon Dioxide Theory of Climatic Change," *Tellus* 8, no. 2 (1956): 140–154, doi: 10.1111/j.2153-3490.1956.tb01206.x.

43. Reid A. Bryson and Thomas J. Murray, *Climates of Hunger* (Madison: University of Wisconsin Press, 1977). Bryson, one of the leading US climate scientists of the time, judged that cooling was the greater threat.

44. US NASA, "Global Land-Ocean Temperature Index," available at http://data.giss.nasa.gov/gistemp/graphs_v3/.

45. Xianyao Chen and Ka-Kit Tung, "Varying Planetary Heat Sink Led to Global-Warming Slowdown and Acceleration," *Science* 345, no. 6199 (2014): 897–899, doi: 10.1126/science.1254937.

46. Yu Kosaka and Shang-Ping Xie, "Recent Global-Warming Hiatus Tied to Equatorial Pacific Surface Cooling," *Nature* 501, no. 7467 (2013): 403–440, doi: 10.1038/nature12534.

47. Gerald A. Meehl, Julie M. Arblaster, John T. Fasullo, Aixue Hu, and Kevin E. Trenberth, "Model-Based Evidence of Deep-Ocean Heat Uptake during Surface-Temperature Hiatus Periods," *Nature Climate Change* 1, no. 7 (2011): 360–364, doi: 10.1038/nclimate1299.

48. Chen and Tung, "Varying Planetary Heat Sink."

49. Camille Parmesan and Gary Yohe, "A Globally Coherent Fingerprint of Climate Change Impact across Natural Systems," *Nature* 421, no. 6918 (2003): 37–42, doi: 10.1038/nature01286.

50. Camille Parmesan, Carlos Duarte, Elvira Poloczanska, Anthony J. Richardson, and Michael C. Singer, "Overstretching Attribution," *Nature Climate Change* 1, no. 1 (2011): 2–4, doi: 10.1038/nclimate1056.

51. "Decades of Data Show Spring Advancing Faster than Experiments Suggest," National Aeronautics and Space Administration (US), May 2, 2012, available at http://www.giss.nasa.gov/research/news/20120502/.

52. Robert M. Barclay, "Variable Variation: Annual and Seasonal Changes in Offspring Sex Ratio in a Bat," *PLoS ONE* 7, no. 5 (2012): e36344, doi: 10.1371/journal.pone.0036344.

53. I-Ching Chen, Jane K. Hill, Ralf Ohlemüller, David B. Roy, and Chris D. Thomas, "Rapid Range Shifts of Species Associated with High Levels of Climate Warming," *Science* 333, no. 6045 (2011): 1024–1026, doi: 10.1126/science.1206432.

54. Markus Huber and Reto Knutti, "Anthropogenic and Natural Warming Inferred from Changes in Earth's Energy Balance," *Nature Geoscience* 5, no. 1 (2012):31–36, doi: 10.1038/ngeo1327.

55. Potsdam Institute for Climate Impact Research and Climate Analytics, *Turn Down the Heat: Climate Extremes, Regional Impacts and the Case for Resilience* (Washington, DC: World Bank, 2013).

56. Carrie A. Schloss, Tristan A. Nunez, and Joshua J. Lawler, "Dispersal Will Limit Ability of Mammals to Track Climate Change in the Western Hemisphere," *Proceedings of the National Academy of Sciences of the United States of America* 109, no. 22 (2012): 8606–8611, doi: 10.1073/pnas.1116791109.

57. Dim Coumou and Stefan Rahmstorf, "A Decade of Weather Extremes," *Nature Climate Change* 2, no. 7 (2012): 491–496.

58. IPCC, *Managing the Risks of Extreme Events and Disasters to Advance Climate Change Adaptation: A special report of the Working Groups I and II of the Intergovernmental Panel on Climate Change* (Cambridge and New York: Cambridge University Press, 2012).

59. Coumou and Rahmstorf, "A Decade of Weather Extremes."

60. Kerry Emanuel, "Global Warming Effects on U.S. Hurricane Damage," *Weather, Climate, and Society* 3 (2011): 261–268, doi: 10.1175/WCAS-D-11-00007.1.

61. Rockström et al., "Safe Operating Space."
62. The global warming effect of black carbon (from diesel engines, the burning of agricultural-residue biomass, and inefficient biomass-burning cookstoves in hundreds of millions of poor, mostly rural, households) is now recognized as significant, especially when taking into account its darkening effect when deposited on reflective ice and snow surfaces (which then also accelerates their melting).
63. This figure is difficult to calculate in any exact sense, since the various greenhouse gases differ not only in their potency but in their rates of decay or removal from the atmosphere. Methane, for example, is short-lived; however, via oxidation it breaks down into carbon dioxide and water vapor—both of them being long-term greenhouse gases.
64. Rajendra K. Pachuri and Andy Reisinger, eds., *Contribution of Working Groups I, II and III to the Fourth Assessment Report of the Intergovernmental Panel on Climate Change* (Geneva: IPCC, 2007). See also IPCC, *Climate Change 2013: The Physical Science Basis: Working Group I Contribution to the Fifth Assessment Report of the Intergovernmental Panel on Climate Change* (Cambridge: Cambridge University Press, 2013). Conveniently, since that one-third reduction cancels out the CO_2-equivalent warming component, atmospheric CO_2 concentration is a good index of the total net warming effect of all human-generated emissions, warming and cooling.
65. Robert K. Kaufmann, Heikki Kauppi, Michael L. Mann, and James H. Stock, "Reconciling Anthropogenic Climate Change with Observed Temperature 1998–2008," *Proceedings of the National Academy of Sciences of the United States of America* 108, no. 29 (2011): 11790–11793, doi: 10.1073/pnas.1102467108.
66. Michael R. Raupach, "What Do Current Emissions Pathways Imply for Future Climate Targets?," *Carbon Management* 2, no. 6 (2011): 625–627.
67. Richard H. Moss, Jae A. Edmonds, Kathy A. Hibbard, Martin R. Manning, Steven K. Rose, Detlef P. van Vuuren, and Timothy R. Carter, "The Next Generation of Scenarios for Climate Change Research and Assessment," *Nature* 463, no. 7282 (2010): 747–756, doi: 10.1038/nature08823. To generate plausible projections of future climate change (conventionally to 2100), climate modelers have typically used a four-step approach during recent decades. First, a plausible profile of trends in socioeconomic factors (population growth, GDP, energy use, agricultural production, etc.) is specified. These range from "business as usual" scenarios, with continuing strong rise in emissions, to scenarios in which mitigation policies achieve substantial emission cuts. That leads to the second step: the modeling of future trends in total global greenhouse gas emissions. In the third step, changes in the composition of the lower atmosphere and hence its overall greenhouse warming potential are estimated. In the fourth step, a global climate model generates estimates of the resultant change in temperature (and rainfall), decade by decade, resulting from each of the scenarios of

future change in atmospheric greenhouse warming potential. A related approach to modeling climate futures, used in the IPCC's Fifth Assessment Report (2013), is to begin at the third step by specifying "representative [emission] concentration pathways" over the course of this century. These "representative [emission] concentration pathways" (or scenarios) are based on a more inclusive approach that incorporates atmospheric, environmental (land surface, vegetation, oceans, and ice-sheets), social, behavioral, and political influences on greenhouse emissions and on the uptake of CO_2 by vegetation on land and sea and by soils (biosequestration). The climatic consequences of the nominated pathways are then estimated.

68. Christian P. Giardina, Creighton M. Litton, Susan E. Crow, and Gregory P. Asner, "Warming-Related Increases in Soil CO_2 Efflux Are Explained by Increased Below-Ground Carbon Flux," *Nature Climate Change* 4, no. 9 (2014): 822–827, doi: 10.1038/nclimate2322.

69. "GISS Surface Temperature Analysis: Global Maps from GHCN v3 Data," National Aeronautics and Space Administration (US), October 5, 2015, available at http://data.giss.nasa.gov/gistemp/maps/.

70. Fei Ji, Zhaohua Wu, Jianping Huang, and Eric P. Chassignet, "Evolution of Land Surface Air Temperature Trend," *Nature Climate Change* 4 (2014): 462–466, doi: 10.1038/nclimate2223.

71. Jennifer A. Francis and Stephen J. Vavrus, "Evidence Linking Arctic Amplification to Extreme Weather in Mid-Latitudes," *Geophysical Research Letters* 39, no. 6 (2012): L06801, doi: 10.1029/2012GL051000.

72. Thomas C. Peterson, Peter A. Stott, and Stephanie Herring, eds., "Explaining Extreme Events of 2011 from a Climate Perspective," *Bulletin of the American Meteorological Society* 93, no. 7 (2012): 1041–1067, doi: 10.1175/BAMS-D-11-00021.1.

Chapter 3

1. Rita Reyburn, Deok Ryun Kim, Michael Emch, Ahmed Khatib, Lorenz Von Seidlein, and Mohammad Ali, "Climate Variability and the Outbreaks of Cholera in Zanzibar, East Africa: A Time Series Analysis," *American Journal of Tropical Medicine and Hygiene* 84, no. 6 (2011): 862–869, doi: 10.4269/ajmh.2011.10-0277.

2. Mercedes Pascual, Xavier Rodó, Stephen P. Ellner, Rita Colwell, and Menno J. Bouma, "Cholera Dynamics and El Niño-Southern Oscillation," *Science* 289, no. 5485 (2000): 1766–1769, doi: 10.1126/science.289.5485.1766.

3. James D. Ford, "Vulnerability of Inuit Food Systems to Food Insecurity as a Consequence of Climate Change: A Case Study from Igloolik, Nunavut," *Regional Environmental Change* 9, no. 2 (2009): 83–100, doi: 10.1007/s10113-008-0060-x.

4. Scot Nickels, Chris Furgal, Mark Buell, and Heather Moquin, *Unikkaaqatigiit—Putting the Human Face on Climate Change: Perspectives from Inuit in Canada* (Ottawa: Inuit Tapiriit Kanatami, Nasivvik Centre

for Inuit Health and Changing Environments at Université Laval and the Ajunnginiq Centre at the National Aboriginal Health Organization, 2005).

5. Patricia Cochran, Orville H. Huntington, Caleb Pungowiyi, Stanley Tom, F. Stuart Chapin III, Henry P. Huntington, Nancy G. Maynard, and Sarah F. Trainor, "Indigenous Frameworks for Observing and Responding to Climate Change in Alaska," *Climatic Change* 120, no. 3 (2013): 557–567, doi: 10.1007/s10584-013-07352.

6. Kirk R. Smith, Alistair Woodward, et al., "Human Health: Impacts, Adaptation and Co-Benefits," in *Climate Change 2014: Impacts, Adaptation, and Vulnerability.* Vol. 1: *Global and Sectoral Aspects: Contribution of Working Group II to the Fifth Assessment Report of the Intergovernmental Panel on Climate Change,* edited by Christopher B. Field, Vicente R. Barros, David Jon Dokken, Katharine J. Mach, Michael D. Mastrandrea, T. Eren Bilir, Monalisa Chatterjee, et al. (Cambridge and New York: Cambridge University Press, 2014), ch. 11.

7. Neil Morisetti, "Climate Change and Resource Security," *British Medical Journal* 344 (2012): e1352, doi: 10.1136/bmj.e1352.

8. Juliet R. Pulliam, Jonathan H. Epstein, Jonathan Dushoff, Sohayati A. Rahman, Michel Bunning, Aziz A. Jamaluddin, Alex D. Hyatt, Hume E. Field, Andrew P. Dobson, and Peter Daszak, "Agricultural Intensification, Priming for Persistence and the Emergence of Nipah Virus: A Lethal Bat-Borne Zoonosis," *Journal of the Royal Society, Interface* 9, no. 66 (2012): 89–101, doi: 10.1098/rsif.2011.0223.

9. Kaw Bing Chua, W. J. Bellini, P. A. Rota, B. H. Harcourt, A. Tamin, S. K. Lam, T. G. Ksiazek, et al., "Nipah Virus: A Recently Emergent Deadly Paramyxovirus," *Science* 288, no. 5470 (2000): 1432–1435, doi: 10.1126/science.288.5470.432.

10. Kaw Bing Chua, Beng Hui Chua, and Chew Wen Wang, "Anthropogenic Deforestation, El Niño and the Emergence of Nipah Virus in Malaysia," *Malaysian Journal of Pathology* 24, no. 1 (2002): 15–21, doi: 10.1.1.510.9697.

11. Cunrui Huang, Adrian G. Barnett, Xiaoming Wang, and Shilu Tong, "Effects of Extreme Temperatures on Years of Life Lost for Cardiovascular Deaths: A Time Series Study in Brisbane, Australia," *Circulation: Cardiovascular Quality and Outcomes* 5, no. 5 (2012): 609–614, doi: 10.1161/circoutcomes.112.965707.

12. Anthony J. McMichael, Paul Wilkinson, R. Sari Kovats, Sam Pattenden, Shakoor Hajat, Ben Armstrong, and Nitaya Vajanapoom, "International Study of Temperature, Heat and Urban Mortality: The 'ISOTHURM' Project," *International Journal of Epidemiology* 37, no. 5 (2008): 1121–1132, doi: 10.1093/ije/dyn086.

13. Tiantian Li, Radley M. Horton, and Patrick L. Kinney, "Projections of Seasonal Patterns in Temperature-Related Deaths for Manhattan, New York," *Nature Climate Change* 3 (2013): 717–721, doi: 10.1038/nclimate1902.

14. Ken Parsons, *Human Thermal Environment: The Effects of Hot, Moderate and Cold Temperatures on Human Health, Comfort and Performance*, 2nd ed. (London: Taylor & Francis, 2003).

15. Patrick L. Kinney, Joel Schwartz, Mathilde Pascal, Elisaveta Petkova, Alain Le Tertre, Sylvia Medina, and Robert Vautard, "Winter Season Mortality: Will Climate Warming Bring Benefits?," *Environmental Research Letters* 10, no. 6 (2015): 064016, doi: 10.1088/1748-9326/10/6/064016.

16. Philip L. Staddon, Hugh E. Montgomery, and Michael H. Depledge, "Climate Warming Will Not Decrease Winter Mortality," *Nature Climate Change* 4, no. 3 (2014): 190–194, doi: 10.1038/nclimate2121.

17. Mark R. Goldstein, Luca Mascitelli, and William B. Grant, "Might Vitamin D Explain the Seasonal Variation of Cardiovascular Disease in Tromsø?," *European Journal of Cardiovascular Prevention & Rehabilitation* 18, no. 4 (2011): 678–679, doi: 10.1177/1741826710389412.

18. Li, Horton, and Kinney, "Projections of Seasonal Patterns."

19. Jean-Marie Robine, Siu Lan K. Cheung, Sophie Le Roy, Herman Van Oyen, Clare Griffiths, Jean-Pierre Michel, and François Richard Herrmann, "Death Toll Exceeded 70,000 in Europe during the Summer of 2003," *Comptes Rendus Biologies* 331, no. 2 (2008): 171–178, doi: 10.1016/j.crvi.2007.12.001.

20. Stéphanie Vandentorren, Florence Suzan, Sylvia Medina, Mathilde Pascal, Adeline Maulpoix, Jean-Claude Cohen, and Martine Ledrans, "Mortality in 13 French Cities during the August 2003 Heat Wave," *American Journal of Public Health* 94, no. 9 (2004): 1518–1520, doi: 10.2105/ajph.94.9.1518.

21. Peter Altman et al., "Killer Summer Heat: Projected Death Toll from Rising Temperatures in America Due to Climate Change," *Natural Resources Defense Council Issues Brief* IB:12-05-C (2012), http://nrdc.org/globalwarming/killer-heat/files/killer-summer-heat-report.pdf.

22. Daniel Oudin Åström, Bertil Forsberg, Kristie L. Ebi, and Joacim Rocklöv, "Attributing Mortality from Extreme Temperatures to Climate Change in Stockholm, Sweden," *Nature Climate Change* 3, no. 12 (2013): 1050–1054, doi: 10.1038/nclimate2022.

23. Antonella Zanobetti, Marie S. O'Neill, Carina J. Gronlund, and Joel D. Schwartz, "Summer Temperature Variability and Long-Term Survival among Elderly People with Chronic Disease," *Proceedings of the National Academy of Sciences of the United States* 109, no. 17 (2012): 6608–6613, doi: 10.1073/pnas.1113070109.

24. Tord Kjellstrom, Ingvar Holmer, and Bruno Lemke, "Workplace Heat Stress, Health and Productivity: An Increasing Challenge for Low- and Middle-Income Countries during Climate Change," *Global Health Action* 2 (2009), doi: 10.3402/gha.v2i0.2047.

25. Rebecca L. Laws, Daniel R. Brooks, Juan José Amador, Daniel E. Weiner, James S. Kaufman, Oriana Ramírez-Rubio, Alejandro Riefkohl, et al., "Changes in Kidney Function among Nicaraguan Sugarcane Workers,"

International Journal of Occupational and Environmental Health 21, no. 3 (2015): 241–250, doi: 10.1179/2049396714Y.0000000102.

26. Catharina Wesseling, Jennifer Crowe, Christer Hogstedt, Kristina Jakobsson, Rebekah Lucas, and David Wegman, eds., *Mesoamerican Nephropathy: Report from the First International Research Workshop on MeN* (San José, Costa Rica: IRET, Universidad Nacional, 2013).

27. Tord Kjellstrom, "Impact of Climate Conditions on Occupational Health and Related Economic Losses: A New Feature of Global and Urban Health in the Context of Climate Change," *Asia-Pacific Journal of Public Health* (2015), doi: 10.1177/1010539514568711.

28. Tord Kjellstrom, Bruno Lemke, and Matthias Otto, "Mapping Occupational Heat Exposure and Effects in South-East Asia: Ongoing Time Trends, 1980–2009, and Future Estimates to 2050," *Industrial Health* 51 (2013): 56–67.

29. John P. Dunne, Ronald J. Stouffer, and Jasmin G. John, "Reductions in Labour Capacity from Heat Stress under Climate Warming," *Nature Climate Change* 3, no. 3 (2013): 563–566, doi: 10.1038/nclimate1827.

30. Andrew E. McKechnie and Blair O. Wolf, "Climate Change Increases the Likelihood of Catastrophic Avian Mortality Events during Extreme Heat Waves," *Biology Letters* 6, no. 2 (2010): 253–256, doi: 10.1098/rsbl.2009.0702.

31. Greg Gordon, A. S. Brown, and T. Pulsford, "A Koala (*Phascolarctos cinereus* Goldfuss) Population Crash during Drought and Heatwave Conditions in South-Western Queensland," *Australian Journal of Ecology* 13, no. 4 (1988): 451–461, doi: 10.1111/j.1442-9993.1988.tb00993.x.

32. André Neveu, "Incidence of Climate on Common Frog Breeding: Long-Term and Short-Term Changes," *Acta Oecologica* 35, no. 5 (2009): 671–678, doi: 10.1016/j.actao.2009.06.012.

33. Veronika Huber, Carola Wagner, Dieter Gerten, and Rita Adrian, "To Bloom or Not to Bloom: Contrasting Responses of Cyanobacteria to Recent Heat Waves Explained by Critical Thresholds of Abiotic Drivers," *Oecologia* 169, no. 1 (2012): 245–256, doi: 10.1007/s00442-011-2186-7.

34. Martin Daufresne, Pierre Bady, and Jean-François Fruget, "Impacts of Global Changes and Extreme Hydroclimatic Events on Macroinvertebrate Community Structures in the French Rhône River," *Oecologia* 151, no. 3 (2007): 544–559, doi: 10.1007/s00442-006-0655-1.

35. Raquel A. Silva, J. Jason West, Yuqiang Zhang, Susan C. Anenberg, Jean-François Lamarque, Drew T. Shindell, William J. Collins, Stig Dalsoren, Greg Faluvegi, and Gerd Folberth, "Global Premature Mortality Due to Anthropogenic Outdoor Air Pollution and the Contribution of Past Climate Change," *Environmental Research Letters* 8, no. 3 (2013): 034005, doi: 10.1088/1748-9326/8/3/034005.

36. Michelle L. Bell, Roger D. Peng, and Francesca Dominici, "The Exposure-Response Curve for Ozone and Risk of Mortality and the Adequacy of Current Ozone Regulations," *Environmental Health Perspectives* 114, no. 4

(2006): 532–536, doi: 10.1289/ehp.8816. See also Michael Jerrett, Richard T. Burnett, C. Arden Pope III, Kazuhiko Ito, George Thurston, Daniel Krewski, Yuanli Shi, Eugenia Calle, and Michael Thun, "Long-Term Ozone Exposure and Mortality," *New England Journal of Medicine* 360 (2009): 1085–1095, doi: 10.1056/NEJMoa0803894.

37. Howard H. Chang, Jingwen Zhou, and Montserrat Fuentes, "Impact of Climate Change on Ambient Ozone Level and Mortality in Southeastern United States," *International Journal of Environmental Research and Public Health* 7, no. 7 (2010): 2866–2880, doi: 10.3390/ijerph7072866. See also Lorenzo M. Polvani, Darryn W. Waugh, Gustavo J. P. Correa, and Seok-Woo Son, "Stratospheric Ozone Depletion: The Main Driver of Twentieth-Century Atmospheric Circulation Changes in the Southern Hemisphere," *Journal of Climate* 24, no. 3 (2011): 795–812, doi: 10.1175/2010JCLI3772.1.

38. Julie Wolf, "Elevated Atmospheric Carbon Dioxide Concentrations Amplify *Alternaria alternate* Sporulation and Total Antigen Production," *Environmental Health Perspectives* 118, no. 9 (2010): 1223–1228, doi: 10.1289/ehp.0901867.

39. Katherine M. Shea, Robert T. Truckner, Richard W. Weber, and David B. Peden, "Climate Change and Allergic Disease," *Journal of Allergy and Clinical Immunology* 122, no. 3 (2008): 443–453, doi: 10.1016/j.jaci.2008.06.032.

40. Eija Yli-Panula, Desta Bey Fekedulegn, Brett James Green, and Hanna Ranta, "Analysis of Airborne *Betula* Pollen in Finland: A 31-Year Perspective," *International Journal of Environmental Health Research and Public Health* 6, no. 6 (2009): 1706–1723, doi: 10.3390/ijerph6061706.

41. Lewis Ziska, Kim Knowlton, Christine Rogers, Dan Dalan, Nicole Tierney, Mary Ann Elder, Warren Filley, et al., "Recent Warming by Latitude Associated with Increased Length of Ragweed Pollen Season in Central North America," *Proceedings of the National Academy of Sciences of the United States of America* 108, no. 10 (2011): 4248–4251, doi: 10.1073/pnas.1014107108.

42. Herminia García-Mozo, Carmen Galán, Purificación Alcázar, C. Díaz de la Guardia, Diego Nieto-Lugilde, Marta Recio, Pablo Hidalgo, Francisco González-Minero, L. Ruiz, and Eugenio Domínguez-Vilches, "Trends in Grass Pollen Season in Southern Spain," *Aerobiologia* 26, no. 2 (2010): 157–169, doi: 10.1007/s10453-009-9153-3.

43. Fay Johnston, Ivan Hanigan, Sarah Henderson, Geoffrey Morgan, and David M. J. S. Bowman, "Extreme Air Pollution Events from Bushfires and Dust Storms and Their Association with Mortality in Sydney, Australia, 1994–2007," *Environmental Research* 111, no. 6 (2011): 811–816, doi: 10.1016/j.envres.2011.05.007.

44. Quirin Schiermeier, "Climate and Weather: Extreme Measures," *Nature* 477 (2011), doi: 10.1038/477148a.

45. Dim Coumou and Alexander Robinson, "Historic and Future Increase in Global Land Area Affected by Monthly Heat Extremes," *Environmental Research Letters* 8 (2013), doi: 10.1088/1748-9326/8/3/034018.

46. Munich Reinsurance, *Topics Geo Natural Catastrophes 2010: Analyses, Assessments, Positions* (Munich: Munchener Ruck, 2010), available at http://preventionweb.net/go/17345.

47. IPCC, *Managing the Risks of Extreme Events and Disasters to Advance Climate Change Adaptation: A Special Report of the Working Groups I and II of the Intergovernmental Panel on Climate Change* (Cambridge and New York: Cambridge University Press, 2012).

48. Smith et al., "Human Health."

49. Cormac Ó Gráda, *Famine: A Short History* (Princeton and Oxford: Princeton University Press, 2009).

50. Josette Sheeran, "The Challenge of Hunger," *The Lancet* 371, no. 9608 (2008): 180–181, doi: 10.1016/S0140-6736(07)61870-4.

51. David B. Lobell, "Prioritizing Climate Change Adaptation Needs for Food Security in 2030," *Science* 319, no. 5863 (2008): 607–610, doi: 10.1126/science.1152339.

52. David S. Battisti and Rosamond L. Naylor, "Historical Warnings of Future Food Insecurity with Unprecedented Seasonal Heat," *Science* 323, no. 5911 (2009): 240–244, doi: 10.1126/science.1164363.

53. Colin D. Butler, "Climate Change, Crop Yields and the Future," *United Nations System Standing Committee on Nutrition News (SCN News)* 38 (2010): 18–25.

54. Julian Cribb, *The Coming Famine: The Global Food Crisis and What We Can Do to Avoid It* (Berkeley: University of California Press, 2010).

55. Simon J. Lloyd, R. Sari Kovats, and Zaid Chalabi, "Climate Change, Crop Yields, and Malnutrition: Development of a Model to Quantify the Impact of Climate Scenarios on Child Malnutrition," *Environmental Health Perspectives* 119, no. 12 (2011): 1817–1823, doi: 10.1289/ehp.1003311.

56. Simon Hales, Sari Kovats, Simon Lloyd, and Diarmid Campbell-Lendrum, eds., *Quantitative Risk Assessment of the Effects of Climate Change on Selected Causes of Death, 2030s and 2050s* (Geneva: World Health Organization, 2014).

57. Michael A. Huston, "Precipitation, Soils, NPP, and Biodiversity: Resurrection of Albrecht's Curve," *Ecological Monographs* 82, no. 3 (2012): 277–296, doi: 10.1890/11-1927.1.

58. Annette Prüss-Ustün, Jamie Bartram, Thomas Clasen, John M. Colford, Oliver Cumming, Valerie Curtis, Sophie Bonjour, et al., "Burden of Disease from Inadequate Water, Sanitation and Hygiene in Low- and Middle-Income Settings: A Retrospective Analysis of Data from 145 Countries," *Tropical Medicine and International Health* 19, no. 8 (2014): 894–905, doi: 10.1111/tmi.12329.

59. Walter W. Immerzeel, Ludovicus P. H. van Beek, and Marc F. P. Bierkens, "Climate Change Will Affect the Asian Water Towers," *Science* 328, no. 5984 (2010): 1382–1385, doi: 10.1126/science.1183188.

60. Mauricio E. Arias, Thomas A. Cochrane, Thanapon Piman, Matti Kummu, Brian S. Caruso, and Timothy J. Killeen, "Quantifying Changes in Flooding and Habitats in the Tonle Sap Lake (Cambodia) Caused by Water Infrastructure Development and Climate Change in the Mekong Basin," *Journal of Environmental Management* 112, no. 15 (2012): 53–66, doi: 10.1016/j.jenvman.2012.07.003.

61. Guillaume Constantin de Magny and Rita R. Colwell, "Cholera and Climate: A Demonstrated Relationship," *Transactions of the American Clinical and Climatological Association* 120 (2009): 119–128.

62. Mercedes Pascual, "Malaria Resurgence in the East African Highlands: Temperature Trends Revisited," *Proceedings of the National Academy of Sciences of the United States of America* 103, no. 15 (2006): 5829–5834, doi: 10.1073/pnas.0508929103.

63. Samir Bhatt, Peter W. Gething, Oliver J. Brady, Jane P. Messina, Andrew W. Farlow, Catherine L. Moyes, John M. Drake, et al., "The Global Distribution and Burden of Dengue," *Nature* 496, no. 7446 (2013): 504–507, doi: 10.1038/nature12060.

64. James Whitehorn and Jeremy Farrar, "Dengue," *British Medical Bulletin* 95, no. 1 (2010): 161–173, doi: 10.1093/bmb/ldq019.

65. Philippe Barbazan, Micheline Guiserix, W. Boonyuan, W. Tuntaprasart, D. Pontier, and J.-P. Gonzalez, "Modelling the Effect of Temperature on Transmission of Dengue," *Medical and Veterinary Entomology* 24, no. 1 (2012): 66–73, doi: 10.1111/j.1365-2915.2009.00848.x.

66. R. Sari Kovats, S. J. Edwards, Shakoor Hajat, Benedict Armstrong, Kristie L. Ebi, and Bettina Menne, "The Effect of Temperature on Food Poisoning: A Time-Series Analysis of Salmonellosis in Ten European Countries," *Epidemiology and Infection* 132, no. 3 (2004): 443–453, doi: 10.1017/S0950268804001992. Comparison of diverse geographically dispersed European urban populations shows that gastroenteritis rates tend to increase by 5–10 percent for each 1°C rise above that particular population's distinctive "threshold" temperature, above which gastroenteritis rates rise more rapidly.

67. Elizabeth J. Carlton, Andrew P. Woster, Peter DeWitt, Rebecca S. Goldstein, and Karen Levy, "A Systematic Review and Meta-Analysis of Ambient Temperature and Diarrhoeal Diseases," *International Journal of Epidemiology* (2015): 1–14, doi: 10.1093/ije/dyv296.

68. Joseph B. McLauchlin, Angelo DePaola, Cheryl A. Bopp, Karen A. Martinek, Nancy P. Napolilli, Christine G. Allison, Shelley L. Murray, Eric C. Thompson, Michele M. Bird, and John P. Middaugh, "Outbreak of *Vibrio parahaemolyticus*," *New England Journal of Medicine* 353, no. 14 (2005): 1463–1470, doi: 10.1056/NEJMoa051594.

69. Lyndon E. Llewellyn, "Revisiting the Association between Sea Surface Temperature and the Epidemiology of Fish Poisoning in the South Pacific: Reassessing the Link between Ciguatera and Climate Change," *Toxicon* 56 no. 5 (2010): 691–697, doi: 10.1016/j.toxicon.2009.08.011.

70. Mark P. Skinner, Tom D. Brewer, Ron Johnstone, Lora E. Fleming, and Richard J. Lewis, "Ciguatera Fish Poisoning in the Pacific Islands (1998 to 2008)," *PLoS Neglected Tropical Diseases* 5, no. 12 (2011): e1416, doi: 10.1371/journal.pntd.0001416.

71. Kathleen A. Alexander, "Climate Change Is Likely to Worsen the Public Health Threat of Diarrheal Disease in Botswana," *International Journal of Environmental Research and Public Health* 10, no. 4 (2013): 1202–1230, doi: 10.3390/ijerph10041202.

72. Martin Lukan, Eva Bullova, and Branislav Petko, "Climate Warming and Tick-borne Encephalitis, Slovakia," *Emerging Infectious Diseases* 16, no. 3 (2010): 524–526, doi: 10.3201/eid1603.081364. The ticks feed on mammals in forest and woodland and the infectious agents, a *spirochete* bacterium that causes Lyme disease and a virus that causes tick-borne encephalitis, circulate naturally between ticks and mammal hosts. The ticks' long coevolutionary cohabiting with the forest mammals has achieved a biological détente, a balance of interests: the blood-feeding ticks are well fed, and their stowaway pathogens cause little harm to the infected mammals. But humans have no such coevolutionary protection, and tick-transmitted infection therefore causes more serious clinical disease.

73. Elisabet Lindgren and Rolf Gustafson, "Tick-borne Encephalitis in Sweden and Climate Change," *The Lancet* 358, no. 9275 (2001): 16–18, doi: 10.1016/S0140-6736(00)05250-8.

74. Kevin D. Lafferty, "The Ecology of Climate Change and Infectious Diseases," *Ecology* 90, no. 4 (2009): 888–900, doi: 10.1890/08-0079.1.

75. Dana Sumilo, Loreta Asokliene, Antra Bormane, Veera Vasilenko, Irina Golovljova, and Sarah E. Randolph, "Climate Change Cannot Explain the Upsurge of Tick-borne Encephalitis in the Baltics," *PLoS ONE* 2, no. 6 (2007): e500, doi: 10.1371/journal.pone.0000500.

76. Craig Baker-Austin, Joaquin A. Trinanes, Nick G. H. Taylor, Rachel Hartnell, Anja Siitonen, and Jaime Martinez-Urtaza, "Emerging Vibrio Risk at High Latitudes in Response to Ocean Warming," *Nature Climate Change* 3, no. 1 (2012): 73–77, doi: 10.1038/nclimate1628.

77. Bethany V. Purse, Phillip S. Mellor, David J. Rogers, Alan R. Samuel, Peter P. C. Mertens, and Matthew Baylis, "Climate Change and the Recent Emergence of Bluetongue in Europe," *Nature Reviews Microbiology* 3, no. 2 (2005): 171–181, doi: 10.1038/nrmicro1090.

78. Sir Leonard Rogers, *Smallpox and Climate in India, Forecasting of Epidemics* (London: H. M. Stationery Office, 1926). Cited in "Editorial: Smallpox and Climate," *American Journal of Public Health* 16, no. 10 (1926): 1027–1029, doi:10.2105/AJPH.16.10.1027.

79. Christopher K. Uejio, Alan Kemp, and Andrew C. Comrie, "Climatic Controls on West Nile Virus and Sindbis Virus Transmission and Outbreaks in South Africa," *Vector-Borne and Zoonotic Diseases* 12, no. 2 (2012): 117–125, doi:10.1089/vbz.2011.0655.

80. Menno Jan Bouma and Christopher Dye, "Cycles of Malaria Associated with El Niño in Venezuela," *Journal of the American Medical Association* 278, no. 21 (1997): 1772–1774, doi: 10.1001/jama.1997.03550210070041.

81. Xavier Rodó, Joan Ballester, Dan Cayan, Marian E. Melish, Yoshikazu Nakamura, Ritei Uehara, and Jane C. Burns, "Association of Kawasaki Disease with Tropospheric Wind Patterns," *Scientific Reports* 1 (2011): 152, doi: 10.1038/srep00152.

82. Rodó et al., "Association of Kawasaki Disease."

83. Jessie M. Creamean, Kaitlyn J. Suski, Daniel Rosenfeld, Alberto Cazorla, Paul J. DeMott, Ryan C. Sullivan, Allen B. White, et al., "Dust and Biological Aerosols from the Sahara and Asia Influence Precipitation in the Western U.S.," *Science* 339, no. 6127 (2013): 1572–1578, doi: 10.1126/science.1227279.

84. Yinjie Yang, Shiho Itahashi, Shin-ichi Yokobori, and Akihiko Yamagishi, "UV-Resistant Bacteria Isolated from Upper Troposphere and Lower Stratosphere," *Biological Sciences in Space* 22, no. 1 (2008): 18–25, doi: 10.2187/bss.22.18. Various ultraviolet-resistant species of bacteria have been identified alive and well in the upper troposphere, despite temperatures of −40°C and high exposure to ultraviolet radiation.

85. Gerald A. Meehl, Aixue Hu, Claudia Tebaldi, Julie M. Arblaster, Warren M. Washington, Haiyan Teng, Benjamin M. Sanderson, Toby Ault, Warren G. Strand, and James B. White III, "Relative Outcomes of Climate Change Mitigation Related to Global Temperature versus Sea-Level Rise," *Nature Climate Change* 2, no. 8 (2012): 576–580, doi: 10.1038/nclimate1529.

86. James E. Hansen and Makiko Sato, "Paleoclimate Implications for Human-Made Climate Change" in *Climate Change: Inferences from Paleoclimate and Regional Aspects*, edited by André Berger, Fedor Mesinger, and Djordje Šijacki (Vienna: Springer, 2012), pp. 21–47, doi: 10.1007/978-3-7091-0973-1_2.

87. Andrea Dutton and Kurk Lambeck, "Ice Volume and Sea Level during the Last Interglacial," *Science* 337, no. 6091 (2012): 216–219, doi: 10.1126/science.1205749.

88. Jon Barnett and W. Neil Adger, "Climate Dangers and Atoll Countries," *Climate Change* 61, no. 3 (2003): 321–337, doi: 10.1023/B:CLIM.0000004559.08755.88. See also Daniel White, Larry Hinzman, Lilian Alessa, John Cassano, Molly Chambers, Kelly Falkner, Jennifer Francis, et al., "The Arctic Freshwater System: Changes and Impacts," *Journal of Geophysical Research: Biogeosciences* 112, no. G4 (2007): G04S54, doi: 10.1029/2006JG000353; Eric D. Larson, Zheng Li, and Robert H. Williams, "Chapter 12—Fossil Energy," in *Global Energy*

Assessment—Towards a Sustainable Future (New York: Cambridge University Press, 2012).

89. Phillip H. Muller, "Marshall Islands Foreign Minister: 'We Are Facing a Climate Disaster,'" *Climate Change News*, July 25, 2014, available at http://www.rtcc.org/2014/03/07/marshall-islands-foreign-minister-we-are-facing-a-climate-disaster/.

90. Melita H. Sunjic, "Top UNHCR Official Warns about Displacement from Climate Change," *UNHCR*, December 9, 2008, available at http://www.unhcr.org/493e9bd94.html.

91. Norman Myers, "Environmental Refugees: A Growing Phenomenon of the 21st Century," *Philosophical Transactions of the Royal Society B: Biological Sciences* 357, no. 1420 (2002): 609–613, doi: 10.1098/rstb.2001.0953.

92. Celia McMichael, Jon Barnett, and Anthony J. McMichael, "An Ill Wind? Climate Change, Migration, and Health," *Environmental Health Perspectives* 120, no. 5 (2012): 646–654, doi: 10.1289/ehp.1104375.

93. Asian Development Bank, *Addressing Climate Change and Migration in Asia and the Pacific* (Manila: Asian Development Bank, 2012).

94. UNHCR. "2015 UNHCR Subregional Operations Profile—East and Horn of Africa," available at http://www.unhcr.org/pages/49e45a846.html.

95. McMichael, Barnett, and McMichael, "An Ill Wind?"

96. The Climate Institute, *A Climate of Suffering* (Sydney: Climate Institute, 2011), available at http://www.climateinstitute.org.au/a-climate-of-suffering.html.

97. Lyndall Strazdins, Sharon Friel, Antony McMichael, Susan Woldenberg Butler, and Elizabeth Hanna, "Climate change and child health in Australia: Likely futures, new inequalities?," *International Public Health Journal* 2, no. 4 (2010): 493-500.

98. Pascal Peduzzi, Bruno Chatenoux, Hy Dao, Andrea De Bono, Christian Herold, Jim Kossin, Frederic Mouton, and Ola Nordbeck, "Global Trends in Tropical Cyclone Risk," *Nature Climate Change* 2, no. 4 (2012): 289–294, doi: 10.1038/nclimate1410.

99. Robert Sallares, Abigail Bouwman, and Cecilia Anderung, "The Spread of Malaria to Southern Europe in Antiquity: New Approaches to Old Problems," *Medical History* 48, no. 3 (2004): 311–328.

100. James Henry Breasted, *The Edwin Smith Surgical Papyrus (Facsimile and Hieroglyphic Transliteration with Translation and Commentary, in Two Volumes)* (Chicago: University of Chicago Press, 1930).

101. Zahi Hawass, Yehia Z. Gad, Somaia Ismail, Rabab Khairat, Dina Fathalla, Naglaa Hasan, Amal Ahmed, et al., "Ancestry and Pathology in King Tutankhamun's Family," *Journal of the American Medical Association* 303, no. 7 (2010): 638–647, doi: 10.1001/jama.2010.121.

102. "Malaria Facts," US Centers for Disease Control and Prevention (CDC), March 4, 2015, available at http://www.cdc.gov/malaria/about/facts.html.

103. Weimin Liu, Yingying Li, Gerald H. Learn, Rebecca S. Rudicell, Joel D. Robertson, Brandon F. Keele, Jean-Bosco N. Ndjango, et al., "Origin of

the Human Malaria Parasite *Plasmodium falciparum* in Gorillas," *Nature* 467, no. 7314 (2010): 420–425, doi: 10.1038/nature09442. More recent plasmodial DNA analyses from fecal samples from wild primates in central Africa point to gorillas as the source of *P. falciparum*. Mosquito-borne yellow fever probably had similar sylvatic origins, as early land-clearing farmers came in contact with primate hosts.

104. Andreas Béguin, Simon Hales, Joacim Rocklöva, Christofer Åströma, Valérie R. Louis, and Rainer Sauerborn, "The Opposing Effects of Climate Change and Socio-economic Development on the Global Distribution of Malaria," *Global Environmental Change* 21 (2011): 1214–1219, doi: 10.1016/j.gloenvcha.2011.06.001.

105. Erin A. Mordecai, Krijn P. Paaijmans, Leah R. Johnson, Christian Balzer, Tal Ben-Horin, Emily de Moor, Amy McNally, et al., "Optimal Temperature for Malaria Transmission Is Dramatically Lower than Previously Predicted," *Ecology Letters* 16, no. 1 (2013): 22–33, doi: 10.1111/ele.12015.

106. John Theilmann and Frances Cate, "A Plague of Plagues: The Problem of Plague Diagnosis in Medieval England," *Journal of Interdisciplinary History* 37, no. 3 (2007): 371–393, doi: 10.1162/jinh.2007.37.3.371.

107. Paul Reiter, "From Shakespeare to Defoe: Malaria in England in the Little Ice Age," *Emerging Infectious Diseases* 6, no. 1 (2000): 1–11, doi: 10.3201/eid0601.000101.

108. Keith R. Briffa, "Annual Climate Variability in the Holocene: Interpreting the Message of Ancient Trees," *Quaternary Science Reviews* 19, no. 1–5 (2000): 87–10, doi: 10.1016/S0277-3791(99)00056-6.

109. The *vivax* parasite needs summer temperatures above 16°C for its sporozoite offspring to mature in the mosquito foregut and become transmissible.

110. Kevin D. Lafferty, "The Ecology of Climate Change and Infectious Diseases," *Ecology* 90, no. 4 (2009): 888–900, doi: 10.1890/08-0079.1.

111. Teresa K. Yamana and Elfatih A. B. Eltahir, "Projected Impacts of Climate Change on Environmental Suitability for Malaria Transmission in West Africa," *Environmental Health Perspectives* 121, no. 10 (2013): 1179–1186, doi: 10.1289/ehp.1206174.

112. M. E. Loevinsohn, "Climatic Warming and Increased Malaria Incidence in Rwanda," *Lancet* 343, no. 889 (1994): 714–718, doi: 10.1016/S0140-6736(94)91586-5.

113. Bouma and Dye, "Cycles of Malaria."

114. Menno J. Bouma, Christopher Dye, and H. J. van der Kaay, "Falciparum Malaria and Climate Change in the Northwest Frontier Province of Pakistan," *American Journal of Tropical Medicine and Hygiene* 55, no. 2 (1996): 131–137.

115. Judith A. Omumbo, Bradfield Lyon, Samuel M. Waweru, Stephen J. Connor, and Madeleine C. Thomson, "Raised Temperatures over

the Kericho Tea Estates: Revisiting the Climate in the East African Highlands Malaria Debate," *Malaria Journal* 10 (2011): 12, doi: 10.1186/1475-2875-10-12.

116. Amir S. Siraj, Mauricio Santos-Vega, Menno Bouma, Damtew Yadeta, Daniel Ruiz Carrascal, and Mercedes Pascual, "Altitudinal Changes in Malaria Incidence in Highlands of Ethiopia and Colombia," *Science* 343, no. 6175 (2014): 1154–1158, doi: 10.1126/science.1244325.

117. Omumbo et al., "Raised Temperatures."

118. David Alonso, Menno J. Bouma, and Mercedes Pascual, "Epidemic Malaria and Warmer Temperatures in Recent Decades in an East African Highland," *Proceedings of the Royal Society B* 278, no. 1712 (2011): 1661–1669, doi: 10.1098/rspb.2010.2020.

119. David Stern, Peter W. Gething, Caroline W. Kabaria, William H. Temperley, Abdisalan M. Noor, Emelda A. Okiro, G. Dennis Shanks, Robert W. Snow, and Simon I. Hay, "Temperature and Malaria Trends in Highland East Africa," *PLoS ONE* 6, no. 9 (2011): e24524, doi: 10.1371/journal.pone.0024524.

120. Béguin et al., "Opposing Effects."

121. Bernard Dixon, "The Future of History," *Lancet Infectious Diseases* 9, no. 6 (2009): 338, doi: 10.1016/S1473-3099(09)70137-9.

122. Robert Dirks, "Famine and Disease," in *The Cambridge World History of Human Disease*, edited by Kenneth F. Kiple (Cambridge: Cambridge University Press, 1993), pp. 157–164.

123. William Checkley, Gillian Buckley, Robert H. Gilman, Ana Mo Assis, Richard L. Guerrant, Saul S. Morris, Kåre Mølbak, Palle Valentiner-Branth, Claudio F. Lanata, and Robert E. Black, "Multi-Country Analysis of the Effects of Diarrhoea on Childhood Stunting," *International Journal of Epidemiology* 37, no. 4 (2008): 816–830.

124. Massimo Livi-Bacci, *Population and History: An Essay on European Demographic History* (Cambridge: Cambridge University Press, 1991), p. 47.

125. Ó Gráda, *Famine*, p. 217.

126. Peter Katona and Judit Katona-Apte, "The Interaction between Nutrition and Infection," *Clinical Infectious Disease* 46, no. 10 (2008): 1582–1588, doi: 10.1086/587658.

127. Andrew B. Appleby, "Epidemics and Famine in the Little Ice Age," *Journal of Interdisciplinary History* 10, no. 4 (1980): 643–663.

128. Appleby, "Epidemics and Famine," p. 654.

129. Sophie E. Moore, Timothy J. Cole, Elizabeth M. E. Poskitt, Bakary J. Sonko, Roger G. Whitehead, Ian A. McGregor, and Andrew M. Prentice, "Season of Birth Predicts Mortality in Rural Gambia," *Nature* 388, no. 6641 (1997): 434; Sophie E. Moore, Timothy J. Cole, Andrew C. Collinson, Elizabeth M. E. Poskitt, Ian A. McGregor, and Andrew M. Prentice, "Prenatal or Early Postnatal Events Predict Infectious Deaths in Young

Adulthood in Rural Africa," *International Journal of Epidemiology* 28, no. 6 (1999): 1088–1095.

130. TDR Thematic Reference Group on Environment, Agriculture and Infectious Diseases of Poverty, *Research Priorities for the Environment, Agriculture and Infectious Diseases of Poverty*, WHO Technical Report Series No. 976 (Geneva: WHO, 2013).

131. Rudolf Brázdil, Miroslav Trnka, Petr Dobrovolný, Kateřina Chromá, Petr Hlavinka, and Zdeněk Žalud, "Variability of Droughts in the Czech Republic, 1881–2006," *Theoretical and Applied Climatology* 97, no. 3 (2009): 297–315, doi: 10.1007/s00704-008-0065-x. The Z-index is not an index of rainfall per se, but of the normalized simple water balance, taking into account both temperature and precipitation. (It is a component of the Palmer drought severity index.)

132. High Level Panel of Experts, *Food Security and Climate Change: A Report by the High Level Panel of Experts on Food Security and Nutrition of the Committee on World Food Security* (Rome: FAO, 2012).

133. Wolfram Schlenker and Michael J. Roberts, "Nonlinear Temperature Effects Indicate Severe Damages to US Crops under Climate Change," *Proceedings of the National Academy of Sciences of the United States of America* 106, no. 37 (2009): 15594–15598, doi: 10.1073/pnas.0906865106.

134. David B. Lobell, Marianne Bänziger, Cosmos Magorokosho, and Bindiganavile Vivek, "Nonlinear Heat Effects on African Maize as Evidenced by Historical Yield Trials," *Nature Climate Change* 1, no. 1 (2011): 42–45, doi: 10.1038/nclimate1043.

135. Walter Mertz, "The Essential Trace Elements," *Science* 213, no. 4514 (1981): 1332–1338, doi: 10.1126/science.7022654.

Chapter 4

1. Alfred Russel Wallace, *The Wonderful Century* (London: Dodd, Mead, 1899), p. 140.

2. Anthony D. Barnosky, Nicholas Matzke, Susumu Tomiya, Guinevere O. U. Wogan, Brian Swartz, Tiago B. Quental, Charles Marshall, et al., "Has the Earth's Sixth Mass Extinction Already Arrived?," *Nature* 471, no. 7336 (2011): 51–57, doi: 10.1038/nature09678.

3. Fred Jourdan, Kip V. Hodges, Bryan Keith Sell, Urs Schaltegger, Michael T. D. Wingate, Lena Zetterstrom Evins, Ulf Söderlund, Peter W. Haines, Dave Phillips, and Tom G. Blenkinsop, "High-Precision Dating of the Kalkarindji Large Igneous Province, Australia, and Synchrony with the Early-Middle Cambrian (Stage 4–5) Extinction," *Geology* 42, no. 6 (2014): 543, doi: 10.1130/G35434.1.

4. J. E. N. Veron, "Mass Extinctions and Ocean Acidification: Biological Constraints on Geological Dilemmas," *Coral Reefs* 27, no. 3 (2008): 459–472, doi: 10.1007/s00338-008-0381-8.

5. Terrence J. Blackburn, Paul E. Olsen, Samuel A. Bowring, Noah M. McLean, Dennis V. Kent, John Puffer, Greg McHone, E. Troy Rasbury, and Mohammed Et-Touhami, "Zircon U-Pb Geochronology Links the End-Triassic Extinction with the Central Atlantic Magmatic Province," *Science* 340, no. 6135 (2013): 941–945, doi: 10.1126/science.1234204.

6. Jessica L. Blois, Phoebe L. Zarnetske, Matthew C. Fitzpatrick, and Seth Finnegan, "Climate Change and Past, Present and Future Biotic Interactions," *Science* 341, no. 6145 (2013): 499–504, doi: 10.1126/science.1237184.

7. Paul G. Harnik, Heike K. Lotze, Sean C. Anderson, Zoe V. Finkel, Seth Finnegan, David R. Lindberg, Lee Hsiang Liow, et al., "Extinctions in Ancient and Modern Seas," *Trends in Ecology and Evolution* 27, no. 11 (2012): 608–617, doi: 10.1016/j.tree.2012.07.010.

8. Zhong-Qiang Chen and Michael J. Benton, "The Timing and Pattern of Biotic Recovery Following the End-Permian Mass Extinction," *Nature Geoscience* 5, no. 6 (2012): 375–383, doi: 10.1038/ngeo1475.

9. Peter J. Mayhew, Gareth B. Jenkins, and Timothy G. Benton, "A Long-Term Association between Global Temperature and Biodiversity, Origination and Extinction in the Fossil Record," *Proceedings of the Royal Society of London B: Biological Sciences* 275, no. 1630 (2008): 47–53, doi: 10.1098/rspb.2007.1302.

10. Paul R. Renne, Alan L. Deino, Frederik J. Hilgen, Klaudia F. Kuiper, Darren F. Mark, William S. Mitchell, Leah E. Morgan, Roland Mundil, and Jan Smit, "Time Scales of Critical Events Around the Cretaceous-Paleogene Boundary," *Science* 339, no. 6120 (2013): 684–687, doi: 10.1126/science.1230492. See also Heiko Pälike, "Impact and Extinction," *Science* 339, no. 6120 (2013): 655–656, doi: 10.1126/science.1233948.

11. Stephen L. Brusatte, Richard J. Butler, Paul M. Barrett, Matthew T. Carrano, David C. Evans, Graeme T. Lloyd, Philip D. Mannion, et al., "The Extinction of the Dinosaurs," *Biological Reviews* 90, no. 2 (2014): 628–642, doi: 10.1111/brv.12128.

12. Barnosky et al., "Has the Earth's Sixth Mass Extinction."

13. James Hansen, Makiko Sato, Pushker Kharecha, David Beerling, Valerie Masson-Delmotte, Mark Pagani, Maureen Raymo, Dana L. Royer, and James C. Zachos, "Target Atmospheric CO_2: Where Should Humanity Aim?" *Open Atmospheric Science Journal* 2 (2008): 217–231, doi: 10.2174/1874282300802010217.

14. Robert A. Rohde, "Ice Age Temperature Changes" (chart), available at http://en.wikipedia.org/wiki/File:Ice_Age_Temperature.png.

15. James C. Zachos, Gerald R. Dickens, and Richard E. Zeebe, "An Early Cainozoic Perspective on Greenhouse Warming and Carbon-Cycle Dynamics," *Nature* 451, no. 7176 (2008): 279–283, doi: 10.1038/nature06588.

16. David Archer, *The Global Carbon Cycle* (Princeton: Princeton University Press, 2010), argues that the world's carbon cycle behaves unpredictably in

different circumstances. During the PETM 55 million years ago, after a huge release of carbon that caused 5°–6°C warming, carbon uptake by natural sinks acted as a *dampener* by restoring the previous "normal" temperature. At other times, the carbon cycle has *amplified* small temperature excursions into major climate events. Humans, says Archer, are now adding a PETM-scale dose of carbon to the atmosphere at a time when the cycle is acting mostly as amplifier.

17. Hansen et al., "Target Atmospheric."

18. Sudhir Kumar, Alan Filipski, Vinod Swarna, Alan Walker, and S. Blair Hedges, "Placing Confidence Limits on the Molecular Age of the Human-Chimpanzee Divergence," *Proceedings of the National Academy of Sciences of the United States of America* 102, no. 52 (2005): 18842–18847, doi: 10.1073/pnas.0509585102.

19. Ann Gibbons, "How a Fickle Climate Made Us Human," *Science* 341, no. 6145 (2013): 474–479, doi: 10.1126/science.341.6145.474.

20. Thure E. Cerling, Kendra L. Chritz, Nina G. Jablonski, Meave G. Leakey, and Fredrick Kyalo Manthi, "Diet of *Theropithecus* from 4 to 1 Ma in Kenya," *Proceedings of the National Academy of Sciences of the United States of America* 110, no. 26 (2013): 10507–10512, doi: 10.1073/pnas.1222571110; Thure E. Cerling, Fredrick Kyalo Manthi, Emma N. Mbua, Louise N. Leakey, Meave G. Leakey, Richard E. Leakey, Francis H. Brown, et al., "Stable Isotope-Based Diet Reconstructions of Turkana Basin Hominins," *Proceedings of the National Academy of Sciences of the United States of America* 110, no. 26 (2013): 10501–10506, doi: 10.1073/pnas.1222568110.

21. James C. Zachos, Nicholas J. Shackleton, Justin S. Revenaugh, Heiko Palike, and Benjamin P. Flower, "Climate Response to Orbital Forcing across the Oligocene-Miocene Boundary," *Science* 292, no. 5515 (2001): 274–278, doi: 10.1126/science.1058288.

22. James E. Hansen, *Storms of My Grandchildren* (New York: Bloomsbury Press, 2009).

23. *Pithecus*, from Greek *pithēkos*, means monkey or ape.

24. Ferran Estebaranz, Jordi Galbany, Laura M. Martínez, Daniel Turbón, and Alejandro Pérez-Pérez, "Buccal Dental Microwear Analyses Support Greater Specialization in Consumption of Hard Foodstuffs for *Australopithecus anamensis*," *Journal of Anthropological Sciences* 90 (2012): 163–185, doi: 10.4436/jass.900060.

25. Shannon P. McPherron, Zeresenay Alemseged, Curtis W. Marean, Jonathan G. Wynn, Denné Reed, Denis Geraads, René Bobe, and Hamdallah A. Béarat, "Evidence for Stone-Tool-Assisted Consumption of Animal Tissues before 3.39 Million Years Ago at Dikika, Ethiopia," *Nature* 466, no. 7308 (2010): 857–860, doi: 10.1038/nature09248.

26. Jonathan G. Wyn, "Influence of Plio-Pleistocene Aridification on Human Evolution: Evidence from Paleosols of the Turkana Basin, Kenya," *American Journal of Physical Anthropology* 123, no. 2 (2004): 106–118, doi: 10.1002/ajpa.10317.

27. Peter B. deMenocal, "African Climate Change and Faunal Evolution during the Pliocene-Pleistocene," *Earth Planetary Science Letters* 220, no. 1–2 (2004): 3–24, doi: 10.1016/S0012-821X(04)00003-2.

28. Sir Joseph Dalton Hooker, director of the Royal Kew Botanical Gardens and leading long-standing supporter of Charles Darwin's theory of evolution, quoted in Iain McCalman, *Darwin's Armada: Four Voyages and the Battle for the Theory of Evolution* (New York: W. W. Norton, 2010), p. 202.

29. Cécile Charrier, Kaumudi Joshi, Jaeda Coutinho-Budd, Ji-Eun Kim, Nelle Lambert, Jacqueline de Marchena, Wei-Lin Jin, et al., "Inhibition of SRGAP2 Function by its Human-specific Paralogs Induces Neoteny during Spine Maturation," *Cell* 149, no. 4 (2012): 923–935, doi: 10.1016/j.cell.2012.03.034.

30. Leslie C. Aiello and Peter Wheeler, "The Expensive Tissue Hypothesis: The Brain and Digestive System in Human and Primate Evolution," *Current Anthropology* 36, no. 2 (1995): 199–221.

31. Richard Wrangham, *Catching Fire: How Cooking Made Us Human* (London: Profile Books, 2010).

32. Clive Finlayson, "Biogeography and Evolution of the Genus *Homo*," *Trends in Ecology and Evolution* 20, no. 8 (2005): 457–463, doi: 10.1016/j.tree.2005.05.019.

33. Leslie C. Aiello and Paul Wheeler, "Neanderthal Thermoregulation and the Glacial Climate," in *Neanderthals and Modern Humans in the European Landscape during the Last Glaciations*, edited by Tjeerd H. van Andel, William Davies, and Leslie Aiello (Cambridge: McDonald Institute for Archaeological Research, 2003), pp. 147–166.

34. Julien Riel-Salvatore, "A Niche Construction Perspective on the Middle-Upper Paleolithic Transition in Italy," *Journal of Archaeological Method and Theory* 17, no. 4 (2010), doi: 10.1007/s10816-010-9093-9.

35. Johannes Krause, Qiaomei Fu, Jeffrey M. Good, Bence Viola, Michael V. Shunkov, Anatoli Derevianko, and Svante Pääbo, "The Complete Mitochondrial DNA Genome of an Unknown Hominin from Southern Siberia," *Nature* 464, no. 7290 (2010): 894–897, doi: 10.1038/nature08976.

36. Michael Marshall, "The Vast Asian Realm of the Lost Humans," *New Scientist*, 28 September 2011, available at http://www.newscientist.com/article/mg21128323.200-the-vast-asi.

37. John R. Stewart and Chris V. Stringer, "Human Evolution Out of Africa: The Role of Refugia and Climate Change," *Science* 335, no. 6074 (2012): 1317–1320, doi: 10.1126/science.1215627.

38. Brenna M. Henn, Christopher R. Gignoux, Matthew Jobin, Julie M. Granka, J. M. MacPherson, Jeffrey M. Kidd, Laura Rodríguez-Botigué, et al., "Hunter-Gatherer Genomic Diversity Suggests a Southern African Origin for Modern Humans," *Proceedings of the National Academy of Sciences of the United States of America* 108, no. 13 (2011): 5154–5162, doi: 10.1073/pnas.1017511108.

39. John J. Shea, "Transitions or Turnovers? Climatically-Forced Extinctions of *Homo sapiens* and Neanderthals in the East Mediterranean Levant," *Quaternary Science Reviews* 27, no. 23–24 (2008): 2253–2270, doi: 10.1016/j.quascirev.2008.08.015.

40. Tatjana Boettger, Elena Yu. Novenko, Andrej A. Velichko, Olga K. Borisova, Konstantin V. Kremenetski, Stefan Knetsch, and Frank W. Junge, "Instability of Climate and Vegetation Dynamics in Central and Eastern Europe during the Final Stage of the Last Interglacial (Eemian, Mikulino) and Early Glaciation," *Quaternary International* 207, no. 1–2 (2009): 137–144, doi: 10.1016/j.quaint.2009.05.006.

41. Ajit Varki, "Dating the Origin of Us," *The Scientist,* November 1, 2013, available at http://www.the-scientist.com/?articles.view/articleNo/38008/title/Dating-the-Origin-of-Us/.

42. Ralf Kittler, Manfred Kayser, and Mark Stoneking, "Molecular Evolution of *Pediculus humanus* and the Origin of Clothing," *Current Biology* 13, no. 16 (2003): 1414–1417, doi: 10.1016/S0960-9822(03)00507-4.

43. Robin A. Weiss, "Apes, Lice and Prehistory," *Journal of Biology* 8, no. 2 (2009): 20, doi: 10.1186/jbiol114.

44. Qiaomei Fu, Alissa Mittnik, Philip L. F. Johnson, Kirsten Bos, Martina Lari, Ruth Bollongino, Chengkai Sun, et al., "A Revised Timescale for Human Evolution Based on Ancient Mitochondrial Genomes," *Current Biology* 23, no. 7 (2013): 553–559, doi: 10.1016/j.cub.2013.02.044.

45. Paul Mellars, Kevin C. Gori, Martin Carr, Pedro A. Soares, and Martin B. Richards, "Genetic and Archaeological Perspectives on the Initial Modern Human Colonization of Southern Asia," *Proceedings of the National Academy of Sciences of the United States of America* 110, no. 26 (2013): 10699–10704, doi: 10.1073/pnas.1306043110.

46. Pedro Soares, Luca Ermini, Noel Thomson, Maru Mormina, Teresa Rito, Arne Röhl, Antonio Salas, Stephen Oppenheimer, Vincent Macaulay, and Martin B. Richards, "Correcting for Purifying Selection: An Improved Human Mitochondrial Molecular Clock," *American Journal of Human Genetics,* 84, no. 6 (2009): 740–759, doi: 10.1016/j.ajhg.2009.05.001.

47. Anne H. Osborne, Derek Vance, Eelco J. Rohling, Nick Barton, Mike Rogerson, and Nuri Fello, "A Humid Corridor across the Sahara for the Migration of Early Modern Humans out of Africa 120,000 Years Ago," *Proceedings of the National Academy of Sciences of the United States of America* 105, no. 43 (2008): 16444–16447, doi: 10.1073_pnas.0804472105.

48. Hugo Reyes-Centeno, Silvia Ghirotto, Florent Détroit, Dominique Grimaud-Hervé, Guido Barbujani, and Katerina Harvati, "Genomic and Cranial Phenotype Data Support Multiple Modern Human Dispersals from Africa and a Southern Route into Asia," *Proceedings of the National Academy of Sciences of the United States of America* 111, no. 20 (2014): 7248–7253, doi: 10.1073/pnas.1323666111.

49. Anton Vaks, Miryam Bar-Matthews, Avner Ayalon, Alan Matthews, Amos Frumkin, Uri Dayan, Ludwik Halicz, Ahuva Almogi-Labin, and Bettina Schilman, "Paleoclimate and Location of the Border between Mediterranean Climate Region and the Saharo-Arabian Desert as Revealed by Speleothems from the Northern Negev Desert, Israel," *Earth and Planetary Science Letters* 249, no. 3–4 (2006): 384–399, doi: 10.1016/j.epsl.2006.07.009.

50. Anders Svensson, M. Bigler, T. Blunier, H. B. Clausen, D. Dahl-Jensen, H. Fischer, S. Fujita, et al., "Direct Linking of Greenland and Antarctic Ice Cores at the Toba Eruption (74 kyr BP)," *Climate of the Past Discussions* 8, no. 6 (2012): 5389–5427, doi: 10.5194/cpd-8-5389-2012.

51. Michael R. Rampino and Stephen Self, "Climate-Volcanism Feedback and the Toba Eruption of ~74,000 Years Ago," *Quaternary Research* 40, no. 3 (1993): 269–280, doi: 10.1006/qres.1993.1081.

52. Alan Robock, Caspar M. Ammann, Luke Oman, Drew Shindell, Samuel Levis, and Georgiy Stenchikov, "Did the Toba Volcanic Eruption of ~74k BP Produce Widespread Glaciation?," *Journal of Geophysical Research* 114, no. D10 (2009): D10107, doi: 10.1029/2008JD011652.

53. Sacha C. Jones, "The Toba Supervolcanic Eruption: Tephra-Fall Deposits in India and Paleoanthropological Implications," in *The Evolution and History of Human Populations in South Asia*, edited by Michael D. Petraglia and Bridget Allchin (Dordrecht: Springer, 2007), pp. 173–200.

54. John Savino and Marie D. Jones, *Supervolcano: The Catastrophic Event that Changed the Course of Human History* (Franklin Lakes: New Page Books, 2007).

55. Jones, "The Toba Supervolcanic Eruption."

56. Stanley H. Ambrose, "Late Pleistocene Human Population Bottlenecks, Volcanic Winter, and Differentiation of Modern Humans," *Journal of Human Evolution* 34, no. 6 (1998): 623–651, doi:10.1006/jhev.1998.0219.

57. F. J. Gathorne-Hardy and W. E. H. Harcourt-Smith, "The Super-Eruption of Toba, Did it Cause a Human Bottleneck?" *Journal of Human Evolution* 45, no. 3 (2003): 227–230, doi: 10.1016/S0047-2484(03)00105-2.

58. Mellars et al., "Genetic and Archaeological Perspectives."

59. Ambrose, "Late Pleistocene Human Population Bottlenecks."

60. George Weber, "Toba–Aftermath: Climate and Environment," George Weber's Lonely Islands, 2007, http://www.andamans.org/toba-aftermath-climate-and-environment/.

61. Nicole Misarti, Bruce P. Finney, James W. Jordan, Herbert D. G. Maschner, Jason A. Addison, Mark D. Shapley, Andrea Krumhardt, and James E. Beget, "Early Retreat of the Alaska Peninsula Glacier Complex and the Implications for Coastal Migrations of First Americans," *Quaternary Science Reviews* 48 (2012): 1–6, doi: 10.1016/j.quascirev.2012.05.014.

62. Referred to as Heinrich events (cooling) and Dansgaard-Oeschger events (warming), respectively.

63. Martin Ziegler, Margit H. Simon, Ian R. Hall, Stephen Barker, Chris Stringer, and Rainer Zahn, "Development of Middle Stone Age Innovation Linked to Rapid Climate Change," *Nature Communications* 4 (2013): 1905, doi: 10.1038/ncomms2897.

64. R. A. Weiss, "The Leeuwenhoek Lecture 2001: Animal Origins of Human Infectious Disease," *Philosophical Transactions of the Royal Society B* 356, no. 1410 (2001): 957–977, doi: 10.1098/rstb.2001.0838.

65. Anthony J. McMichael, "Zoonoses," in *Proceedings, International Conference on Climate, Environment and Infectious Diseases*, Monaco, March 2012.

66. Ro McFarlane, Adrian Sleigh, and Anthony J. McMichael, "Synanthropy of Wild Mammals as a Determinant of Emerging Infectious Diseases in the Asian-Australasian Region," *EcoHealth* 9, no. 1 (2012): 24–35, doi: 10.1007/s10393-012-0763-9.

67. Stephen G. Webb, *Palaeopathology of Aboriginal Australians: Health and Disease across a Hunter-Gatherer Continent* (Cambridge: Cambridge University Press, 2009), p. 24.

68. Mike Smith, *The Archaeology of Australia's Deserts* (Cambridge: Cambridge University Press, 2013), p. 110.

69. R. G. Kimber, "Hunter-Gatherer Demography: The Recent Past in Central Australia," in *Hunter-Gatherer Demography: Past and Present* (Oceania Monograph 39), edited by Betty Meehan and Neville G. White (Sydney: University of Sydney, 1990), pp. 160–170.

70. Peter B. Thorley, "Shifting Location, Shifting Scale: A Regional Landscape Approach to the Prehistoric Archaeology of the Palmer River Catchment, Central Australia," PhD diss., Northern Territory University, 1998, p. 42.

71. Webb, *Palaeopathology of Aboriginal Australians*, p. 173.

72. A. W. G. Pike, D. L. Hoffmann, Marcos García-Diez, Paul B. Pettitt, José Javier Alcolea-González, Rodrigo De Balbin-Behrmann, C. González-Sainz, et al., "U-Series Dating of Paleolithic Art in 11 Caves in Spain," *Science* 336, no. 6087 (2012): 1409–1413, doi: 10.1126/science.1219957.

73. Robert L. Cieri, Steven E. Churchill, Robert G. Franciscus, Jingzhi Tan, and Brian Hare, "Craniofacial Feminization, Social Tolerance, and the Origins of Behavioral Modernity," *Current Anthropology* 55, no. 4 (2014): 419–443, doi: 10.1086/677209.

74. Rachel E. Wood, Cecilio Barroso-Ruíz, Miguel Caparrós, Jesús F. Jordá Pardo, Bertila Galván Santos, and Thomas F. G. Higham, "Radiocarbon Dating Casts Doubt on the Late Chronology of the Middle to Upper Palaeolithic transition in Southern Iberia," *Proceedings of the National Academy of Sciences of the United States of America* 110, no. 8 (2013): 2781–2786, doi: 10.1073/pnas.1207656110.

75. Clive Finlayson, Francisco Giles Pacheco, Joaquín Rodríguez-Vidal, Darren A. Fa, José María Gutierrez López, Antonio Santiago Pérez, Geraldine Finlayson, et al., "Late Survival of Neanderthals at the Southernmost Extreme of Europe," *Nature* 443, no. 7113 (2006): 850–853, doi: 10.1038/nature05195.

76. P. C. Tzedakis, K. A. Hughen, I. Cacho, and K. Harvati, "Placing Late Neanderthals in a Climatic Context," *Nature* 449, no. 7159 (2007): 206–208, doi: 10.1038/nature06117.

77. Love Dalén, Ludovic Orlando, Beth Shapiro, Mikael Brandström Durling, Rolf Quam, M. Thomas P. Gilbert, J. Carlos Díez Fernández-Lomana, Eske Willerslev, Juan Luis Arsuaga, and Anders Götherström, "Partial Genetic Turnover in Neanderthals: Continuity in the East and Population Replacement in the West," *Molecular Biology and Evolution* 29, no. 8 (2012): 1893–1897, doi: 10.1093/molbev/mss074.

78. Krause et al., "The Complete Mitochondrial DNA Genome."

79. Sriram Sankararaman, Swapan Mallick, Michael Dannemann, Kay Prüfer, Janet Kelso, Svante Pääbo, Nick Patterson, and David Reich, "The Genomic Landscape of Neanderthal Ancestry in Present-Day Humans," *Nature* 507, no. 7492 (2014): 354–357, doi: 10.1038/nature12961.

80. Universität Bonn, " 'Immune Gene' in Humans Inherited from Neanderthals, Study Suggests," *ScienceDaily*, November 22, 2013, http://www.sciencedaily.com/releases/2013/11/131122084405.htm.

81. Emilia Huerta-Sánchez, Xin Jin, Asan, Zhuoma Bianba, Benjamin M. Peter, Nicolas Vinckenbosch, Yu Liang, et al., "Altitude Adaptation in Tibetans Caused by Introgression of Denisovan-like DNA," *Nature* 512, no. 7513 (2014): 194–197, doi: 10.1038/nature13408.

82. Laurent Abi-Rached, Matthew J Jobin, Subhash Kulkarni, Alasdair McWhinnie, Klara Dalva, Loren Gragert, Farbod Babrzadeh, et al., "The Shaping of Modern Human Immune Systems by Multiregional Admixture with Archaic Humans," *Science* 334, no. 6052 (2011): 89–94, doi: 10.1126/science.1209202.

83. Varki, "Dating the Origin of Us."

84. Katerina Harvati, "Neanderthals," *Evolution: Education and Outreach* 3, no. 3 (2010): 367–376, doi: 10.1007/s12052-010-0250-0.

85. Clive Finlay and José S. Carrion, "Rapid Ecological Turnover and its Impact on Neanderthal and Other Human Populations," *Trends in Ecology and Evolution* 22, no. 4 (2008): 213–222, doi: 10.1016/j.tree.2007.02.001.

86. Graham W. Prescott, David R. Williams, Andrew Balmford, Rhys E. Green, and Andrea Manica, "Quantitative Global Analysis of the Role of Climate and People in Explaining Late Quaternary Megafaunal Extinctions," *Proceedings of the National Academy of Sciences of the United States of America* 109, no. 12 (2012): 4527–4531, doi: 10.1073/pnas.1113875109.

87. Stephen Wroe, Judith H. Field, Michael Archer, Donald K. Grayson, Gilbert J. Price, Julien Louys, J. Tyler Faith, Gregory E. Webb, Iain Davidson, and Scott D. Mooney, "Climate Change Frames Debate over the Extinction of Megafauna in Sahul (Pleistocene Australia–New Guinea)," *Proceedings of the National Academy of Sciences of the United States of America* 110, no. 22 (2013): 8777–8781, doi: 10.1073/pnas.1302698110.

88. Eske Willerslev, John Davison, Mari Moora, Martin Zobel, Eric Coissac, Mary E. Edwards, Eline D. Lorenzen, et al., "Fifty Thousand Years of Arctic

Vegetation and Megafaunal Diet," *Nature* 506, no. 7486 (2014): 47–51, doi: 10.1038/nature12921.

89. Mark N. Cohen, *The Food Crisis in Prehistory: Overpopulation and the Origins of Agriculture* (New Haven: Yale University Press, 1979).

90. Ofer Bar-Yosef, "The Natufian Culture in the Levant: Threshold to the Origins of Agriculture," *Evolutionary Anthropology* 6, no. 5 (1998): 159–177.

91. Gordon Hillman, Robert Hedges, Andrew Moore, Susan Colledge, and Paul Pettitt, "New Evidence of Lateglacial Cereal Cultivation at Abu Hureyra on the Euphrates," *The Holocene* 11, no. 4 (2001): 383–393, doi: 10.1191/095968301678302823.

92. "A map of the Levant with Natufian regions across present-day Israel, Palestine, and a long arm extending into Lebanon and Syria," Wikipedia, https://en.wikipedia.org/wiki/Natufian_culture#/media/File:NatufianSpread.svg.

93. Angela E. Close, "*Plus Ça Change*: The Pleistocene-Holocene Transition in Northeast Africa," in *Humans at the End of the Ice Age*, edited by Lawrence Guy Straus, Berit Valentin Eriksen, Jon M. Erlandson, and David R. Yesner (New York: Plenum Press, 1996), pp. 43–60.

94. The Younger Dryas gets its name from sedimentary pollen evidence that the little arctic flowering plant *Dryas octopetala* was able to extend its range south to newly cooled Denmark. This had previously happened more briefly around 14,000 BP during, yes, the Older Dryas.

95. James T. Teller, David W. Leverington, and Jason D. Mann, "Freshwater Outbursts to the Oceans from Glacial Lake Agassiz and Their Role in Climate Change During the Last Deglaciation," *Quaternary Science Reviews* 21, no. 8–9 (2002): 879–887, doi: 10.1016/S0277-2791(01)00145-7. See also Julian B. Murton, Mark D. Bateman, Scott R. Dallimore, James T. Teller, and Zhirong Yang, "Identification of Younger Dryas Outburst Flood Path from Lake Agassiz to the Arctic Ocean," *Nature* 464, no. 7289 (2010): 740–743, doi: 10.1038/nature08954.

96. Malcolm A. LeCompte, Albert C. Goodyear, Mark N. Demitroff, Dale Batchelor, Edward K. Vogel, Charles Mooney, Barrett N. Rock, and Alfred W. Seidel, "Independent Evaluation of Conflicting Microspherule Results from Different Investigations of the Younger Dryas Impact Hypothesis," *Proceedings of the National Academy of Sciences of the United States of America* 109, no. 44 (2012): E2960–E2969, doi: 10.1073/pnas.1208603109.

97. Yingzhe Wu, Mukul Sharma, Malcolm A. LeCompte, Mark N. Demitroff, and Joshua D. Landis, "Origin and Provenance of Spherules and Magnetic Grains at the Younger Dryas Boundary," *Proceedings of the National Academy of Sciences of the United States of America* 110, no. 38 (2013): E3557–E3566, doi: 10.1073/pnas.1304059110

98. Close, "*Plus Ça Change.*"

99. Peter J. Mitchell, Royden Yates, and John E. Parkington, "At the Transition: The Archaeology of the Pleistocene-Holocene Boundary in Southern

Africa," in *Humans at the End of the Ice Age*, edited by Lawrence Guy Straus et al. (New York: Plenum Press, 1996), pp. 15–41.

100. Charles C. Mann, "The Birth of Religion," *National Geographic*, June 2011, available at http://ngm.nationalgeographic.com/2011/06/gobekli-tepe/mann-text/1.

101. Graeme Barker, *The Agricultural Revolution in Prehistory: Why Did Foragers Become Farmers?* (Oxford: Oxford University Press, 2006).

102. Scott C. Meeks and David G. Anderson, "Evaluating the Effect of the Younger Dryas on Human Population Histories in the Southeastern United States," in *Hunter-Gatherer Behavior: Human Response during the Younger Dryas*, edited by Metin I. Eren (Walnut Creek: Left Coast Press, 2012), pp. 111–138.

103. The dates for Paleo-Indian migration are subject to continuing research.

104. Wu et al., "Origin and Provenance."

105. James P. Kennett and Allen West, "Biostratigraphic Evidence Supports Paleoindian Population Disruption at Approximately 12.9 ka," *Proceedings of the National Academy of Sciences of the United States of America* 105, no. 50 (2008): E110, doi: 10.1073/pnas.0809004196.

106. George Willcox, Ramon Buxo, and Linda Herveux, "Late Pleistocene and Early Holocene Climate and the Beginnings of Cultivation in Northern Syria," *The Holocene* 19, no. 1 (2009): 151–158, doi: 10.1177/0959683608098961.

107. Jared Diamond, *Guns, Germs and Steel: The Fate of Human Societies* (New York: W. W. Norton, 1997).

108. Jules Janick, "Ancient Egyptian Agriculture and the Origins of Horticulture," *ISHS Acta Horticulturae* 582 (2002): 23–39, doi: 10.17660/ActaHortic.2002.582.1.

109. Sylvain Glémin and Thomas Bataillon, "A Comparative View of the Evolution of Grasses under Domestication," *New Phytologist* 183, no. 2 (2009): 273–290, doi: 10.1111/j.1469-8137.2009.02884.x.

110. Elizabeth A. Kellogg, "Evolutionary History of the Grasses," *Plant Physiology* 125, no. 3 (2001): 1198–1205, doi: 10.1104/pp.125.3.1198.

111. Clark S. Larsen, "The Agricultural Revolution as Environmental Catastrophe: Implications for Health and Lifestyle in the Holocene," *Quaternary International* 150, no. 1 (2006): 12–20, doi: 10.1016/j.quaint.2006.01.004.

112. Amanda Mummert, Emily Esche, Joshua Robinson, and George J. Armelagos, "Stature and Robusticity during the Agricultural Transition: Evidence from the Bioarchaeological Record," *Economics & Human Biology* 9, no. 3 (2011): 284–301, doi: 10.1016/j.ehb.2011.03.004.

113. Webb, *Palaeopathology of Aboriginal Australians*, p. 279.

114. Webb, *Palaeopathology of Aboriginal Australians*, p. 280.

115. Roberts, Charlotte, "Infectious disease in biocultural perspective: past, present and future work in Britain," in Margaret Cox and Simon Mays

(eds.), *Human Osteology: In Archaeology and Forensic Science* (New York: Cambridge University Press, 2000), 145–62.

116. Melinda A. Zeder, "Domestication and Early Agriculture in the Mediterranean Basin: Origins, Diffusion, and Impact," *Proceedings of the National Academy of Sciences of the United States of America* 105, no. 33 (2008): 11597–11604, doi: 10.1073/pnas.0801317105.

117. Jared E. Decker, Stephanie D. McKay, Megan M. Rolf, JaeWoo Kim, Antonio Molina Alcalá, Tad S. Sonstegard, Olivier Hanotte, et al., "Worldwide Patterns of Ancestry, Divergence, and Admixture in Domesticated Cattle," *PLoS Genetics* 10, no. 3 (2014): e1004254, doi: 10.1371/journal.pgen.1004254.

118. McFarlane, Sleigh, and McMichael, "Synanthropy of Wild Mammals."

119. George Willcox, "The Roots of Cultivation in Southwestern Asia," *Science* 341, no. 6141 (2013): 39–40, doi: 10.1126/science.1240496.

120. Panel on Climate Variability on Decade-to-Century Time Scales, National Research Council, *Decade-to-Century-Scale Climate Variability and Change: A Science Strategy* (Washington, DC: National Academies Press, 1998).

121. Joan Feynman and Alexander Ruzmaikin, "Climate Stability and the Development of Agricultural Societies," *Climatic Change* 84, no. 3 (2007): 295–311, doi: 10.1007/s10584-007-9248-1.

122. Jared Diamond and Peter Bellwood, "Farmers and Their Languages: The First Expansions," *Science* 300, no. 5619 (2003): 597–603, doi: 10.1126/science.1078208.

123. Mark Cohen, "Introduction: Rethinking the Origins of Agriculture," *Current Anthropology* 50, no. 5 (2009): 591–595, doi: 10.1086/603548.

124. D. J. Cohen, "The Origin of Domesticated Cereals and the Pleistocene–Holocene Transition in Eastern Asia," *Review of Archaeology* 19, no. 2 (1998): 22–29.

125. Feynman and Ruzmaikin, "Climate Stability."

126. Feynman and Ruzmaikin, "Climate Stability."

127. Diamond, *Guns, Germs*.

Chapter 5

1. William Burroughs, ed., *Climate: Into the 21st Century* (Cambridge: Cambridge University Press, 2003), p. 66: Around 14,000 years ago, solar radiation increased marginally in response to critical changes in Earth's orbital and rotational parameters—enough to draw the ITCZ further north. At 65°N, slight variations (up to 9 percent) in incoming solar radiation, due to combinations of these orbital/rotational (Milankovitch) cycles, are the primary influence on the advance or retreat of ice-sheets. Immediately after the Last Glacial Maximum, 18,000 years ago, the ice-sheets began retreating—indicating a period of increasing insolation at that high northern latitude. Several millennia later, the increase was sufficient to induce a northern shift in the ITCZ.

2. Raymond S. Bradley, "Holocene Perspectives on Future Climate Change," in *Nature Climate Variability and Global Warming: A Holocene Perspective,* edited by Richard W. Battarbee and Heather A. Binney (Oxford: Wiley Blackwell, 2008), p. 255.

3. US National Climatic Data Center, "The Mid-Holocene 'Warm Period'", August 20, 2008, http://www.ncdc.noaa.gov/paleo/globalwarming/holocene.html.

4. Barry Cunliffe, *Europe Between the Oceans, 9000 BC–1000 AD* (London: Yale University Press, 2008).

5. Luigi Luca Cavalli-Sforza and Francesco Cavalli-Sforza, *The Great Human Diaspora: The History of Diversity and Evolution,* trans. Sarah Thorne (New York: Perseus Books, 1996).

6. Andrew Curry, "Archaeology: The Milk Revolution," *Nature* 500, no. 7460 (2013): 20–22, doi: 10.1038/500020a.

7. Cavalli-Sforza and Cavalli-Sforza, *Great Human Diaspora.*

8. Yuval Itan, Adam Powell, Mark A. Beaumont, Joachim Burger, and Mark G. Thomas, "The Origins of Lactase Persistence in Europe," *PLoS Computational Biology* 5, no. 8 (2009): e1000491, doi: 10.1371/journal.pcbi.1000491.

9. Nina G. Jablonski, "The Evolution of Human Skin and Skin Color," *Annual Review in Anthropology* 3 (2004): 585–623, doi: 10.1146/annurev.anthro.33070203.143955.

10. Jonathan Kingdon, *Lowly Origin: Where, When and Why Our Ancestors First Stood Up* (Princeton: Princeton University Press, 2004).

11. Anthony J. McMichael, *Human Frontiers, Environments and Disease* (Cambridge: Cambridge University Press, 2001).

12. Jared E. Decker, Stephanie D. McKay, Megan M. Rolf, JaeWoo Kim, Antonio Molina Alcalá, Tad S. Sonstegard, Olivier Hanotte, et al., "Worldwide Patterns of Ancestry, Divergence, and Admixture in Domesticated Cattle," *PLoS Genetics* 10, no. 3 (2014): e1004254, doi: 10.1371/journal.pgen.1004254.

13. Alessia Ranciaro, Michael C. Campbell, Jibril B. Hirbo, Wen-Ya Ko, Alain Froment, Paolo Anagnostou, Maritha J. Kotze, et al., "Genetic Origins of Lactase Persistence and the Spread of Pastoralism in Africa," *American Journal of Human Genetics* 94, no. 3 (2014): 496–510, doi:10.1016/j.ajhg.2014.02.009.

14. Julie Dunne, Richard Evershed, Mélanie Salque, Lucy Cramp, Silvia Bruni, Kathleen Ryan, Stefano Biagetti, and Savino Di Lernia, "First Dairying in Green Saharan Africa in the Fifth Millennium BC," *Nature* 486, no. 7403 (2012): 390–394, doi: 10.1038/nature11186.

15. Dunne et al., "First Dairying."

16. Burroughs, *Climate,* pp. 223–225. Note, too, that analysis of mineral-dust sediment in the Atlantic seabed off the North African coast shows a great reduction in the amount of Saharan dust during the period 12,000 to 6,000 years ago, when the Sahara was mostly well watered.

17. Burroughs, *Climate,* p. 66.

18. Masatoshi Yoshino, "Climatic Change and Ancient Civilizations," in *Encyclopaedia of World Climatology*, edited by John E. Oliver (Dordrecht: Springer, 2005), pp. 192–199.

19. Friedhelm Steinhilber, Jose A. Abreu, Jürg Beer, Irene Brunner, Marcus Christl, Hubertus Fischer, Ulla Heikkilä, et al., "9,400 Years of Cosmic Radiation and Solar Activity from Ice Cores and Tree Rings," *Proceedings of the National Academy of Sciences of the United States of America* 109, no. 16 (2012): 5967–5971, doi: 10.1073/pnas.1118965109.

20. Barbara Barich, *People, Water, and Grain: The Beginnings of Domestication in the Sahara and the Nile Valley* (Rome: L'Erma di Bretschneider, 1998).

21. Paul C. Sereno, Elena A. A. Garcea, Hélène Jousse, Christopher M. Stojanowski, Jean-François Saliège, Abdoulaye Maga, Oumarou A. Ide, et al., "Lakeside Cemeteries in the Sahara: 5000 Years of Holocene Population and Environmental Change," *PLoS ONE* 3, no. 8 (2008): e2995, doi: 10.1371/journal.pone.0002995.

22. Hélène Jousse, "What Is the Impact of Holocene Climatic Changes on Human Societies? Analysis of West African Neolithic Populations Dietary Customs," *Quaternary International* 151, no. 1 (2006): 63–73, doi: 10.1016/j.quaint.2006.01.015.

23. Michael Dee, David Wengrow, Andrew Shortland, Alice Stevenson, Fiona Brock, Linus Girdland Flink, and Christopher Bronk Ramsey, "An Absolute Chronology for Early Egypt Using Radiocarbon Dating and Bayesian Statistical Modelling," *Proceedings of the Royal Society A: Mathematical, Physical and Engineering Sciences* 469, no. 2159 (2013): 20130395, doi: 10.1098/rspa.2013.0395.

24. d-maps.com, Fertile Crescent (maps), http://d-maps.com/pays.php?lib=fertile_crescent_maps&num_pay=303&lang=en.

25. Hans J. Nissen, *The Early History of the Ancient Near East, 9000–2000*, trans. Elizabeth Lutzeier with Kenneth J. Northcott (Chicago and London: University of Chicago Press, 1988). Nissen notes that most of latter-day Sumeria, extending toward Babylon itself, was under seawater during this Chalcolithic (carbonate sedimentation) period.

26. Kurt Lambeck, "Shoreline Reconstructions for the Persian Gulf since the Last Glacial Maximum," *Earth and Planetary Science Letter* 142, no. 1–2 (1996): 43–57, doi: 10.1016/0012-821X(96)00069-6.

27. Harvey Weiss, "Beyond the Younger Dryas: Collapse as Adaptation to Abrupt Climate Change in Ancient West Asia and the Ancient Eastern Mediterranean," in *Environmental Disasters and the Archaeology of Human Response*, edited by Garth Bawdon and Richard M. Reycraft (Albuquerque: Maxwell Museum of Anthropology, 2000), pp. 75–98.

28. Burroughs, *Climate*, p. 66.

29. John L. Brooke, *Climate Change and the Course of Global History: A Rough Journey* (Cambridge: Cambridge University Press, 2014).

30. Nissen, *Early History of the Ancient Near East*. See also Gwendolyn Leick, *Mesopotamia: The Invention of the City* (London: Penguin, 2002).

31. Clark S. Larsen, "Animal Source Foods and Human Health during Evolution," *Journal of Nutrition* 133, no. 11 (2003): 3893S–3897S.

32. Lawrence J. Angel, "Health as a Crucial Factor in the Changes from Hunting to Developed Farming in the Eastern Mediterranean," in *Paleopathology at the Origins of Agriculture*, edited by Mark N. Cohen and George J. Armelagos (London: Academic Press, 1984), pp. 51–73.

33. Vincenzo Formicola and Brigitte M. Holt, "Resource Availability and Stature Decrease in Upper Palaeolithic Europe," *Journal of Anthropological Sciences* 85 (2007): 147–155.

34. Stephen J. Corbett, Anthony J. McMichael, and Andrew M. Prentice, "Type 2 Diabetes, Cardiovascular Disease and the Evolutionary Paradox of the Polycystic Ovary Syndrome: A Fertility First Hypothesis," *American Journal of Human Biology* 21, no. 5 (2009): 587–598, doi: 10.1002/ajhb.20937.

35. Thomas McKeown, *The Origins of Human Disease* (Oxford: Basil Blackwell, 1988), p. 51.

36. Jessica M. C. Pearce-Duvet, "The Origin of Human Pathogens: Evaluating the Role of Agriculture and Domestic Animals in the Evolution of Human Disease," *Biological Reviews of the Cambridge Philosophical Society* 81, no. 3 (2006): 369–382, doi: 10.1017/S1464793106007020.

37. Robin A. Weiss and Anthony J. McMichael, "Social and Environmental Risk Factors in the Emergence of Infectious Diseases," *Nature Medicine* 10, no. 12 supplement (2004): S70–S76, doi: 10.1038/nm1150.

38. A view that was explicit in Abdel R. Omran, "The Epidemiologic Transition: A Theory of the Epidemiology of Population Change," *Millbank Quarterly* 49, no. 4 (1971): 509–538, doi: 10.1111/j.1468-0009.2005.00398.x.

39. Thomas Hobbes, *Leviathan: or The Matter, Forme and Power of a Common Wealth Ecclesiasticall and Civill* (London: Andrew Crooke, 1651), p. i.xviii.1.

40. William H. McNeill, *Plagues and Peoples* (New York: Anchor, 1976), gave the first systematic, and eminently readable, account of the sequence of historical surges in infectious disease entry and spread in human populations.

41. Robin A. Weiss, "The Leeuwenhoek Lecture 2001. Animal Origins of Human Infectious Disease," *Philosophical Transactions B: Biological Sciences* 356, no. 1410 (2001): 957–977, doi: 10.1098/rstb.2001.0838.

42. Evilena Anastasiou, Kirsi O. Lorentz, Gil J. Stein, and Piers D. Mitchell, "Prehistoric Schistosomiasis Parasite Found in the Middle East," *The Lancet Infectious Diseases* 14, no. 7 (2014): 553–554, doi: 10.1016/S14733099(14)70794-7.

43. Kirsten I. Bos, Kelly M. Harkins, Alexander Herbig, Mireia Coscolla, Nico Weber, Iñaki Comas, Stephen A. Forrest, et al., "Pre-Columbian Mycobacterial Genomes Reveal Seals as a Source of New World Human Tuberculosis," *Nature* 514, no. 7523 (2014): 494–497, doi: 10.1038/nature13591.

44. R. Brosch, S. V. Gordon, M. Marmiesse, P. Brodin, C. Buchrieser, K. Eiglmeier, T. Garnier, et al., "A New Evolutionary Scenario for the *Mycobacterium tuberculosis* Complex," *Proceedings of the National Academy of Sciences of the United States of America* 99, no. 6 (2002): 3684–3689, doi: 10.1073/pnas.052548299.

45. Helen D. Donoghue, "Human Tuberculosis—An Ancient Disease Elucidated by Ancient Microbial Biomolecules," *Microbes and Infections* 11, no. 14–15 (2009): 1156–1163, doi: 10.1016/j.micinf.2009.08.008.

46. Weiss, "Leeuwenhoek Lecture."

47. Ro McFarlane, Adrian Sleigh, and Anthony J. McMichael, "Synanthropy of Wild Mammals as a Determinant of Emerging Infectious Diseases in the Asian-Australasian Region," *EcoHealth* 9, no. 1 (2012): 24–35, doi: 10.1007/s10393-012-0763-9; Eric P. Hoberg and Daniel R. Brooks, "Evolution in Action: Climate Change, Biodiversity Dynamics and Emerging Infectious Disease," *Philosophical Transactions of the Royal Society of London B: Biological Sciences* 370, no. 1665 (2015): 20130553, doi: 10.1098/rstb.2013.0553.

Chapter 6

1. The Mesopotamian climate was the changeable product of several meteorological systems that interfaced and interacted at this geographic crossroads. The region's climate was driven by Eurasian continental circulation patterns that were variously driven from the North Atlantic, the rhythmic summertime monsoonal activity from the Indian Ocean, the variable southern intrusions of Russian boreal air masses during winter, sporadic strong El Niño teleconnections from the Pacific, and indolent longer-term north-south shifts in the rain-bearing Inter-Tropical Convergence Zone.

2. Simone Riehl, Konstantin E. Pustovoytov, Heike Weippert, Stefan Klett, and Frank Hole, "Drought Stress Variability in Ancient Near Eastern Agricultural Systems Evidenced by $\delta^{13}C$ in Barley Grain," *Proceedings of the National Academy of Sciences of the United States of America* (PNAS) 111, no. 34 (2014): 12348–12353, doi: 10.1073/pnas.1409516111. (When barley is water-stressed while growing, the proportion of heavier, and very stable, carbon isotopes in its cells is raised.)

3. William J. Burroughs, *Climate Change in Prehistory: The End of the Reign of Chaos* (Cambridge: Cambridge University Press, 2005), p. 50. The Siberian High was centered on the Baikal region of Siberia, but extended widely through much of Central Asia as well as northern Arctic Siberia. During times of weakened northern atmospheric circulation, it extended its icy hand southward.

4. Adapted from Riehl et al., "Drought Stress Variability."

5. Tony McMichael.

6. Peter B. deMenocal, "Cultural Responses to Climate Change during the Late Holocene," *Science* 292, no. 5517 (2001): 667–673, doi: 1126/science.1059287.

7. Ellery Frahm and Joshua M. Feinberg, "Environment and Collapse: Eastern Anatolian Obsidians at Urkesh (Tell Mozan, Syria) and the Third-Millennium Mesopotamian Urban Crisis," *Journal of Archaeological Science* 40, no. 4 (2012): 1866–1878, doi: 10.1016/j.jas.2012.11.026.

8. Riehl et al., "Drought Stress Variability."

9. Arne Wossink, *Challenging Climate Change: Competition and Cooperation among Pastoralists and Agriculturalists in Northern Mesopotamia (c. 3000–1600 bc)* (Leiden: Sidestone Press, 2009), p. 148.

10. H. M. Cullen, P. B. deMenocal, S. Hemming, G. Hemming, F. H. Brown, T. Guilderson, and F. Sirocko, "Climate Change and the Collapse of the Akkadian Empire: Evidence from the Deep Sea," *Geology* 28, no. 4 (2000): 379–382, doi: 10.1130/00917613(2000)28<379:ccatco>2.0.co;2.

11. deMenocal, "Cultural Responses."

12. William R. Thompson, "Complexity, Diminishing Marginal Returns, and Serial Mesopotamian Fragmentation," *Journal of World-Systems Research* 10, no. 3 (2004): 613–652, doi: 10.5195/jwsr.2004.288, quoted in "Sumerian Civilization," *New World Encyclopedia,* October 27, 2015, available at http://www.newworldencyclopedia.org/entry/Sumerian_Civilization.

13. Magnus Widdell, "Historical Evidence for Climate Instability and Environmental Catastrophes in Northern Syria and the Jazira: The Chronicle of Michael the Syrian," *Environment and History* 13, no. 1 (2007): 47–70, doi: 10.3197/096734007779748255.

14. "Hammurabi's Babylonia," Wikimedia Commons, http://en.wikipedia.org/wiki/Hammurabi.

15. Yongjin Wang, Hai Cheng, R. Lawrence Edwards, Yaoqi He, Xinggong Kong, Zhisheng An, Jiangying Wu, Megan J. Kelly, Carolyn A. Dykoski, and Xiangdong Li, "The Holocene Asian Monsoon: Links to Solar Changes and North Atlantic Climate," *Science* 308, no. 5723 (2005): 854–857, doi: 10.1126/science.1106296.

16. Luigi Luca Cavalli-Sforza and Francesco Cavalli-Sforza, *The Great Human Diaspora: The History of Diversity and Evolution*, trans. Sarah Thorne (New York: Perseus, 1996).

17. Textile Display Area, Asian Civilisations Museum, Singapore, 2012.

18. Adapted from Liviu Giosan, Peter D. Clift, Mark G. Macklin, Dorian Q. Fuller, Stefan Constantinescu, Julie A. Durcan, Thomas Stevens, et al., "Fluvial Landscapes of the Harappan Civilization," *Proceedings of the National Academy of Sciences of the United States of America* 109, no. 29 (2012): E1688–E1694, doi: 10.1073/pnas.1112743109.

19. Hubert H. Lamb, *Climate, History and the Modern World*, 2d ed. (London and New York: Routledge, 1995), p. 2.

20. Giosan et al., "Fluvial Landscapes."

21. Yama Dixit, David A. Hodell, and Cameron A. Petrie, "Abrupt Weakening of the Summer Monsoon in Northwest India ~4100 Yr Ago," *Geology* 42, no. 4 (2014): 2014129, doi: 10.1130/G35236.1.

22. Gwen R. Schug, "Infection, Disease, and Biosocial Processes at the End of the Indus Civilization," *PLoS ONE 8*, no. 12 (2013): e84814, doi: 10.1371/journal.pone.0084814.

23. David Kaniewski, Elise Van Campo, Joël Guiot, Sabine Le Burel, Thierry Otto, and Cecile Baeteman, "Environmental Roots of the Late Bronze Age Crisis," *PLoS ONE 8*, no. 8 (2013): e71004, doi: 10.1371/journal.pone.0071004.

24. David Kaniewski, Etienne Paulissen, Elise Van Campo, Harvey Weiss, and Joachim Bretschneider, "Late Second–Early First Millennium BC Abrupt Climate Changes in Coastal Syria and Their Possible Significance for the History of the Eastern Mediterranean," *Quaternary Research* 74, no. 2 (2010): 207–215, doi: 10.1016/j.yqres.2010.07.010.

25. Andrea Salimbeti, "The Greek Age of Bronze: Sea Peoples," updated October 28, 2015, http://www.salimbeti.com/micenei/sea.htm.

26. David Abulafia, *The Great Sea: A Human History of the Mediterranean* (London: Allen Lane, 2011).

27. Kenneth J. Hsu, "Sun, Climate, Hunger, and Mass Migration," *Science in China Series D: Earth Sciences* 41, no. 5 (1998): 449–472, doi: 10.1007/BF02877737.

28. Eric H. Cline, *1177 BC: The Year Civilization Collapsed* (Princeton: Princeton University Press, 2014).

29. Amos Nur and Eric H. Cline, "Poseidon's Horses: Plate Tectonics and Earthquake Storms in the Late Bronze Age Aegean and Eastern Mediterranean," *Journal of Archaeological Science* 27, no. 1 (2000): 43–63, doi: 10.1006/jasc.1999.0431.

30. Cline, *1177 BC.*

31. Rhys Carpenter, *Discontinuity in Greek Civilization* (Cambridge: Cambridge University Press, 1966), quoted in Brian Fagan, *The Long Summer: How Climate Changed Civilization* (New York: Basic Books, 2004), p. 182.

32. Lamb, *Climate, History*, p. 149.

33. Stefanie Jacomet, Michel Magny, Conradin A. Burga, "Klima- und Seespiegelschwankungen im Verlauf des Neolithikums und ihr Auswirkungen auf die Besiedlung der Seeufer," in *Die Schweiz von Paläolithikum bis zum frühen Mittelalter*, Vol. 2: *Neolithikum*, edited by Werner E. Stöckli, Urs Niffeler, and Eduard Gross-Klee (Basel: Schweizerische Gesellschaft für Ur- und Frühgeschichte, 1995), pp. 53–58.

34. Cunliffe, *Europe Between the Oceans*, p. 349.

Chapter 7

1. John L. Brooke, *Climate Change and the Course of Global History: A Rough Journey*, Studies in Environment and History (New York: Cambridge University Press, 2014), p. 325.

2. Nicola Terrenato, "The Essential Countryside of the Roman World," in *Classical Archaeology*, edited by Susan E. Alcock and Robin Osborne (Malden: Blackwell, 2007), p. 142.

3. Brooke, *Climate Change*, p. 325.

4. The centuries of warm and stable conditions that extended a Mediterranean-type climate into Europe during the Classical Optimum are commonly labelled the Roman Warm Period.

5. Wallace S. Broecker and Aaron E. Putnam, "Hydrologic Impacts of Past Shifts of Earth's Thermal Equator Offer Insight into Those to Be Produced by Fossil Fuel CO_2," *Proceedings of the National Academy of Sciences of the United States of America* 110, no. 42 (2013): 16710–16715, doi: 10.1073/pnas.1301855110. This northern extension resembles several previous similar latitude excursions, when the quasi-equatorial Inter-Tropical Convergence Zone (ITCZ) strays further north, associated with warming, and pushing the higher-latitude climate zones north as well.

6. Brian Fagan, *The Long Summer: How Climate Changed Civilization* (New York: Basic Books, 2004), pp. 190–212.

7. Carole L. Crumley, "Alternative Forms of Social Order," in *Heterarchy, Political Economy, and the Ancient Maya: The Three Rivers Region of the East-Central Yucatan*, edited by Vernon L. Scarborough, Fred Valdez, and Nicholas P. Dunning (Tucson: University of Arizona Press, 2003), pp. 136–165.

8. Peter Heather, *The Fall of the Roman Empire: A New History* (London: Pan Macmillan, 2005).

9. Bryan Ward-Perkins, *The Fall of Rome and the End of Civilization* (Oxford: Oxford University Press, 2005).

10. Edward Gibbon, *The History of the Decline and Fall of the Roman Empire* (London: Everyman's Library, 2010).

11. In this section on climatic conditions during the rise and decline of the Roman Empire, I have drawn on a number of published papers. In particular, I have referred to the comprehensive review by Michael McCormick, Ulf Büntgen, Mark A. Cane, Edward R. Cook, Kyle Harper, Peter Huybers, Thomas Litt, et al., "Climate Change during and after the Roman Empire: Reconstructing the Past from Scientific and Historical Evidence," *Journal of Interdisciplinary History* 43, no. 2 (2012): 169–220, doi: 10.1162/JINH_a_00379, which provides an integrated account of climatic trends and changes in the Western and Eastern Roman Empire(s) during the period 100 B.C.E. to 800 C.E. That synthesis is based on eleven different paleo-climatic reconstructions published over the past decade or so. Other useful sources include Ulf Büntgen, Willy Tegel, Kurt Nicolussi, Michael McCormick, David Frank, Valerie Trouet, Jed O. Kaplan, et al., "2500 Years of European Climate Variability and Human Susceptibility," *Science* 331, no. 6017 (2011): 578–582, and Hubert H. Lamb, *Climate, History and the Modern World* (London: Routledge, 1995).

12. Büntgen et al., "2500 Years."

13. Lamb, *Climate, History*, pp. 156–157.

14. Carole L. Crumley, "Celtic Settlement Before the Conquest: The Dialectics of Landscape and Power," in *Regional Dynamics: Burgundian Landscapes*

in Historical Perspective, edited by Carole L. Crumley and William H. Maquandt (San Diego: Academic Press, 1987), pp. 237–264.

15. Carole L. Crumley, "The Ecology of Conquest: Contrasting Agropastoral and Agricultural Societies' Adaptation to Climate Change," in *Historical Ecology: Cultural Knowledge and Changing Landscapes*, edited by Carole L. Crumley (Santa Fe: School of American Research Press, 1994).

16. McCormick et al., "Climate Change."

17. Heather, *The Fall*, pp. 112–114. See also McCormick et al., "Climate Change": refers to the declining economic conditions and political stability of Rome's northern European provinces, while the southern provinces remained stable.

18. Dionysios Ch. Stathakopoulos, *Famine and Pestilence in the Late Roman and Early Byzantine Empire: A Systematic Survey of Subsistence Crises and Epidemics*, Birmingham Byzantine and Ottoman Monographs 9 (Aldershot and Burlington, VT: Ashgate, 2004), pp. 36–39.

19. Wolfgang Behringer, *A Cultural History of Climate*, trans. Patrick Camiller (Cambridge: Polity Press, 2010), p. 65.

20. Stathakopoulos, *Famine and Pestilence*, p. 92.

21. Dionysios Ch. Stathakopoulos, "Plagues of the Roman Empire," in *Encyclopedia of Pestilence, Plagues and Pandemics*, edited by Joseph P. Byrne (Westport: Greenwood Press, 2008), pp. 536–538.

22. R. J. Littman and M. L. Littman, "Galen and the Antonine Plague," *American Journal of Philology* 94, no. 3 (1973): 243–255.

23. Stathakopoulos, "Plagues of the Roman Empire."

24. Donald A. Henderson, "Smallpox: Clinical and Epidemiologic Features," *Emerging Infectious Diseases* 5, no. 4 (1999): 537–539, doi: 10.3201/eid0504.990415; also Steadman Upham, "Smallpox and Climate in the American Southwest," *American Anthropologist* 88, no. 1 (1986): 115–128, doi: 10.1525/aa.19866.99.1.02a00080, and Hiroshi Nishiura and Tomoko Kashiwagi, "Smallpox and Season: Reanalysis of Historical Data," *Interdisciplinary Perspectives on Infectious Diseases* 2009 (2009): 591935, doi: 10.1155/2009/591935.

25. Stathakopoulos, "Plagues of the Roman Empire."

26. Pontius, "The Life and Passion of Cyprian, Bishop and Martyr," trans. Ernest Wallis, in *The Ante-Nicene Fathers*, Vol. 5, edited by Alexander Roberts and James Donaldson (Edinburgh: T & T Clark, 1885), quoted in "Plague of Cyprian," Wikipedia, accessed December 6, 2015, http://en.wikipedia.org/wiki/Plague_of_Cyprian#cite_note-Furuse2010-7.

27. Joseph A. Tainter, *The Collapse of Civilisations* (Cambridge: Cambridge University Press, 1988).

28. William H. McNeill, *Plagues and Peoples* (New York: Anchor Books, 1976).

29. McCormick et al., "Climate Change."

30. Fagan, *The Long Summer*.

31. Adapted from Andreas Kunze, "Distribution Map of Europe," Wikimedia Commons, available at https://commons.wikimedia.org/wiki/File:Distribution_map_of_Europe_blank_crop.svg.

32. Heather, *The Fall.*

33. William Rosen, *Justinian's Flea: Plague, Empire and the Birth of Europe* (New York: Viking, 2006). Rosen provides a popular-science summary of ancient and modern estimates of the human toll of this pandemic, along with a well-informed, detailed discussion of the molecular biology of the bacterium and the infectious process.

34. David M. Wagner, Jennifer Klunk, Michaela Harbeck, Alison Devault, Nicholas Waglechner, Jason W. Sahl, Jacob Enk, et al., "*Yersinia pestis* and the Plague of Justinian 541–543 AD: A Genomic Analysis," *The Lancet Infectious Diseases* 14, no. 4 (2014): 319–326, doi: 10.1016/S1473-3099(13)70323-2.

35. Wendy Orent, *Plague: The Mysterious Past and Mystifying Future of the World's Most Dangerous Disease* (New York: Free Press, 2004).

36. Ancient historical records freely used words that can be translated "plague" generically, as in the Antonine and Cyprian plagues. Likewise "pox" and "fever." The bubonic form of the infection, now known to be caused by *Yersinia pestis*, is sometimes referred to as the Plague (capitalized). Confusion has often arisen in the historical interpretation of these "plague" epidemics.

37. Jean-Daniel Vigne and Frédérique Audoin-Rouzeau, "La colonisation de l'Europe occidentale par le Rat noir, contraintes méthodologiques, appel à collaborations," *Nouvelles de l'Archéologie* 47 (1992): 42–44.

38. Eva Panagiotakopulu, "Pharaonic Egypt and the Origins of Bubonic Plague," *Journal of Biogeography* 31, no. 2 (2004): 269–275.

39. Procopius, *Histories of the Wars*, trans. H. B. Dewing, Loeb Classical Library (Cambridge: Harvard University Press, 1981), quoted in Cheston B. Cunha and Burke A. Cunha, "Great Plagues of the Past and Remaining Questions," in *Paleomicrobiology: Past Human Infections,* edited by Didier Raoult and Michel Drancourt (Berlin: Springer-Verlag, 2008), pp. 1–20.

40. Book of Revelation 14:19, quoted in William Rosen, *Justinian's Flea.*

41. This was a novel setting in which a major epidemic of infection was the cause, not the sequel, of a food shortage and famine. Rats consume grain voraciously, and spoil twice as much again in the process. Further, when food is abundant their breeding accelerates, with new pregnancies occurring while current litters are still being suckled—biological mass production, capitalizing on good dietary times.

42. Giovanna Morelli, Yajun Song, Camila J. Mazzoni, and Mark Eppinger, "*Yersinia pestis* Genome Sequencing Identifies Patterns of Global Phylogenetic Diversity," *Nature Genetics* 42, no. 12 (2010): 1140–1143, doi: 10.1038/ng.705; also Ingrid Wiechmann and Gisela Grupe, "Detection of *Yersinia pestis* DNA in Two Medieval Skeletal Finds from Aschheim (Upper Bavaria, 6th century A.D.)," *American Journal of Physical Anthropology* 126, no. 1 (2004): 48–55, doi: 10.1002/ajpa.10276.

43. Wagner et al., "*Yersinia pestis.*"

44. Christine A. Smith, "Plague in the Ancient World: A Study from Thucydides to Justinian," *Student Historical Journal* XXVIII (1996–1997), available at http://www.loyno.edu/~history/journal/1996-7/1996-7.htm. See also Panagiotakopulu, "Pharaonic Egypt."

45. Michael McCormick, "Toward a Molecular History of the Justinianic Pandemic," in *Plague and the End of Antiquity: The Pandemic of 541–750*, edited by Lester K. Little (Cambridge: Cambridge University Press, 2007), pp. 290–312.

46. Costas Tsiamis, Effie Poulakou-Rebelakou, and Eleni Petridou, "The Red Sea and the Port of Clysma: A Possible Gate of Justinian's Plague," *Gesnerus* 66, no. 2 (2009): 209–217.

47. Ole J. Benedictow, "*Yersinia pestis*, the Bacterium of Plague, Arose in East Asia: Did It Spread Westwards via the Silk Roads, the Chinese Maritime Expeditions of Zheng He or over the Vast Eurasian Populations of Sylvatic (Wild) Rodents?," *Journal of Asian History* 47, no. 1 (2013): 1–32, doi: 10.13173/jasiahist.47.1.0001.

48. McCormick, "Toward a Molecular History."

49. McNeill, *Plagues and Peoples*, p. 109.

50. This fragmentary document by Procopius is reportedly preserved in the Collection of Oribasius: 164-1 Lib. xliv.cap. 17, in *Oeuvres d'Oribase*, edited by Ulco Cats Bussemaker and ⌈Charles Daremberg, Vol. 3 (Paris: Imprimerie Nationale, 1851): 607.

51. Panagiotakopulu, "Pharaonic Egypt." The Nile rat (*Arvicanthis niloticus*, a natural reservoir of bubonic plague), black rat, rat flea, and human flea were all evident in excavated samples, suggesting that plague has been a long-term resident in the Nile Valley.

52. *The Papyrus Ebers*, trans. Bendix Ebbell (Copenhagen: Ejnar Munksgaard, 1937).

53. Procopius BP 2.22.6, p. 250.13–18, quoted in McCormick, "Toward a Molecular History," p. 303.

54. Source: Aksum 300–700 c.e. http:// images.classwell.com/ mcd_ xhtml_ ebooks/ 2005_ world_ history/ images/ mcd_ awh2005_ 0618376798_ p226_ f1.jpg; Silk Road, Wikimedia Commons, https://commons.wikimedia.org/ wiki/File:Silk_route_copy.jpg.

55. Scholasticus Evagrius, *Ecclesiastical History Book 4*, trans. E. Walford (London: S. Bagster and Sons, 1846).

56. McCormick, "Toward a Molecular History," p. 304. The Kushites (in today's Sudan) and Himyarites (in southern Yemen) were part of the kingdom of Aksum.

57. Wiechmann and Grupe, "Detection of *Yersinia pestis* DNA."

58. Morelli et al., "*Yersinia pestis*"; see also Mark Achtman, Kerstin Zurth, Giovanna Morelli, Gabriela Torrea, Annie Guiyoule, and Elisabeth Carniel, "*Yersinia pestis*, the Cause of Plague, Is a Recently Emerged Clone of

Yersinia pseudotuberculosis," *Proceedings of the National Academy of Sciences of the United States of America* 96, no. 24, 1999: 14043–14048, doi: 10.1073/pnas.96.24.14043.

59. Dionysios Stathakopoulos, "Plague of Justinian; First Pandemic," in *Encyclopedia of Pestilence, Pandemics and Plagues*, edited by Joseph P. Byrne (Westport: Greenwood Press, 2008), pp. 532–535.

60. David Keys, *Catastrophe: An Investigation into the Origins of the Modern World* (New York: Ballantine, 1999). See also the related proposition in Rosen, *Justinian's Flea*.

61. Tsiamis et al., "The Red Sea and the Port of Clysma." This land-and-sea route was presumably usually subject to climatic constraints on flea and bacterial survival because of excessive and drying temperatures.

62. Wagner et al., "*Yersinia pestis*."

63. Keys, *Catastrophe*.

64. Lars Berg Larsen, Bo Møllesøe Vinther, Keith R. Briffa, Tom M. Melvin, Henrik Brink Clausen, Phil D. Jones, Marie Louise S. Andersen, et al., "New Ice Core Evidence for a Volcanic Cause of the A.D. 536 Dust Veil," *Geophysical Research Letters* 35, no. 4 (2008): L04708, doi: 10.1029/2007GL032450.

65. Procopius, *Histories of the Wars*, 4.14.5.

66. Richard B. Stothers, "The Great Tambora Eruption in 1815 and Its Aftermath," *Science* 224, no. 4654 (1984): 1191–1198, doi: 10.1126/science.224.4654.1191.

67. Other evidence from Greenland ice-cores points to a comet shower at around that time. A research team from Columbia University, New York, has identified a precise layer of tiny spherules of condensed rock vapor in the ice cores, suggestive of the type of aerosolized terrestrial debris caused by multiple comet impacts. Dallas H. Abbot, "Comet Smashes Triggered Ancient Famine," *New Scientist*, January 7, 2009.

68. Larsen et al., "New Ice Core Evidence."

69. Bo Gräslund, "Fimbulvintern, Ragnarök och klimatkrisen år 536–537 e. Kr.," *Saga och sed* (2007): 93–123, quoted in Daniel Löwenborg, "Excavating the Digital Landscape: GIS Analyses of Social Relations in Central Sweden in the 1st Millennium AD," PhD diss., University of Uppsala, 2010.

70. Daniel Löwenborg, "Landscapes of Death: GIS Modelling of a Dated Sequence of Prehistoric Cemeteries in Västmanland, Sweden," *Antiquity* 83, no. 322 (2009): 1134–1143, doi: 10.1017/S0003598X00099415. That study is explained in its wider context in Löwenborg, *Excavating the Digital Landscape*.

71. Harold W. Brown and Franklin A. Neva, *Basic Clinical Parasitology* (New York: Appleton-Century Crofts, 1975).

72. Kenneth L. Gage, Thomas R. Burkot, Rebecca J. Eisen, and Edward B. Hayes, "Climate and Vectorborne Diseases," *American Journal of Preventative Medicine* 35, no. 5 (2008): 436–450, doi: 10.1016/j.amepre.2008.08.030.

73. Gage et al., "Climate and Vector-Borne Diseases."

74. Ralph St. John Brooks, "The Influence of Saturation Deficiency and of Temperature on the Course of Epidemic Plague," *Journal of Hygiene, London* 15, no. S1 (1917): 881–899.

75. A. W. Bacot and C. J. Martin, "The Respective Influences of Temperature and Moisture upon the Survival of the Rat Flea (*Xenopsylla cheopis*) Away from its Host," *Journal of Hygiene, London* 23, no. 1 (1924): 98–105.

76. McNeill, *Plagues and Peoples*, p. 262.

77. Rosen, *Justinian's Flea*, citing a meticulous demographer.

78. Ulf Buntgen, Vladimir S. Myglan, Fredrik Charpentier Ljungqvist, Michael McCormick, Nicola Di Cosmo, Michael Sigl, Johann Jungclaus, et al., "Cooling and Societal Change during the Late Antique Little Ice Age from 536 to around 660 AD," *Nature Geoscience* 9 (2016): 231–236, doi: 10.1038/ngeo2652; Wagner et al., "*Yersinia pestis.*"

79. Jared Diamond, *Collapse: How Societies Choose to Fail or Succeed* (New York: Penguin, 2005), p. 160, points out that this is a misleading descriptor. During the rainy season (May–October), the region is humid, wet tropical forest, but during the dry half-year it is really a seasonal desert. Further, there is a fivefold difference in annual rainfall: the northern, Yucatan region receives much less of the annual monsoon rains than does the more densely forested highland area in the southwest, in today's upper Central America.

80. Adapted from Arthur Demarest, *Ancient Maya: The Rise and Fall of a Civilisation* (Cambridge: Cambridge University Press, 2005); B. L. Turner II and Jeremy A. Sabloff, "Classic Period Collapse of the Central Maya Lowlands: Insights about Human-Environment Relationships for Sustainability," *Proceedings of the National Academy of Sciences of the United States of America* 109, no. 35 (2012): 13908–13914, doi: 10.1073/pnas.1210106109.

81. Robert J. Sharer and Loa P. Traxler, *The Ancient Maya* (Stanford: Stanford University Press, 2006).

82. Richardson B. Gill, *The Great Maya Droughts: Water, Life, and Deaths* (Albuquerque: University of New Mexico Press, 2000), p. 255.

83. Vernon Scarborough, *Flow of Power: Ancient Water Systems and Landscapes* (Santa Fe: SAR Press, 2003).

84. Douglas J. Kennett, Sebastian F. M. Breitenbach, Valorie V. Aquino, Yemane Asmerom, Jaime Awe, James U. L. Baldini, Patrick Bartlein, et al., "Development and Disintegration of Maya Political Systems in Response to Climate Change," *Science* 338, no. 6108 (2012): 788–791, doi: 10.1126/scienece.1226299.

85. Turner and Sabloff, "Classic Period."

86. David A. Hodell, "Possible Role of Climate in the Collapse of Classic Maya Civilization," *Nature* 375, no. 6530 (1995): 391–394, doi: 10.1038/375391a0.

87. Turner and Sabloff, "Classic Period."

88. Nicholas P. Dunning, Timothy P. Beach, and Sheryl Luzzadder-Beach, "Kax and Kol: Collapse and Resilience in Lowland Maya Civilization,"

Proceedings of the National Academy of Sciences of the United States of America
109, no. 10 (2012): 3652–3657, doi: 10.1073/pnas.1114838109.

89. Joseph Tainter provides a succinct definition of "collapse" as a fundamental
    and pronounced decline in sociopolitical complexity taking place within
    two or three generations. See Joseph A. Tainter, "Problem Solving:
    Complexity, History, Sustainability," *Population and Environment* 22, no.
    1 (2000): 3–41. Karl Butzer points out that the concept of collapse "has
    intuitive appeal but ambiguous meaning, and has been applied to states,
    nations, or complex societies, in the sense that such entities rise and
    flourish, but eventually disintegrate and fail," in Karl W. Butzar, "Collapse,
    Environment, and Society," *Proceedings of the National Academy of Sciences
    of the United States of America* 109, no. 10 (2012): 3632–3639, doi: 10.1073/
    pnas.1114845109.

90. Lori E. Wright and Christine D. White, "Human Biology in the Classic
    Maya Collapse: Evidence from Paleopathology and Paleodiet," *Journal of
    World Prehistory* 10, no. 2 (1996): 147–198.

91. Diamond, *Collapse*, p. 170.

92. Richardson B. Gill, Paul A. Mayewski, Johan Nyberg, Gerald H. Haug, and
    Larry C. Peterson, "Drought and the Maya Collapse," *Ancient Mesoamerica*
    18, no. 2 (2007): 283–302, doi: 10.1017/S0956536107000193.

93. Gerald H. Haug, Detlef Gunther, Larry C. Peterson, Daniel M. Sigman,
    Konrad A. Hughen, and Beat Aeschlimann, "Climate and the Collapse of
    Maya Civilization," *Science* 299, no. 5613 (2003): 1731–1735, doi: 10.1126/
    science.1080444.

94. Turner and Sabloff, "Classic Period." The same research team gleaned
    additional information from an archaeological hieroglyphic chronology
    of social stress, conflict, warfare, symbolic carving activity, and the
    construction of edifices. Considered together, the stalagmite chemistry and
    the archaeological records showed that declines in rainfall, and the three
    evident periods of drought, coincided with social, cultural, and political
    setbacks during the Terminal Classic period.

95. D. W. Stahle, J. Villanueva Diaz, Dorian J. Burnette, J. Paredes, R. R.
    Heim, Falko K. Fye, Rodolfo Acuna Soto, Matthew D. Therrell, Malcolm
    K. Cleaveland, and Daniel K. Stahle, "Major Mesoamerican Droughts of
    the Past Millennium," *Geophysical Research Letters* 38 (2011): L05703, doi:
    10.1029/2010GL046472.

96. Adapted from Kennett et al., "Development and Disintegration."

97. B. I. Cook, K. J. Anchukaitis, J. O. Kaplan, M. J. Puma, M. Kelley, and
    D. Gueyffier, "Pre-Columbian Deforestation as an Amplifier of Drought in
    Mesoamerica," *Geophysical Research Letters* 39 (2012): L16706, doi: 10.1029/
    2012GL052565.

98. Martin Medina-Elizalde and Eelco J. Rohling, "Collapse of Classic Maya
    Civilization Related to Modest Reduction in Precipitation," *Science* 335,
    no. 6071 (2012): 956–959, doi: 10.1126/science.1216629.

99. Gill, *The Great Maya Droughts*.
100. Turner and Sabloff, "Classic Period."
101. Vernon L. Scarborough, Nicholas P. Dunning, Kenneth B. Tankersley, Christopher Carr, Eric Weaver, Liwy Grazioso, Brian Lane, et al., "Water and Sustainable Land Use at the Ancient Tropical City of Tikal, Guatemala," *Proceedings of the National Academy of Sciences of the United States of America* 109, no. 31 (2012): 12408–12413, doi: 10.1073/pnas.1202881109.
102. Wright and White, "Human Biology."
103. Mark Golitko, James Meierhoff, Gary M. Feinman, and Ryan Williams, "Complexities of Collapse: The Evidence of Maya Obsidian as Revealed by Social Network Graphical Analysis," *Antiquity* 86, no. 332 (2012): 507–523, doi: 10.1017/S0003598X00062906.
104. Gill, *The Great Maya Droughts*, p. 100.
105. Nicholas E. Graham, Malcolm K. Hughes, Caspar M. Ammann, Kim M. Cobb, Martin P. Hoerling, Douglas J. Kennett, and James P. Kennett, "Tropical Pacific—Mid-Latitude Teleconnections in Medieval Times," *Climate Change* 83, no. 1 (2007): 241–285, doi: 10.1007/s10584-007-9239-2.
106. Edward R. Cook, Richard Seager, Mark A. Cane, and David W. Stahle, "North American Drought: Reconstructions, Causes, and Consequences," *Earth Science Reviews* 81, no. 1–2 (2007): 93–134, doi: 10.1016/j.earscirev.2006.12.002.
107. Based on Figure 9.b of Larry V. Benson, Michael S. Berry, Edward A. Jolie, Jerry D. Spangler, David W. Stahle, and Eugene M. Hattori, "Possible Impacts of Early-11th-, Middle-12th-, and Late-13th-Century Droughts on Western Native Americans and the Mississippian Cahokians," *Quaternary Science Reviews* 26, no. 3–4 (2007): 336–350, doi: 10.1016/j.quascirev.2006.08.001.
108. Adapted from Cook et al., "North American Drought."
109. Cook et al., "North American Drought."
110. Fagan, *The Long Summer*. This section about the Anasazi draws on that part of Brian Fagan's book, which contains many interesting and evocative descriptions of how and why they lived across the variable terrain of the Four Corners region—and dealt with variable rainfall.
111. Clark Spencer Larsen, *Bioarchaeology: Interpreting Behaviour from the Human Skeleton* (Cambridge: Cambridge University Press, 1999), p. 35. Human bones and teeth provide much information about the health and way of life of past populations. These tissues provide a cumulative record of disease, physiological stress, trauma, activity patterns, diet, nutrition, and many other factors that constitute the life history of both the individual and the population.
112. Timothy A. Kohler and Kelsey M. Reese, "Long and Spatially Variable Neolithic Demographic Transition in the North American Southwest," *Proceedings of the National Academy of Sciences of the United States of America* 111, no. 28 (2014): 10101–10106, doi: 10.1073/pnas.1404367111.

113. Diamond, *Collapse*.
114. Diamond, *Collapse*, pp. 151–153.
115. Fagan, *The Long Summer*, p. 228.
116. Aiguo Dai, "Drought Under Global Warming: A Review," *Wiley Interdisciplinary Reviews: Climate Change* 2, no. 1 (2011): 45–65, doi: 10.1002/wcc.81.
117. Benson et al., "Possible Impacts."

Chapter 8

1. Franck Lavigne, Jean-Philippe Degeai, Jean-Christophe Komorowski, Sébastien Guillet, Vincent Robert, Pierre Lahitte, Clive Oppenheimer, et al., "Source of the Great A.D. 1257 Mystery Eruption Unveiled, Samalas Volcano, Rinjani Volcanic Complex, Indonesia," *Proceedings of the National Academy of Sciences of the United States of America* 110, no. 42 (2013): 16742–16747, doi: 10.1073/pnas.1307520110.
2. Richard B. Stothers, "Climatic and Demographic Consequences of the Massive Volcanic Eruption of 1258," *Climatic Change* 45, no. 2 (2000): 361–374, doi: 10.1023/A:1005523330643.
3. Peter B. deMenocal, "Cultural Responses to Climate Change during the Late Holocene," *Science* 292, no. 5517 (2001): 667–673, doi: 10.1126/science.1059287.
4. Sami Solanki, "Solar Variability and Climate Change: Is There a Link?" *Astronomy and Geophysics* 43, no. 5 (2002): 5.9–5.13, doi: 10.1046/j.1468-4004.2002.43509.x.
5. Temperature graph adapted from Ulf Büntgen et al., "2500 years of European Climate Variability and Human Susceptibility," *Science* 331, no. 6017 (2011): 578–582, doi: 10.1126/science.1197175. Solar activity graph: "Solar cycle," Wikipedia, http://en.wikipedia.org/wiki/Solar_variation.
6. Ulf Büntgen, Tomáš Kyncl, Christian Ginzler, David S. Jacks, Jan Esper, Willy Tegel, Karl-Uwe Heussner, and Josef Kyncl, "Filling the Eastern European Gap in Millennium-long Temperature Reconstructions," *Proceedings of the National Academy of Sciences of the United States of America* 110, no. 5 (2013): 1773–1778, doi: 10.1073/pnas.1211485110.
7. Jean M. Grove, "Climatic Change in Northern Europe over the Last Two Thousand Years and Its Possible Influence on Human Activity," in *Climate Development and History of the North Atlantic Realm: Hanse Conference on Climate History*, edited by Gerald Wefer, Wolfgang H. Berger, and Karl-Ernst Behre (Berlin: Springer, 2002), pp. 313–326.
8. Hubert Lamb, *Climate, History and the Modern World* (London: Routledge, 1995).
9. Lamb, *Climate, History*.
10. A. G. Dawson, K. Hickey, P. A. Mayewski, and A. Nesje, "Greenland (GISP2) Ice Core and Historical Indicators of Complex North Atlantic

Climate Changes during the Fourteenth Century," *The Holocene* 17, no. 4 (2007): 427–434, doi: 10.1177/0959683607077010.

11. Bruce M. S. Campbell, "Nature as Historical Protagonist: Environment and Society in Pre-Industrial England," *Economic History Review* 6, no. 2 (2010): 281–314, doi: 10.111/j.1468-0289.2009.00492.x.

12. Campbell, "Nature as Historical Protagonist."

13. Adapted from Campbell, "Nature as Historical Protagonist."

14. Campbell, "Nature as Historical Protagonist."

15. Wolfgang Behringer, *A Cultural History of Climate* (London: Polity Press, 2010).

16. William C. Jordan, *The Great Famine: Northern Europe in the Early Fourteenth Century* (Princeton: Princeton University Press: 1996).

17. Behringer, *Cultural History of Climate*.

18. Jordan, *The Great Famine*, pp. 127–150.

19. Umberto Eco, *The Name of the Rose* (London: Vintage Classics, 2004).

20. Eco, *Name of the Rose*, p. 179.

21. Jordan, *The Great Famine*.

22. Nils C. Stenseth, Bakyt B. Atshabar, Mike Begon, Steven R. Belmain, Eric Bertherat, Elisabeth Carniel, Kenneth L. Gage, Herwig Leirs, and Lila Rahalison, "Plague: Past, Present, and Future," *PLoS Medicine* 5, no. 1 (2008): 9–13, doi: 10.1371/journal.pmed.0050003.

23. Giovanni Boccaccio, *The Decameron*, Vol. 1, trans. by Richard Aldington (New York: Dell Laurel, 1930).

24. Stenseth et al., "Plague."

25. Ole Jørgen Benedictow, *The Black Death 1346–1353: The Complete History* (Woodbridge: Boydell Press, 2004).

26. Benedictow, *The Black Death*.

27. Sheldon Watts, *Epidemics and History: Disease, Power and Imperialism* (New Haven: Yale University Press, 1997).

28. Ann G. Carmichael, "Universal and Particular: The Language of Plague, 1348–1500," *Medical History* 52, no. S27 (2008): 17–52, doi: 10.1017/S0025727300072070.

29. Ulf Büntgen, Christian Ginzler, Jan Esper, Willy Tegel, and Anthony J. McMichael, "Digitizing Historical Plague," *Clinical Infectious Diseases* 55, no. 11 (2012): 1586–1588, doi: 10.10933/cid/cis723.

30. Susan Scott and Christopher Duncan, *Return of the Black Death: The World's Greatest Serial Killer* (London: John Wiley, 2004).

31. Phyllis Pobst, "Should We Teach That the Cause of the Black Death Was Bubonic Plague?" *History Compass* 11, no. 10 (2013): 808–820, doi: 10.1111/hic3.12081.

32. K. Birkelbach, "Black Death: Modern Medical Debate," in *Pestilence, Pandemics and Plagues*. edited by Joseph P. Byrne (Westport: Greenwood Press), pp. 72–74.

33. Susan Scott and Christopher J. Duncan, *Biology of Plagues: Evidence from Historical Populations* (Cambridge: Cambridge University Press, 2001).

34. Saravanan Ayyadurai, Florent Sebbane, Didier Raoult, and Michel Drancourt, "Body Lice, *Yersinia pestis* Orientalis, and Black Death," *Emerging Infectious Diseases* 16, no. 5 (2010): 892–893, doi: 10.3201/eid1605.091280.

35. Hans Zinsser, *Rats, Lice and History* (New Brunswick and London: Transaction, 2008).

36. Michel Drancourt, Michel Signoli, La Vu Dang, Bruno Bizot, Véronique Roux, Stéfan Tzortzis, and Didier Raoult, "*Yersinia pestis* Orientalis in Remains of Ancient Plague Patients," *Emerging Infectious Disease* 13, no. 2 (2007): 332–333, doi: 10.3201/eid1302.060197. The graves were from 1590 in Lambec and 1722 in Marseilles.

37. Stephanie Haensch, Raffaella Bianucci, Michel Signoli, Minoarisoa Rajerison, Michael Schultz, Sacha Kacki, Marco Vermunt, et al., "Distinct Clones of *Yersinia pestis* Caused the Black Death," *PLoS Pathogens* 6, no. 10 (2010): e1001134. doi: 10.1371/journal.ppat.1001134.

38. Giovanna Morelli, Yajun Song, Camila J. Mazzoni, and Mark Eppinger, "*Yersinia pestis* Genome Sequencing Identifies Patterns of Global Phylogenetic Diversity," *Nature Genetics* 42, no. 12 (2010): 1140–1143, doi: 10.1038/ng.705.

39. Kirsten I. Bos, Verena J. Schuenemann, G. Brian Golding, Hernán A. Burbano, Nicholas Waglechner, Brian K. Coombes, Joseph B. McPhee, et al., "A Draft Genome of *Yersinia pestis* from Victims of the Black Death," *Nature* 478, no. 7370 (2011): 506–510, doi: 10.1038/nature10549.

40. Gordon Manley, "Central England Temperatures: Monthly Means 1659 to 1973," *Quarterly Journal of the Royal Meteorological Society* 100, no. 425 (1974): 389–405, doi: 10.1002/qj.49710042511.

41. Alan D. Dyer, "The Influence of Bubonic Plague in England 1500–1667," *Medical History* 22, no. 3 (1978): 308–326, doi: 10.1017/S0025727300032932.

42. Kenneth L. Gage and Michael Y. Kosoy, "Natural History of Plague: Perspectives from More than a Century of Research," *Annual Review of Entomology* 50 (2005): 505–528, doi: 10.1146/annurev.ento.50.0671803.130337. See also Gage et al., "Climate and Vectorborne Diseases."

43. Robert R. Parmenter, Ekta Pratap Yadav, Cheryl A. Parmenter, Paul Ettestad, and Kenneth L. Gage, "Incidence of Plague Associated with Increased Winter-Spring Precipitation in New Mexico," *American Journal of Tropical Medicine and Hygiene* 61, no. 5 (1999): 814–821.

44. Russell E. Enscore, "Modeling Relationships Between Climate and the Frequency of Human Plague Cases in the Southwestern United States, 1960–1997," *American Journal of Tropical Medicine and Hygiene* 66, no. 2 (2002): 186–196.

45. Tamara Ben Ari, Alexander Gershunov, Kenneth L. Gage, Tord Snäll, Paul Ettestad, Kyrre L. Kausrud, and Nils Chr. Stenseth, "Human Plague in the USA: The Importance of Regional and Local Climate," *Biology Letters* 4, no.

6 (2008): 737–740, doi: 10.1098/rsbl.2008.0363. Increased rainfall resulting from shifts in the Pacific Decadal Oscillation and from the La Niña phase of ENSO appear to be important.

46. Lei Xu, Qiyong Liu, Leif Chr. Stige, Tamara Ben Ari, Xiye Fang, Kung-Sik Chan, Shuchun Wang, Nils Chr. Stenseth, and Zhibin Zhang, "Nonlinear Effect of Climate on Plague during the Third Pandemic in China," *Proceedings of the National Academy of Sciences of the United States of America* 108, no. 25 (2011): 10214–10219, doi: 10.1073/pnas.1019486108.

47. Xu et al., "Nonlinear Effect."

48. Fa-Hu Chen, Jian-Hui Chen, Jonathan Holmes, Ian Boomer, Patrick Austin, John B. Gates, Ning-Lian Wang, Stephen J. Brooks, and Jia-Wu Zhang, "Moisture Changes over the Last Millennium in Arid Central Asia: A Review, Synthesis and Comparison with Monsoon Region," *Quaternary Science Reviews* 29, no. 7–8 (2010): 1055–1068, doi: 10.1016/j.quascirev.2010.01.005.

49. Kallie Szczepanski, "Black Death in Asia: Bubonic Plague," About Education, updated October 27, 2015, http://asianhistory.about.com/od/asianenvironmentalhistory/p/Black-Death-In-Asia-Bubonic-Plague.htm.

50. Linné K. Kausrud, Mike Begon, Tamara Ben Ari, Hildegunn Viljugrein, Jan Esper, Ulf Büntgen, Herwig Leirs, et al., "Modeling the Epidemiological History of Plague in Central Asia: Palaeoclimatic Forcing on a Disease System Over the Past Millennium," *BMC Biology* 8 (2012): 112–116, doi: 10.1186/1741-7007-8-112.

51. Nils Chr. Stenseth, Noelle I. Samia, Hildegunn Viljugrein, Kyrre Linné Kausrud, Mike Begon, Stephen Davis, Herwig Leirs, et al., "Plague Dynamics Are Driven by Climate Variation," *Proceedings of the National Academy of Sciences of the United States of America* 103, no. 35 (2006): 13110–13115, doi: 10.1073/pnas.0602447103.

52. Kausrud et al., "Modeling the Epidemiological History."

53. Kausrud et al., "Modeling the Epidemiological History."

54. Kausrud et al., "Modeling the Epidemiological History." See also Jin-Qi Fang and Guo Liu, "Relationship between Climatic Change and the Nomadic Southward Migrations in Eastern Asia during Historical Times," *Climate Change* 22, no. 2 (1992): 151–168, doi: 10.1007/BF00142964.

55. Lamb, *Climate, History*, p. 200.

56. David J. Barker, "The Fetal and Infant Origins of Adult Disease," *British Medical Journal* 301, no. 6761, 1990: 1111; David J. Barker, "Fetal Origins of Coronary Heart Disease," *British Medical Journal* 311, no. 6998, 1995: 171–174; David J. Barker, Johan G. Eriksson, Tom Forsén, and Clive Osmond, "Fetal Origins of Adult Disease: Strength of Effects and Biological Basis," *International Journal of Epidemiology* 31, no. 6 (2002): 1235–1239, doi: 10.1093/ije/31.6.1235.

57. Lamb, *Climate, History*, pp. 200–201.

58. William F. Ruddiman, "The Anthropogenic Greenhouse Era Began Thousands of Years Ago," *Climatic Change* 61, no. 3 (2003): 261–293, doi:

10.1023/B:CLIM.0000004577.17928.fa. See also William F. Ruddiman, "How Did Humans First Alter Global Climate?" *Scientific American* 292, no. 3 (2005): 46–53, doi: 10.10338/scientificamerican0305-46.

59. William F. Ruddiman, *Plows, Plagues and Petroleum: How Humans Took Control of Climate* (Princeton: Princeton University Press, 2005).

60. Ruddiman, *Plows, Plagues*.

61. Paul Slack, "The Disappearance of Plague: An Alternative View," *Economic History Review* 34, no. 3 (1981): 469–476, doi: 10.1111/j.1468-0289.1981. tb02081.x.

62. W. G. Hoskins, "Harvest Fluctuations and English Economic History, 1480–1619," *Agricultural History Review* 12, no. 1 (1964): 28–46. See also W. G. Hoskins, BBC broadcast, November 24, 1964, quoted in Lamb, *Culture, History*, p. 399.

63. Daniel Schaller, *Theologischer Heroldt* (Magdeburg, 1595), quoted in Behringer, *A Cultural History*, p. 93.

64. Geoffrey Parker, "Crisis and Catastrophe: The Global Crisis of the Seventeenth Century Reconsidered," *American Historical Review* 113, no. 4 (2008): 1053–1079.

65. David D. Zhang, Harry F. Lee, Cong Wang, Baosheng Li, Qing Pei, Jane Zhang, and Yulun An, "The Causality Analysis of Climate Change and Large-Scale Human Crisis," *Proceedings of the National Academy of Sciences of the United States of America*, 108, no. 42 (2011): 17296–17301, doi: 10.1073/pnas.1104268108.

66. Thomas Hobbes, *Leviathan; or, The Matter, Forme, and Power of a Commonwealth, Ecclesiasticall and Civill*, edited by Richard Tuck (Cambridge: Cambridge University Press, 1996), p. 89; originally published 1651.

67. Parker, "Crisis and Catastrophe," p. 1061.

68. Voltaire [François-Marie Arouet], *Essai sur les moeurs et l'esprit des nations*, 2 vols. (Paris: Garnier, 1963), originally published 1756, quoted in Parker, "Crisis and Catastrophe," p. 1064.

69. Parker, "Crisis and Catastrophe," p. 1065.

70. Parker, "Crisis and Catastrophe," p. 1073.

71. John Keay, *China: A History* (London: Harper, 2008), p. 417.

72. Caiming Shen, Wei-Chyung Wang, Zhixin Hao, and Wei Gong, "Exceptional Drought Events over Eastern China during the Last Five Centuries," *Climate Change* 85 (2007): 453–471, doi: 10.1007/s10584-007-9283-y.

73. Helen Dunstan, "The Late Ming Epidemics: A Preliminary Survey," *Ch'ing-Shih wen-t'i* 3, no. 3 (1975): 1–59.

74. Ka-wai Fan, "Climatic Change and Dynastic Cycles in Chinese History: A Review Essay," *Climate Change* 101, no. 3 (2010): 565–573, doi: 10.1007/s10584-009-9702-3.

75. Parker, "Crisis and Catastrophe."

76. William H. McNeill, *Plagues and Peoples* (Garden City: Anchor, 1976), pp. 259–269. Appendix: Epidemics in China (list compiled in 1940 by

J. H. Cha from the original two volumes of Ch'en Kao-yung's *Chung-kuo li-tai t'ien-tsai jen-huo piao*).

77. Anthony J. McMichael, "Insights from Past Millennia into Climatic Impacts on Human Health and Survival," *Proceedings of the National Academy of Sciences of the United States of America* 109, no. 13 (2012): 4730–4737, doi: 10.1073/pnas.1120177109.

78. Shengsheng Gong, "Changes of the Temporal-spatial Distribution of Epidemic Disasters in 770BC–AD1911 China," *Acta Geographica Sinica* 58, no. 6 (2003): 870–878.

79. David D. Zhang, C. Y. Jim, George C-S Lin, Yuan-Qing He, James J. Wang, and Harry F. Lee, "Climatic Change, Wars and Dynastic Cycles in China over the Last Millennium," *Climatic Change* 76, no. 3 (2006): 459–477, doi: 10.1007/s10584-005-9024-z.

80. David D. Zhang, Jane Zhang, Harry F. Lee, and Yuan-qing He, "Climate Change and War Frequency in Eastern China over the Last Millennium," *Human Ecology* 35, no. 4 (2007): 403–414, doi: 10.1007/s10745-007-9115-8.

81. Zhang et al., "Climatic Change, Wars and Dynastic Cycles."

82. Zhang et al., "Climate Change and War Frequency in Eastern China."

83. Fan, "Climate Change and Dynastic Cycles."

84. Xunming Wang, "Climate, Desertification, and the Rise and Collapse of China's Historical Dynasties," *Human Ecology* 38, no. 1 (2010): 157–172, doi: 10.1007/s10745-009-9298-2.

85. Gergana Yancheva, Norbert R. Nowaczyk, Jens Mingram, Peter Dulski, Georg Schettler, Jörg F. W. Negendank, Jiaqi Liu, Daniel M. Sigman, Larry C. Peterson, and Gerald H. Haug, "Influence of the Intertropical Convergence Zone on the East Asian Monsoon," *Nature* 445, no. 7123 (2007): 74–77, doi: 10.1038/nature05431.

86. Edward R. Cook, Kevin J. Anchukaitis, Brendan M. Buckley, Rosanne D. D'Arrigo, Gordon C. Jacoby, and William E. Wright, "Asian Monsoon Failure and Megadrought," *Science* 328, no. 5977 (2010): 486–489, doi: 10.1126/science.1185188.

87. Shen et al., "Exceptional Drought."

88. Zhang et al., "Climate Change and War Frequency in Eastern China."

89. T. Jiang, *Recent History of Chinese Population* (Hangzhou: Hangzhou University Press, 1993), quoted in Zhang et al., "Climatic Change, Wars and Dynastic Cycles."

90. Patrick D. Nunn, *Climate, Environment and Society in the Pacific during the Last Millennium*, Developments in Earth and Environmental Sciences 6 (Amsterdam and Oxford: Elsevier Science, 2007).

91. Janet Davidson, *The Prehistory of New Zealand* (Auckland: Longman Paul, 1984).

92. B. Foss Leach, "Prehistoric Communities in Palliser Bay, New Zealand," PhD diss., University of Otago, 1976, available at http://hdl.handle.net/10523/499.

93. Robert Fogel, *The Escape from Hunger and Premature Death, 1700–2100: Europe, America and the Third World* (Cambridge: Cambridge University Press, 2004).

94. Paul Kennedy, *The Rise and Fall of the Great Powers* (New York: Random House, 1987), p. 677.

95. John Dexter Post, *Food Shortage, Climatic Variability, and Epidemic Disease in Preindustrial Europe: The Mortality Peak in the Early 1740s* (Ithaca: Cornell University Press, 1985).

96. Helen A. Fletcher, Helen D. Donoghue, John Holton, Ildikó Pap, and Mark Spigelman, "Widespread Occurrence of *Mycobacterium tuberculosis* DNA from 18th–19th Century Hungarians," *American Journal of Physical Anthropology* 120, no. 2 (2003): 144–152, doi: 10.1002/ajpa.10114.

97. Richard H. Grove, "Revolutionary Weather: The Climatic and Economic Crisis of 1788–1795 and the Discovery of El Niño," in *Sustainability or Collapse? An Integrated History and Future of People on Earth*, edited by Robert Costanza, Lisa J. Graumlich, and William L. Steffen (Cambridge: MIT Press, 2007), pp. 151–167.

98. John Withington, *A Disastrous History of the World: Chronicles of War, Earthquakes, Plague and Flood* (London: Piatkus, 2008), p. 14.

99. Grove, "Revolutionary Weather."

100. Richard H. Grove, "Global Impact of the 1789–93 El Niño," *Nature* 393, no. 6683 (1998): 318–319, doi: 10.1038/30636.

101. William J. Dawson, "Wolfgang Amadeus Mozart—Controversies Regarding His Illnesses and Death: A Bibliographic Review," *Medical Problems of Performing Artists* 25, no. 2 (2010): 49–53.

102. L. Karhausen, "A Selection of Diagnostic Hypotheses Purporting to Explain Mozart's Terminal Illness," December 16, 2010, http://karhausenlmd.blogspot.com.au/.

103. Richard H. Zegers, Andreas Weigl, and Andrew Steptoe, "The Death of Wolfgang Amadeus Mozart: An Epidemiologic Perspective," *Annals of Internal Medicine* 151, no. 4 (2009): 274–278, doi: 10.7326/0003-4819-151-4-200908280-00010.

104. Charles Gibson, *The Aztecs Under Spanish Rule: A History of the Indians of the Valley of Mexico* (Stanford: Stanford University Press, 1964).

105. Grove, "Revolutionary Weather," p. 157.

106. Kenneth R. Foster, Mary F. Jenkins, and Anna C. Toogood, "The Philadelphia Yellow Fever Epidemic of 1793," *Scientific American* 279, no. 2 (1998): 68–74.

107. Grove, "Revolutionary Weather."

108. Henry F. Diaz and Gregory J. McCabe, "A Possible Connection between the 1878 Yellow Fever Epidemic in the Southern United States and the 1877–78 El Niño Episode," *Bulletin of the American Meteorological Society* 80, no. 1 (1999): 21–27, doi: 10.1175/1520-0477(1999)080<0021:APCBTY>2.0.CO;2.

109. Kevin D. Lafferty, "The Ecology of Climate Change and Infectious Disease," *Ecology* 90, no. 4 (2009): 888–900, doi: 10.1890/08-0079.1.

110. Robert Hughes, *The Fatal Shore: A History of the Transportation of Convicts to Australia, 1787–1868* (London and Sydney: Collins Harvill, 1987).

111. Joelle Gergis, David J. Karoly, and Rob J. Allan, "A Climate Reconstruction of Sydney Cove, New South Wales, Using Weather Journal and Documentary Data, 1788–1791," *Australian Meteorological and Oceanographic Journal* 58, no. 2 (2009): 83–98.

112. B. Gandevia and J. Cobley, "Mortality at Sydney Cove, 1788–1792," *Australian and New Zealand Journal of Medicine* 4, no. 2 (1974): 111–125.

113. David Collins, *An Account of the English colony in New South Wales* (London: Cadell and David, 1798), quoted in Gergis, Karoly, and Allan, "A Climate," p. 94.

114. Watkin Tench, *A Complete Account of the Settlement at Port Jackson* (London: Nicol and Sewell, 1793).

115. Grove, "Revolutionary Weather."

116. Kevin Hamilton and Rolando R. Garcia, "El Niño/Southern Oscillation Events and Their Associated Midlatitude Teleconnections 1531–1841," *Bulletin of the American Meteorological Association* 67, no. 11 (1986): 1354–1361, doi: 10.1175/1520-0477(1986)067<1354:ENOEAT>2.0.CO;2.

117. Gandevia and Cobley, "Mortality at Sydney Cove."

118. Gandevia and Cobley, "Mortality at Sydney Cove." These figures exclude deaths from physical trauma.

119. Stephen Nicholas and Richard H. Steckel, "Heights and Living Standards of English Workers during the Early Years of Industrialisation, 1770–1815," *Journal of Economic History* 51, no. 4 (1991): 937–957, doi: 10.1017/S0022050700040171.

120. Ivan Hanigan, Colin D. Butler, Philip N. Kokic, and Michael F. Hutchinson, "Suicide and Drought in New South Wales, Australia, 1970–2007," *Proceedings of the National Academy of Sciences of the United States of America* 109, no. 35 (2012): 13950–13955, doi: 10.1073/pnas.1112965109.

121. Bertrand Timbal et al., *Understanding the Anthropogenic Nature of the Observed Rainfall Decline Across South-Eastern Australia*, CAWCR No. 026 (Melbourne: Centre for Australian Weather and Climate Research, 2010).

Chapter 9

1. Gillen D. Wood, *Tambora: The Eruption that Changed the World* (Princeton and Oxford: Princeton University Press, 2014).

2. Rebecca Lines-Kelly, "Environmental Agriculture: History Reconstruction Confirms Changes," *Agriculture Today,* July 2012, available at http://www.dpi.nsw.gov.au/content/archive/agriculture-today-stories/ag-today-archive/july-2012/history-reconstruction-confirms-changes. See also SEARCH: South Eastern Australian Recent Climate History, "Unearthing Australia's Climate History," December 17, 2009, http://climatehistory.com.au/.

3. Dorothea Mackellar's much-loved poem, from 1912, "My Country," available at http://www.dorotheamackellar.com.au/archive/mycountry.htm.
4. Hanigan et al., "Suicide and Drought in New South Wales."
5. Michael E. Mann, Jose D. Fuentes, and Scott Rutherford, "Underestimation of Volcanic Cooling in Tree-Ring-Based Reconstructions of Hemispheric Temperatures," *Nature Geoscience* 5, no. 3 (2012): 202–205, doi: 10.1038/ngeo1394.
6. Clive Oppenheimer, "Climatic, Environmental and Human Consequences of the Largest Known Historic Eruption: Tambora Volcano (Indonesia) 1815," *Progress in Physical Geography* 27, no. 2 (2003): 230–259, doi: 10.1191/0309133303pp379ra.
7. Jihong Cole-Dai, David G. Ferris, Alyson L. Lanciki, Joël Savarino, Mélanie Baroni, and Mark H. Thiemens, "Cold Decade (AD 1810–1819) Caused by Tambora (1815) and Another (1809) Stratospheric Volcanic Eruption," *Geophysical Research Letters* 36, no. 22 (2009): L22703, doi: 10.1029/2009GL040882.
8. Oppenheimer, "Climatic, Environmental."
9. Mann et al., "Underestimation of Volcanic Cooling."
10. Those eruptions were clustered in time around the seismically volatile year 1811, when, in America, a series of earthquakes caused a temporary reversal of flow of the Mississippi River. "A spirit of change, and a restlessness seemed to pervade the very inhabitants of the forest," wrote the London-born traveler Charles Joseph La Trobe, describing disoriented squirrels plunging, lemming-like, into the Ohio River.
11. Oppenheimer, "Climatic, Environmental."
12. Anthony J. McMichael, "Insights from Past Millennia into Climatic Impacts on Human Health and Survival," *Proceedings of the National Academy of Sciences of the United States of America* 109, no. 13 (2012): 4730–4737, doi: 10.1073/pnas.1120177109.
13. McMichael, "Insights from Past Millennia."
14. Clive Oppenheimer, *Eruptions that Shook the World* (Cambridge: Cambridge University Press, 2011).
15. Oppenheimer, "Climatic, Environmental."
16. Chester Dewey, "Results of Meteorological Observations, made at Williamstown, Massachusetts," *Memoirs of the American Academy of Arts and Sciences* 4, parts 1–2 (1816): 388–389.
17. Cormac Ó Gráda, *Famine: A Short History* (Princeton and Oxford: Princeton University Press, 2009), p. 106.
18. Ann G. Carmichael, "Infectious Diseases and Human Agency: An Historical Overview," in *Interactions between Global Change and Human Health*, edited by M. O. Andreae, U. Confalonieri, and A. J. McMichael (Vatican City: Pontificia Academia Scientiarum, 2006), pp. 3–46.
19. Mercedes Pascual, Menno J. Bouma, and Andrew P. Dobson, "Cholera and Climate: Revisiting the Quantitative Evidence," *Microbes and Infection* 4, no. 2 (2002): 237–245, doi: 10.1016/S1286-4579(01)01533-7.

20. Rita R. Colwell, "Global Climate and Infectious Disease: The Cholera Paradigm," *Science* 274, no. 5295 (1996): 2025–2031, doi: 10.1126/science.274.5295.2025.

21. Dorothy H. Crawford, *Deadly Companions: How Microbes Shaped Our History* (Oxford: Oxford University Press, 2009).

22. Michael G. Mulhall, *The Dictionary of Statistics* (London, G. Routledge and Sons, 1892), p. 64.

23. Brian Fagan, *Floods, Famines, and Emperors: El Niño and the Fate of Civilizations* (New York: Basic Books, 1999), pp. 234–243.

24. Mike Davis, *Late Victorian Holocausts: El Niño Famines and the Making of the Third World* (London: Verso, 2001).

25. G. Gong, Q. Ge, and K. Xu, "Influence of Climatic Changes on Agriculture," in *Historical Climatic Changes in China*, edited by P. Zhang (Jinan: Shandong Science and Technology Press, 1996), pp. 406–425. Harvest yields during the cold 1840–1890 period were 10–25 percent lower than in the warmer 1730–1770 period.

26. Xu et al., "Nonlinear Effect."

27. Nils Chr. Stenseth, "Plague Dynamics Are Driven by Climate Variation," *Proceedings of the National Academy of Sciences of the United States of America* 103, no. 35 (2006): 13110–13115, doi: 10.1073/pnas.0602447103.

28. Xu et al., "Nonlinear Effect."

29. Xu et al., "Nonlinear Effect."

30. Gage et al., "Climate and Vectorborne Diseases."

31. Terence Hull, "Plague in Java," in *Death and Disease in Southeast Asia: Explorations in Social, Medical and Demographic History*, edited by Norman G. Owen (Singapore: Oxford University Press for Asian Studies Association of Australia, 1987), pp. 210–234.

32. Anthony J. McMichael, "Paleoclimate and Bubonic Plague: A Forewarning of Future Risk?" *BMC Biology* 8 (2010): 108, doi: 10.1186/1741-7007-8-108.

33. D. C. Cavanaugh and J. E. Williams, "Plague: Some Ecological Interrelationships," in *Fleas*, edited by Robert Traub and Helle Starcke (Rotterdam: Taylor and Francis, 1980).

34. Amartya Sen, *Poverty and Famines: An Essay on Entitlement and Deprivation* (Oxford: Oxford University Press, 1981).

35. Ó'Gráda, *Famine*. The Bengali *aman* rice crops in late 1942 were thought to be at risk, and this was the autumn–winter harvest that provided most of the year's output.

36. When I visited Calcutta as a student traveler in 1963, the impressive Howrah Bridge spanning the aromatic Hooghly River, a two-span iron-and-steel suspension bridge built in the year of the Bengal Famine, was an inviting subject for photography. But I was quickly waved away by agitated officials. In 1963, exactly 20 years after the Bengal Famine and that earlier fear of a wartime Japanese invasion, India was briefly fearful of a flare-up of the 1962 Sino-Indian border war—and tried to prohibit photos that might assist Chinese bomber pilots.

37. See also Chapter 12: the 1930s drought was predictable, though not in its severity, in light of the longstanding cyclical recurrence of droughts in this region every 20–22 years.

38. Cormac Ó Gráda, "Famines Past, Famine's Future," *Development and Change* 42, no. 1 (2011): 49–69, doi: 10.1111/j.1467-7660.2010.01677.x.

39. Frank Dikötter, *Mao's Great Famine: The History of China's Most Devastating Catastrophe, 1958–1962* (New York: Walker, 2010).

40. United Nations, Population Division of Department of Economic and Social Affairs, *World Population Prospects: The 2012 Revision* (New York: United Nations, 2012).

41. David B. Lobell, Marianne Bänziger, Cosmos Magorokosho, and Bindiganavile Vivek, "Nonlinear Heat Effects on African Maize as Evidenced by Historical Yield Trials," *Nature Climate Change* 1, no. 1 (2011): 42–45, doi: 10.1038/nclimate1043.

42. Michael Buerk, "Extent of Ethiopia Famine Revealed," BBC News, October 23, 1984, http://news.bbc.co.uk/2/hi/8315248.stm.

43. Abraham Vergese's novel *Cutting for Stone* (New York: Vintage Books, 2010) tells how the emperor's dog was much better fed than most Ethiopians.

44. Ó'Gráda, *Famine*, pp. 254–255.

45. Daniel Goodkind and Loraine West, "The North Korean Famine and Its Demographic Impact," *Population and Development Review* 27, no. 2 (2001): 219–238, doi: 10.1111/j.1728-4457.2001.00219.x.

46. Ó'Gráda, *Famine*, pp. 255–256.

47. For example, K. Saxton, A. Falconi, S. Goldman-Mellor, and R. Catalano, "No Evidence of Programmed Late-Life Mortality in the Finnish Famine Cohort," *Journal of Developmental Origins of Health and Disease* 4, no. 1 (2013): 30–34, doi: 10.1017/S2040174412000517.

48. Yonghong Wang, Xiaolin Wang, Yuhan Kong, John H. Zhang, and Qing Zeng, "The Great Chinese Famine Leads to Shorter and Overweight Females in Chongqing Chinese Population after 50 Years," *Obesity* 18, no. 3 (2010): 588–592, doi: 10.1038/oby.2009.296.

49. Winnie Fung and Wei Ha, "Intergenerational Effects of the 1959–61 China Famine," in *Risks, Shocks, and Human Development: On the Brink*, edited by Ricardo Fuentes-Nieva and Papa A. Seck (London: Palgrave Macmillan, 2010), pp. 222–254.

50. Pascual et al., "Cholera and Climate."

51. Luigi Vezzulli, Ingrid Brettar, Elisabetta Pezzati, Philip C. Reid, Rita R. Colwell, Manfred G. Höfle, and Carla Pruzzo, "Long-Term Effects of Ocean Warming on the Prokaryotic Community: Evidence from the Vibrios," *ISME Journal—Multidisciplinary Journal of Microbial Ecology* 6 (2012): 21–30, doi: 10.1038/ismej.2011.89.

52. A. S. Siraj et al., "Altitudinal Changes in Malaria Incidence."

53. Kostas Danis, Annick Lenglet, Maria Tseroni, Agoritsa Baka, Sotiris Tsiodras, and Stefanos Bonovas, "Malaria in Greece: Historical and Current Reflections on a Re-Emerging Vector Borne Disease," *Travel Medicine and Infectious Disease* 11, no. 1 (2013): 8–14, doi: 10.1016/j.tmaid.2013.01.001.

54. Hubert H. Lamb, *Climate, History and the Modern World* (London and New York: Routledge, 1995), p. 315.

55. Dim Coumou and Stefan Rahmstorf, "A Decade of Weather Extremes," *Nature Climate Change* 2, no. 7 (2012): 491–496, doi: 10.1038/nclimate1452.

56. Munich Re, *Group Annual Report* (Sydney: Munich Re, 2011), available at http://www.munichre.com/site/corporate/get/documents/mr/assetpool .shared/Documents/0_Corporate%20Website/_Publications/302-07342_ en.pdf.

57. Coumou and Rahmstorf, "A Decade."

58. Aslak Grinsted, John C. Moore, and Svetlana Jevrejeva, "Homogeneous Record of Atlantic Hurricane Surge Threat Since 1923," *Proceedings of the National Academy of Sciences of the United States of America* 109, no. 48 (2012): 19601–19605, doi: 10.1073/pnas.1209542109.

59. Joan Brunkard, Gonza Namulanda, and Raoult Ratard, "Hurricane Katrina Deaths, Louisiana, 2005," *Disaster Medicine and Public Health Preparedness* 2, no. 4 (2008): 215–223, doi: 10.1097/DMP.0b013e31818aaf55.

60. John Manuel, "The Long Road to Recovery: Environmental Health Impacts of Hurricane Sandy," *Environmental Health Perspectives* 121, no. 5 (2013): A152–A159, doi: 10.1289/ehp.121-a152.

61. Smith et al., "Human Health."

62. Gulrez Shah Azhar, Dileep Mavalankar, Amruta Nori-Sarma, Ajit Rajiva, Priya Dutta, Anjali Jaiswal, Perry Sheffield, Kim Knowlton, and Jeremy J. Hess, "Heat-Related Mortality in India: Excess All-Cause Mortality Associated with the 2010 Ahmedabad Heat Wave," *PLoS ONE* 9, no. 3 (2014): e91831, doi: 10.371/journal.pone.0091831.

63. Charmian M. Bennett, Keith B. G. Dear, Anthony J. McMichael, "Shifts in the seasonal distribution of deaths in Australia, 1968–2007," *International Journal of Biometeorology* 58, Issue 5 (2014): 835–842.

Chapter 10

1. Raymond S. Bradley, *Paleoclimatology: Reconstructing Climates of the Quaternary*, 2nd ed. (San Diego: Harcourt, 2008), p. 264.

2. Franz-Xaver Neubert, Rogier B. Mars, Adam G. Thomas, Jerome Sallet, and Matthew F. S. Rushworth, "Comparison of Human Ventral Frontal Cortex Areas for Cognitive Control and Language with Areas in Monkey Frontal Cortex," *Neuron* 81, no. 3 (2014): 700–713, doi: 10.1016/j.neuron.2013.11.012.

3. John Robert McNeill, "Diamond in the Rough: Is There a Genuine Environmental Threat to Security? A Review Essay," *International Security* 30, no. 1 (2005): 178–195.

4. Lee R. Kump, "The Last Great Global Warming," *Scientific American* 305 (2011): 56–61.
5. World Bank, *Turn Down the Heat: Why a 4°C Warmer World Must Be Avoided* (Washington, DC: World Bank, 2012), available at http://documents.worldbank.org/curated/en/2012/11/17097815/turn-down-heat-4°c-warmer-world-must-avoided.
6. Niall Ferguson, *Civilization: The Six Killer Apps of Western Power* (London: Penguin, 2012).
7. Ó Gráda, *Famine*, p. 31.
8. Ó Gráda, *Famine*, pp. 14–16.
9. Department of Defense, *Quadrennial Defense Review Report* (Washington, DC: US Department of Defense, 2010).
10. Bernice Lee, Felix Preston, Jaakko Kooroshy, Rob Bailey, and Glada Lahn, *Resources Future* (London: Chatham House, 2012), p. 75.
11. Jeffrey Mazo, *Climate Conflict: How Global Warming Threatens Security and What to Do About It* (London: Routledge, 2010).
12. Kurt M. Campbell, ed., *Climatic Cataclysm: The Foreign Policy and National Security Implications of Climate Change* (Washington, DC: Brookings Institution Press, 2008).
13. Peter Schwartz and Doug Randall, *An Abrupt Climate Change Scenario and Its Implications for United States National Security* (Washington, DC: Environmental Media Services, 2003).
14. Solomon Hsiang, Marshall Burke, and Edward Miguel, "Quantifying the Influence of Climate on Human Conflict," *Science* 341, no. 6151 (2013): 1235367, doi: 10.1126/science.1235367.
15. Hsiang, Burke, and Miguel, "Quantifying the Influence of Climate."
16. Civil war was classified as a conflict in which at least 25 combat deaths occurred.
17. Ó'Gráda, *Famine*.
18. Zhang et al., "Causality Analysis of Climate Change."
19. Patricia M. Lambert, "Patterns of Violence in Prehistoric Hunter-Gatherer Societies of Coastal Southern California," in *Troubled Times: Violence and Warfare in the Past*, edited by Debra L. Martin and David W. Frayer (Langhorne: Gordon and Breach, 1997), p. 376.
20. Thomas Homer-Dixon, "Environmental Scarcities and Violent Conflict: Evidence from Cases," *International Security* 19 (1994): 5–40. See also Colin H. Kahl, *States, Scarcity and Civil Strife in the Developing World* (Princeton: Princeton University Press, 2006).
21. Hsiang, Burke, and Miguel, "Quantifying the Influence of Climate."
22. Gwen Robbins Schug, K. Elaine Blevins, Brett Cox, Kelsey Gray, and V. Mushrif-Tripathy, "Infection, Disease, and Biosocial Processes at the End of the Indus Civilization," *PLoS ONE* 8, no. 12 (2013): e84814, doi: 10.1371/journal.pone.0084814.

23. Robert F. Worth, "Earth Is Parched Where Syrian Farms Thrived," *New York Times,* October 13, 2010.

24. Shahrzad Mohtadi, "Climate Change and the Syrian Uprising," *Bulletin of the Atomic Scientists,* August 16, 2012, http://thebulletin.org/climate-change-and-syrian-uprising; Colin P. Kelley, Shahrzad Mohtadi, Mark A. Cane, Richard Seager, and Yochanan Kushnir, "Climate Change in the Fertile Crescent and Implications of the Recent Syrian Drought," *Proceedings of the National Academy of Sciences of the United States of America* 112, no. 11 (2015): 3241–3246, doi: 10.1073/pnas.1421533112.

25. Worth, "Earth Is Parched."

26. Devin C. Bowles, Colin D. Butler, and Sharon Friel, "Climate Change and Health in Earth's Future," *Earth's Future* 2 (2014): 60–67, doi: 10.1002/2013EF000177.

27. Jennifer Leaning and Debarati Guha-Sapir, "Natural Disasters, Armed Conflict, and Public Health," *New England Journal of Medicine* 369 no. 19 (2013): 1836–1842, doi: 10.1056/NEJMra1109877.

28. Jarvis Lionel, Hugh Montgomery, Neil Morisetti, and Ian Gilmore, "Climate Change, Ill Health, and Conflict," *British Medical Journal* 342 (2011): d1819, doi: 10.1136/bmj.d1819.

29. Colin D. Butler, "Do We Face a Third Revolution in Human History? If So, How Will Public Health Respond?" *Journal of Public Health* 30, no. 4 (2008): 364–365, doi: 10.1093/pubmed/fdn082.

30. Gwynne Dyer, *The Fight for Survival as the World Overheats* (Toronto: Random House Canada, 2008).

31. Thomas F. Homer-Dixon, *Environment, Scarcity, and Violence* (Princeton: Princeton University Press, 1999).

32. Michael T. Klare, *Resource Wars: The New Landscape of Global Conflict* (New York: Metropolitan, 2001).

33. Clionadh Raleigh and Henrik Urdal, "Climate Change, Environmental Degradation and Armed Conflict," *Political Geography* 26, no. 6 (2007): 674–694, doi: 10.1016/j.polgeo.2007.06.005.

34. Celia McMichael, Jon Barnett, and Anthony J. McMichael, "An Ill Wind? Climate Change, Migration, and Health," *Environmental Health Perspectives* 120, no. 5 (2012): 646–654, doi: 10.1289/ehp.1104375.

35. Asian Development Bank, *Addressing Climate Change and Migration in Asia and the Pacific* (Mandaluyong City, Philippines: Asian Development Bank, 2012).

36. Barbara Tuchman, *A Distant Mirror: The Calamitous 14th Century* (New York: Random House, 1978): xix–xx.

37. Geoffrey Parker, "Crisis and Catastrophe: The Global Crisis of the Seventeenth Century Reconsidered," *American Historical Review* 113, no. 4 (2008): 1053–1079, doi: 10.1086/ahr.113.4.1053.

38. Hubert H. Lamb, *Climate, History and the Modern World* (London and New York: Routledge, 1995), p. 3.

39. Michel P. Coleman, "A Plague Epidemic in Voluntary Quarantine," *International Journal of Epidemiology* 15, no. 3 (1986): 379–385, doi: 10.1093/ije/15.3.379. See also Geraldine Brooks, *Year of Wonders: A Novel of the Plague* (London: Viking, 2001).

40. Brooks, *Year of Wonders*.

41. Tord Kjellstrom, Alistair Woodward, Laila Gohar, Jason Lowe, Bruno Lemke, Lauren Lines, David Briggs, Chris Freyberg, Matthias Otto, and Olivia Hyatt, "The Risk of Heat Stress to People," in *Climate Change: A Risk Assessment*, edited by James Hynard and Tom Rodger (The Foreign and Commonwealth Office, UK), available at http://www.csap.cam.ac.uk/projects/climate-change-risk-assessment/.

42. Jean-Marie Robine, Siu Lan K. Cheung, Sophie Le Roy, Herman Van Oyen, Claire Griffiths, Jean-Pierre Michel, and François Richard Herrmann, "Death Toll Exceeded 70,000 in Europe during the Summer of 2003," *Comptes Rendus Biologies* 331, no. 2 (2008): 171–178, doi: 10.1016/j.crvi.2007.12.001.

43. Simone Russo, Alessandro Dosio, Rune G. Graversen, Jana Sillmann, Hugo Carrao, Martha B. Dunbar, Andrew Singleton, Paolo Montagna, Paulo Barbola, and Jürgen V. Vogt, "Magnitude of Extreme Heat Waves in Present Climate and Their Projection in a Warming World," *Journal of Geophysical Research: Atmospheres* 119, no. 22 (2014): 12500–12512, doi: 10.1002/2014JD022098.

44. Mark Zastrow, "Speedy Study Claims Climate Change Doubled Chances of European Heatwave," *Nature (News)*, July 13, 2015, doi: 10.1038/nature.2015.17940.

45. Parker, "Crisis and Catastrophe."

46. Zhang et al., "Causality Analysis of Climate Change."

47. Stephen V. Boyden, *The Biology of Civilisation* (Sydney: UNSW Press, 2004).

48. Anthony J. McMichael, *Human Frontiers, Environments and Disease: Past Patterns, Future Uncertainties* (Cambridge: Cambridge University Press, 2001).

49. Robert Beaglehole, Ruth Bonita, Richard Horton, Majid Ezzati, Neeraj Bhala, Mary Amuyunzu-Nyamongo, Modi Mwatsama, and K. Srinath Reddy, "Measuring Progress on NCDs: One Goal and Five Targets," *The Lancet* 380, no. 9850 (2012): 1283–1285, doi: 10.1016/S0140-6736(122)61692-4.

50. OECD, *Future Global Shocks: Improving Risk Governance* (Paris: Organisation for Economic Co-operation and Development, 2012).

51. OECD, *Future Global Shocks*.

52. Kathryn J. Bowen, Fiona Miller, Va Dany, Anthony J. McMichael, and Sharon Friel, "Enabling Environments? Insights into the Policy Context for Climate Change and Health Adaptation Decision-Making in Cambodia," *Climate and Development* 5, no. 4 (2013): 277–287, doi: 10.1080/17565529.2013.833077.

53. William H. McNeill, *Plagues and Peoples* (New York: Anchor Books, 1977).

54. Celia McMichael et al., "An Ill Wind?"

55. Patrick D. Nunn, "Climate, Environment and Society in the Pacific during the Last Millennium," in *Developments in Earth and Environmental Sciences* (Amsterdam: Elsevier, 2007).

56. John R. McNeill, "Of Rats and Men: A Synoptic Environmental History of the Island Pacific," *Journal of World History* 5, no. 2 (1994): 299–349. See also John R. McNeill, "Islands in the Rim: Ecology and History in and around the Pacific, 1521–1996," in *Pacific Centuries: Pacific and Pacific Rim History since the Sixteenth Century*, edited by D. O. Flynn, L. Frost, and A. J. H. Latham (London: Routledge, 1999), pp. 70–84. Both quoted in Nunn, "Climate, Environment, and Society."

Chapter 11

1. René Dubos, "Trend is Not Destiny," *Engineering and Science* 34, no. 3 (1971): 5–10, available at http://resolver.caltech.edu/CaltechES:34.3.dubos.

2. Joseph Tainter, *The Collapse of Complex Societies* (Cambridge and New York: Cambridge University Press, 1988).

3. William Ophuls, *Immoderate Greatness: Why Civilizations Fail* (Charleston: Creative Space Publishing, 2012).

4. Safa Motesharrei, Jorge Rivas, and Eugenia Kalnay, "Human and Nature Dynamics (HANDY): Modeling Inequality and Use of Resources in the Collapse or Sustainability of Societies," *Ecological Economics* 101 (2014): 90–102, doi: 10.1016/j.edolecon.2014.02.014.

5. Neubert et al., "Comparison of Human Ventral Frontal Cortex Areas."

6. George Marshall, *Carbon Detox: Your Step-by-Step Guide to Getting Real about Climate Change* (London: Octopus, 2007).

7. Clive Hamilton, *Requiem for a Species* (Crows Nest, Australia: Allen & Unwin, 2010).

8. Andrew T. Guzman, *Overheated: The Human Cost of Climate Change* (Oxford: Oxford University Press, 2013).

9. William E. Rees, "Carrying Capacity and Sustainability: Waking Malthus' Ghost," in *Introduction to Sustainable Development*, edited by David V. J. Bell and Yuk-kuen Annie Cheung (Oxford: Encyclopedia of Life Support Systems, 2009).

10. Paul R. Ehrlich and Anne H. Ehrlich, "Can a Collapse of Global Civilization Be Avoided?" *Proceedings of the Royal Society B* 280, no. 1754 (2013): 20122845, doi: 10.1098/rspb.2012.2845.

11. Martin Rees, *Our Final Century: Will the Human Race Survive the Twenty-first Century?* (London: William Heinemann, 2003).

12. René Dubos, *A God Within* (London: Angus and Robertson, 1973), p. 12.

13. Clive Hamilton, *Growth Fetish* (Crows Nest, Australia: Allen and Unwin, 2003).

14. J. R. McNeill, *Something New Under the Sun: An Environmental History of the Twentieth-Century World* (New York: W. W. Norton, 2000), p. 336.

15. McNeill, *Something New.*
16. Herman E. Daly, *Beyond Growth: The Economics of Sustainable Development* (Boston: Beacon, 1996).
17. Herman Daly, "The Negative Natural Interest Rate and Uneconomic Growth," Center for the Advancement of the Steady State Economy, January 2014, http://steadystate.org/the-negative-natural-interest-rate-and-uneconomic-growth/.
18. Adam Smith, *An Enquiry into the Nature and Causes of the Wealth of Nations* (London: W. Strahan and T. Cadell, 1776).
19. John Stuart Mill, *Principles of Political Economy* (London: John William Parker, 1848).
20. When the Dalai Lama was asked what surprised him most about humanity, he replied: "Man sacrifices his health in order to make money. Then he sacrifices money to recuperate his health. And then he is so anxious about the future that he does not enjoy the present; the result being that he does not live in the present or the future; he lives as if he is never going to die, and then dies having never really lived."
21. Paul C. Roberts, *The Failure of Laissez Faire Capitalism and Economic Dissolution of the West* (Atlanta: Clarity Press, 2013).
22. Ian Dunlop, "Planet's Future Is Upon Us," *Canberra Times*, September 6, 2013, available at http://www.canberratimes.com.au/comment/planets-future-is-upon-us-20130905-2t7pk.html#ixzz2eNpShG8C. Dunlop is a Club of Rome member.
23. Nicholas Stern, *The Stern Review: The Economics of Climate Change* (London: UK Government, 2006).
24. Ángel Gurria, "Charting Progress, Building Visions, Improving Life," Third OECD World Forum on Statistics, Knowledge and Policy, Busan, South Korea, October 27–30, 2009.
25. Joseph Stiglitz, "Progress, What Progress?" *OECD Observer* 272, March 2009, http://www.oecdobserver.org/news/archivestory.php/aid/2793/Progress,_what_progress_.html.
26. American Psychological Association (APA), "Happier Consumers Can Lead to Healthier Environment, Research Reveals," Science Daily, August 9, 2014, http://www.sciencedaily.com/releases/2014/08/140809141434.htm.
27. Ida Kubiszewski, Robert Costanza, Carol Franco, Philip Lawn, John Talberth, Tim Jackson, and Camille Aylmer, "Beyond GDP: Measuring and Achieving Global Genuine Progress," *Ecological Economics* 93 (2013): 57–68, doi: 10.1016/j.ecolecon.2013.04.019.
28. Partha Dasgupt, *Human Well-Being and the Natural Environment* (Oxford: Oxford University Press, 2001).
29. Richard Heinberg, *The End of Growth: Adapting to Our New Economic Reality* (Gabriola Island: New Society Publishers, 2011).
30. Robert B. Richardson, ed., *Building a Green Economy: Perspectives from Ecological Economics* (East Lansing: Michigan State University Press, 2013).

31. Mike Salvaris, "Measuring the Kind of Australia We Want: The Australian National Development Index, the Gross Domestic Product and the Global Movement to Redefine Progress," *Australian Economic Review* 46, no. 1 (2013): 79–91, doi: 10.1111/j.1467-8462.2013.00711.x.

32. UNU-IHDP and UNEP, *Inclusive Wealth Report 2012: Measuring Progress toward Sustainability* (Bonn: UNU-IHDP, 2012).

33. Robert Constanza, Ida Kubiszewski, Enrico Giovannini, Hunter Lovins, Jacqueline McGlade, Kate E. Pickett, Kristín Vala Ragnarsdóttir, Debra Roberts, Roberto De Vogli, and Richard Wilkinson, "Development: Time to Leave GDP Behind," *Nature* 505, no. 7483 (2014): 283–285, doi: 10.1038/505283a.

34. Kubiszewski et al., "Beyond GDP." The Genuine Progress Indicator (GPI) data in this analysis were available from 17 countries, accounting for almost 60 percent of total global GDP. All estimates were in 2005 US$.

35. But see Thomas Princen, Michael Maniates, and Ken Conca, eds., *Confronting Consumption* (Cambridge and London: MIT Press, 2002), p. 318, for a brief discussion about the "pernicious idea of consumer sovereignty and its impact on policymaking and public perception."

36. John F. Kennedy, Yale University Commencement (June 11, 1962), available at http://millercenter.org/president/speeches/speech-3370.

37. Daniel Gilbert, "If Only Gay Sex Caused Global Warming," *Los Angeles Times*, July 2, 2006, available at http://articles.latimes.com/2006/jul/02/opinion/op-gilbert2.

38. John Cook, "The Scientific Guide to Global Warming Skepticism," *Skeptical Science*, December 8, 2010, http://www.skepticalscience.com/The-Scientific-Guide-to-Global-Warming-Skepticism.html.

39. Scientific skepticism (a central element in testing evidence, as in Karl Popper's criterion of falsifiability) differs fundamentally from the cynicism of much "denialism" (which dismisses evidence, however compelling, in order to achieve confusion, inaction, or protection of vested interest).

40. Thomas Picketty, *Capital in the 21st Century*, trans. Arthur Goldhammer (Cambridge and London: Harvard University Press, 2014).

41. David G. Victor, *Why Do Smart People Disagree about Facts? Some Perspectives on Climate Denialism* (La Jolla: Laboratory on International Law and Regulation UC San Diego, 2014), available at http://ilar.ucsd.edu/assets/001/505666.pdf.

42. Emma Tonkin, "Deep Down We Know," quoted in Libby Skeels, Benjamin Nisenbaum, Carol Ride, Sue Pratt, and Bronwyn Wauchope, *Let's Speak about Climate Change* (Melbourne: Psychology for a Safe Climate, 2013), available at http://media.wix.com/ugd/59da79_4172ae06668b49978140408bae365688.pdf.

43. Paul Gilding, *The Great Disruption: Why the Climate Crisis Will Bring On the End of Shopping and the Birth of a New World* (Sydney: Bloomsbury, 2011).

44. Max Planck, *Scientific Autobiography*, trans. Frank Gaynor (New York: Philosophical Library, 1949), pp. 33–34.

45. Barbara Kingsolver, *Flight Behavior* (New York: Harper, 2012), p. 283. See also Patrick A. Guerra and Steven M. Reppert, "Coldness Triggers Northward Flight in Remigrant Monarch Butterflies," *Current Biology* 23, no. 5 (2013): 419–423, doi: 10.1016/j.cub.2013.01.052.

46. Matthew Feinberg and Robb Willer, "Apocalypse Soon? Dire Messages Reduce Beliefs in Global Warming by Contradicting Just-World Beliefs," *Psychological Science* 22, no. 1 (2011): 34–38, doi: 10.1177/0956797610391911.

47. Cardinal George Pell, Catholic archbishop of Sydney, Australia, wrote in an opinion piece in the (right-wing) daily newspaper *The Australian* (May 10, 2006): "Pagan emptiness and fears about nature have led to hysteric and extreme claims about global warming. In the past pagans sacrificed animals and even humans in vain attempts to placate capricious and cruel gods. Today they demand a reduction in carbon dioxide emissions."

48. Bible, Gen. 8 and 9.

49. German Advisory Council on Global Change, *Solving the Climate Dilemma: The Budget Approach* (Berlin: WBGU, 2009), available at http://www .preventionweb.net/go/11474.

50. Pope Francis, "Apostolic Exhortation *Evangelii Gaudium*," November 24, 2013, available at http://w2.vatican.va/content/francesco/en/apost_ exhortations/documents/papa-francesco_esortazione-ap_20131124_evangelii- gaudium.html.

51. Credit Suisse, *Global Wealth Report 2013* (Zurich: Credit Suisse, 2013), available at https://publications.credit-suisse.com/tasks/render/file/ ?fileID=BCDB1364-A105-0560-1332EC9100FF5C83.

52. Picketty, *Capital in the 21st Century*.

53. Smith et al., "Human Health."

54. Junfeng Zhang and Kirk R. Smith, "Household Air Pollution from Coal and Biomass Fuels in China: Measurements, Health Impacts, and Interventions," *Environmental Health Perspectives* 115, no. 6 (2007): 848–855, doi: 10.1289/ehp.9479.

55. Rex Tillerson, CEO ExxonMobil Corp, quoted in Michael Babad, "ExxonMobil CEO: 'What Good Is It to Save the Planet if Humanity Suffers?" *Toronto Globe and Mail*, May 30, 2013, available at http:// www.theglobeandmail.com/report-on-business/top-business-stories/ exxon-mobil-ceo-what-good-is-it-to-save-the-planet-if-humanity-suffers/ article12258350/.

56. Andy Haines, George Alleyne, Ilona Kickbusch, and Carlos Dora, "From the Earth Summit to Rio+20: Integration of Health and Sustainable Development," *The Lancet* 379, no. 9832 (2012): 2189–2197, doi: 10.1016/ S0140-6736(12)60779-X.

57. Joseph E. Stiglitz, Amartya Sen, and Jean-Paul Fitoussi, *Report by the Commission on the Measurement of Economic Performance and Social Progress* (Paris: OECD, 2009).

58. Richard Munang, Jesica Andrews, Keith Alverson, and Desta Mebratu, "Harnessing Ecosystem-Based Adaptation to Address the Social Dimensions of Climate Change," *Environment: Science and Policy for Sustainable Development* 56, no. 1 (2014): 18, doi: 10.1080/00139157.2014.861676.

59. John Williams, "Strategic Thinking on Environmental Policy in Australia," *Sustainable Population Australia Inc. Newsletter* 113 (2013): 3, https://www .population.org.au/sites/default/files/newsletters/nl201312_113.pdf.

60. Jan Kunnas, Eoin McLaughlin, Nick Hanley, David Greasley, Les Oxley, and Paul Warde, "Counting Carbon: Historic Emissions from Fossil Fuels, Long-run Measures of Sustainable Development and Carbon Debt," *Scandinavian Economic History Review* 62, no. 3 (2014): 243–265, doi: 10.1080/03585522.2014.896284. The authors argue that an enlightened step in that direction would be an act of extraordinary international enlightenment in which the poor countries cancel the historical global carbon debt incurred by long-emitting industrialized rich countries and rich countries cancel the development assistance financial debts being carried by poor countries. The arithmetic would not be exact, and would upset obsessional accountants—but that is hardly the world's main worry at the moment.

61. James Lovelock, *The Revenge of Gaia* (London: Allen Lane, 2007), p. 10.

62. Stephen Gill, ed., *Global Crises and the Crisis of Global Leadership* (Cambridge: Cambridge University Press, 2011).

63. Frank Biermann, Philipp Pattberg, Harro van Asselt, and Fariborz Zelli, "The Fragmentation of Global Governance Architectures: A Framework for Analysis," *Global Environmental Politics* 9, no. 4 (2009): 14–40.

64. Harro van Asselt, *The Fragmentation of Global Climate Governance* (Cheltenham and Northampton: Edward Elgar, 2014).

65. Paul G. Harris, *What's Wrong with Climate Policy, and How to Fix It* (Cambridge: Polity, 2013).

66. W. Neil Adger, "Climate Change, Human Well-Being and Insecurity," *New Political Economy* 13, no. 2 (2010): 275–292, doi: 10.1080/13563460903290912.

67. Frank Biermann, Kenneth Abbott, Steinar Andresen, Karin Bäckstrand, Steven Bernstein, Michele M. Betsill, Harriet Bulkeley, et al., "Navigating the Anthropocene: Improving Earth System Governance," *Science* 335, no. 6074 (2012): 1306–1307, doi: 10.1126/science.1217255.

68. Quirin Schiermeier, "Combined Climate Pledges of 146 Nations Fall Short of 2°C Target," *Nature (News),* October 30, 2015, doi: 10.1038/ nature.2015.18693.

69. Justin Gillis, "Climate Accord Is a Healing Step, if Not a Cure," *New York Times,* December 12, 2015, available at http://www.nytimes.com/2015/12/13/ science/earth/climate-accord-is-a-healing-step-if-not-a-cure.html?_r=0.

70. Jeff Tollefson, "Is the 2°C World a Fantasy?" *Nature* 527, no. 7579 (2015): 436–438, doi: 10.1038/527436a.

71. Sewell Chan, "Paris Accord Considers Climate Change as a Factor in Mass Migration," *New York Times,* December 12, 2015, available at http://www.nytimes.com/2015/12/13/world/europe/paris-accord-considers-climate-change-as-a-factor-in-mass-migration.html.

72. Chan, "Paris Accord."

73. Cameron Muir, "Powering Asia: The Battle between Energy and Food," *Griffith Review* 49 (2015), https://griffithreview.com/articles/powering-asia/.

74. Gilding, *The Great Disruption.*

75. See Libby Robin and Cameron Muir, "Slamming the Anthropocene: Performing Climate Change in Museums," *reCollections* 10, no. 1 (2015), http://recollections.nma.gov.au/issues/volume_10_number_1/papers/slamming_the_anthropocene.

76. Mark Stafford Smith and Julian Cribb, *Dry Times: Blue Print for a Red Land* (Collingwood: CSIRO Publishing, 2009), chapter 9.

77. Rebecca J. McLain, Patrick T. Hurley, Marla R. Emery, and Melissa R. Poe, "Gathering 'Wild' Food in the City: Rethinking the Role of Foraging in Urban Ecosystem Planning and Management," *International Journal of Justice and Sustainability* 19, no. 2 (2014): 220–240, doi: 10.1080/13549839.2013.841659.

78. Michele M. Betsill and Harriet Bulkeley, "Cities and the Multilevel Governance of Climate Change," *Global Governance Journal* 12, no. 2 (2006): 141–159.

79. UNISDR, *Global Assessment Report 2011: Revealing Risk, Redefining Development* (Geneva: UN International Strategy for Disaster Reduction, 2011), http://www.preventionweb.net/english/hyogo/gar/2011/en/home/index.html. (In the past 30 years the proportion of people living in flood-prone river basins has approximately doubled, as has the proportion living on cyclone-exposed coastlines.)

80. Simone Tilmes, John Fasullo, Jean-François Lamarque, Daniel R. Marsh, Michael Mills, Kari Alterskjær, Helene Muri, et al., "The Hydrological Impact of Geoengineering in the Geoengineering Model Intercomparison Project," *Journal of Geophysical Research: Atmosphere*s 118, no. 19 (2013): 11036–11058, doi: 10.1002/jgrd.50868.

81. Mike Raupach, "Earth System Science at a Crossroads," *Global Change* 79 (2012): 22–25, available at http://www.igbp.net/news/features/features/earthsystemscienceatacrossroads.5.19b40be31390c033ede80001358.html.

82. William Ophuls, "Rousseau, Not Calvin," *The Good Society* 11, no. 3 (2002): 97–98, doi: 10.1353/gso.2003.0015.

83. The Great Transition Initiative (GTI), Boston, MA, http://www.greattransition.org/.

84. Stephen Boyden, *The Biology of Civilisation: Understanding Human Nature as a Force in Nature* (Sydney: UNSW Press, 2004).

85. F. Stuart Chapin III, Steward T. A. Pickett, Mary E. Power, Robert B. Jackson, David M. Carter, and Clifford Duke, "Earth Stewardship: A Strategy for Social-Ecological Transformation to Reverse Planetary Degradation," *Journal of Environmental Studies and Science* 1, no. 1 (2011): 44–53, doi: 10.1007/s13412-011-0010-7.

86. Hui Pan and Yong-Wei Zhang, "GaN/ZnO Superlattice Nanowires as Photocatalyst for Hydrogen Generation: A First-Principles Study on Electronic and Magnetic Properties," *Nano Energy* 1, no. 3 (2012): 488–493, doi: 10.1016/j.nanoen.2012.03.001.

87. Steven C. Sherwood and Matthew Huber, "An Adaptability Limit to Climate Change Due to Heat Stress," *Proceedings of the National Academy of Sciences of the United States of America* 107, no. 21 (2010): 9552–9555, doi: 10.1073/pnas.0913352107.

88. Colin D. Butler, John Powles, and Anthony J. McMichael, "Human Disease: Effects of Economic Development," in *Encyclopedia of Life Sciences* (Chichester: John Wiley & Sons, 2012), doi: 10.1002/9780470015902 .a0003292.pub2.

89. David Griggs, Mark Stafford-Smith, Owen Gaffney, Johan Rockström, Marcus C. Öhman, Priya Shyamsundar, Will Steffen, Gisbert Glaser, Norichika Kanie, and Ian Noble, "Policy: Sustainable Development Goals for People and Planet," *Nature* 495, no. 7441 (2013): 305–307, doi: 10.1038/ 495305a.

90. Michael J. Russell, Wolfgang Nitschke, and Elbert Branscomb, "The Inevitable Journey to Being," *Philosophical Transactions of the Royal Society B: Biological Sciences* 368, no. 1622 (2013): 20120254, doi: 10.1098/ rstb.2012.0254.

91. See Stanley Salthe and Gary Fuhrman, "The Cosmic Bellows: The Big Bang and the Second Law," *Cosmos and History: The Journal of Natural and Social Philosophy* 1, no. 2 (2005): 295–318.

92. Aixue Hu, Yangyang Xu, Claudia Tebaldi, Warren M. Washington, and Veerabhadran Ramanathan, "Mitigation of Short-Lived Climate Pollutants Slows Sea-Level Rise," *Nature Climate Change* 3, no. 8 (2013): 730–734, doi: 10.1038/nclimate1869.

93. Simon Szreter, "Industrialization and Health," *British Medical Bulletin* 69, no. 1 (2004): 75–86, doi: 10.1093/bmb/ldh005.

94. H. Charles J. Godfray, John R. Beddington, Ian R. Crute, Lawrence Haddad, David Lawrence, James F. Muir, Jules Pretty, Sherman Robinson, Sandy M. Thomas, and Camilla Toulmin, "Food Security: The Challenge of Feeding 9 Billion People," *Science* 327, no. 5967 (2010): 812–818, doi: 10.1126/science.1185383.

95. Cribb, *The Coming Famine*.

96. IAASTD, *Agriculture at a Crossroads: The Synthesis Report. Science and Technology* (Washington, DC: International Assessment of Agricultural Knowledge, Science and Technology for Development, 2008), available at http://www.agassessment.org.

97. Erle C. Ellis, Jed O. Kaplan, Dorian Q. Fuller, Steve Vavrus, Kees Klein Goldewijk, and Peter H. Verbur, "Used Planet: A Global History," *Proceedings of the National Academy of Sciences of the United States of America* 110, no. 20 (2013): 7978–7985, doi: 10.1073/pnas.1217241110.

98. Anthony J. McMichael, John W. Powles, Colin D. Butler, and Ricardo Uauy, "Food, Livestock Production, Energy, Climate Change, and Health," *The Lancet* 370, no. 9594 (2007): 1253–1263, doi: 10.1016/S0140-6736(07)61256-2; Philip McMichael, "Agro-Fuels, Food Security, and the Metabolic Rift," *Kurswechsel* 3 (2008): 14–22.

99. Philip McMichael, "Food Regime Crisis and Revaluing the Agrarian Question," in *Rethinking Agricultural Policy Regimes: Food Security, Climate Change and the Future Resilience of Global Agriculture*, edited by R. Almas and H. Campbell (Bingley: Emerald Books, 2012); McMichael et al., "Food, Livestock Production."

100. FAO press release (Rome, October 4, 2013), issued after a meeting between La Via Campesina (based in South America, representing small-hold farmers) and the FAO, available at http://viacampesina.org/en/index.php/main-issues-mainmenu-27/food-sovereignty-and-trade-mainmenu-38/1497.

101. World Bank, *World Development Report 2008: Agriculture for Development* (Washington, DC: The World Bank, 2007), http://go.worldbank.org/2DNNMCBGI0.

102. Philip McMichael, "A Food Regime Genealogy," *Journal of Peasant Studies* 36, no. 1 (2009): 139–169, doi: 10.1080/03066150902820354.

103. Philip D. McMichael and Mindi Schneider, "Food Security Politics and the Millennium Development Goals," *Third World Quarterly* 32, no. 1 (2011): 119–139, doi: 10.1080/01436597.2011.543818.

104. H. Valin, P. Havlík, A. Mosnier, M. Herrero, E. Schmid, and M. Obersteiner, "Agricultural Productivity and Greenhouse Gas Emissions: Trade-Offs or Synergies between Mitigation and Food Security?," *Environmental Research Letters* 8, no. 3 (2013): 035019, doi: 10.1088/1748-9326/8/3/035019.

105. G. Philip Robertson, Katherine L. Gross, Stephen K. Hamilton, Douglas A. Landis, Thomas M. Schmidt, Sieglinde S. Snapp, and Scott M. Swinton, "Farming for Ecosystem Services: An Ecological Approach to Production Agriculture," *BioScience* (2014), doi: 10.1093/biosci/biu037.

106. Gebisa Ejeta, "African Green Revolution Needn't Be a Mirage," *Science* 327, no. 5967 (2010): 831–832, doi: 10.1126/science.1187152.

107. Godfray et al., "Food Security."

108. See University of Nottingham, "World-Changing Technology Enables Crops to Take Nitrogen from the Air," *Science News*, July 25, 2013, http://www.sciencedaily.com/releases/2013/07/130725125024.htm.

109. American Society for Microbiology, "Report Proposes Microbiology's Grand Challenge to Help Feed the World," *Science News*, August 27, 2013, http://www.sciencedaily.com/releases/2013/08/130827204536.htm.

110. Stephen Gardiner, *A Perfect Moral Storm: The Ethical Tragedy of Climate Change* (Oxford: Oxford University Press, 2011).

111. Picketty, *Capital in the 21st Century*.

112. Gardiner, *Perfect Moral Storm*.

113. Stern, *The Stern Review*.

114. Alfred Russel Wallace, quoted in Iain McCalman, *Darwin's Armada* (London: Simon and Schuster, 2009), p. 240.

INDEX

Page references for figures are indicated by *f*, for tables by *t*, and for boxes by *b*.